Distributed Energy Resources Management 2018

Distributed Energy Resources Management 2018

Special Issue Editors

Pedro Faria
Zita Vale

MDPI • Basel • Beijing • Wuhan • Barcelona • Belgrade

MDPI

Special Issue Editors
Pedro Faria
GECAD
Portugal

Zita Vale
Engineering Institute
Portugal

Editorial Office
MDPI
St. Alban-Anlage 66
4052 Basel, Switzerland

This is a reprint of articles from the Special Issue published online in the open access journal *Energies* (ISSN 1996-1073) from 2018 to 2020 (available at: https://www.mdpi.com/journal/energies/special_issues/distributed_energy_resources_management2018).

For citation purposes, cite each article independently as indicated on the article page online and as indicated below:

LastName, A.A.; LastName, B.B.; LastName, C.C. Article Title. *Journal Name* **Year**, *Article Number*, Page Range.

ISBN 978-3-03928-170-1 (Pbk)
ISBN 978-3-03928-171-8 (PDF)

Contents

About the Special Issue Editors

Pedro Faria completed his Ph.D in 2016 and works in the field of power systems, with a focus on energy markets, smart grids, and demand response, and has published 1 patent and over 150 papers. His current work includes renewable-based distributed generation, energy storage, and electric vehicles. In these fields, optimization, clustering, and classification methods have been applied to real and simulated environment problems. He has been developing business models for the modeling, aggregation, and remuneration of consumers participating in electricity markets and in demand response programs. He has also worked in the real-time simulation of power and energy systems. He has participated in a number of relevant national and international research projects, contributing to models and their implementation, testing, demonstration, and piloting.

Zita Vale is a full professor at the Polytechnic Institute of Porto, Portugal. She received her diploma in Electrical Engineering in 1986 and her Ph.D in 1993, both from the University of Porto. Zita Vale works in the area of power and energy systems, with special interest in the application of artificial intelligence techniques. She has been involved in more than 50 funded projects related to the development and use of knowledge-based systems, multi-agent systems, genetic algorithms, neural networks, particle swarm intelligence, constraint logic programming and data mining. The main application fields of these projects include smart grids, accommodating an intensive use of renewable energy sources, distributed energy resources (DER) and distributed generation (DG). She addresses the management of energy resources, the impact of DER on electrical networks, the negotiation of DER in electricity markets, demand response, storage, energy management in buildings, and electrical vehicles, including the ones with gridable capability (V2G). Her work also has applications in electricity markets—addressing contracts, prices and tariffs, decision-support for market participants, aggregation, ancillary services, and wholesale and local market simulation—and incontrol centers, namely through intelligent alarm processing, intelligent interfaces and intelligent tutors. She has published over 800 works, including more than 100 papers in international scientific journals, and more than 500 papers in international scientific conferences.

![energies logo] *energies*

MDPI

Editorial

Distributed Energy Resources Management 2018

Pedro Faria [1,2,*] and Zita Vale [2]

[1] GECAD—Research Group on Intelligent Engineering and Computing for Advanced Innovation and
     Development, Polytechnic of Porto, 4200-072 Porto, Portugal
[2] Polytechnic of Porto, Rua Dr. Antonio Bernardino de Almeida, 431, 4200-072 Porto, Portugal; zav@isep.ipp.pt
*   Correspondence: pnf@isep.ipp.pt; Tel.: +351-228-340-511; Fax: +351-228-321-159

Received: 5 November 2019; Accepted: 23 December 2019; Published: 30 December 2019

1. Introduction

The Special Issue "Distributed Energy Resources Management 2018" includes 13 papers [1–13], and is a continuation of the Special Issue "Distributed Energy Resources Management" published in https://www.mdpi.com/journal/energies/special_issues/distributed_energy_resources_management2018. The success of the previous Special Issue, for which the summary is published in [14], shows the unquestionable relevance of distributed energy resources in the operation of power and energy systems at both the distribution level, and at the wide power system level. Improving the management of distributed energy resources makes possible the accommodation of a higher penetration of intermittent distributed generation and electric vehicle charging stations. Demand response (DR) programs, particularly the ones with a distributed nature, allow for consumers to contribute to increased system efficiency while receiving benefits.

This Special Issue addresses the management of distributed energy resources with a focus on methods and techniques to achieve optimized operation, to aggregate resources within the scope of virtual power players and other types of aggregators, and to remunerate them. The integration of distributed resources in electricity markets is also addressed as an enabler for their increased and efficient use.

The topics of this Special Issue include the following:

- DR;
- electricity markets;
- renewable energy integration;
- real-time simulation;
- smart grids; and
- virtual power players.

13 research papers have been published in this Special Issue, with the following statistics:

- Submissions: 16; published: 13; rejected: 3.
- Average paper processing time: 43.98 days.
- Authors' geographical distribution: Belgium (1), China (1), Korea (1), Portugal (5), Slovakia (1), Spain (4), The Netherlands (1).

2. Published Papers Highlights

This paper provides a summary of the Special Issue of *Energies* covering the published articles [1–13], which address several topics related to distributed energy resources management. Table 1 identifies the most relevant topics in each publication of the 2018 Edition. These topics have been selected by the Editors as the most relevant among the 13 papers published.

Table 1. Topics covered in each publication.

Topic	Publication												
	[1]	[2]	[3]	[4]	[5]	[6]	[7]	[8]	[9]	[10]	[11]	[12]	[13]
Demand response (DR)	x			x	x		x				x	x	x
Distributed generation	x	x	x		x		x	x	x	x			x
Electricity markets and aggregation	x	x	x	x	x							x	
End-users and their comfort	x			x	x	x		x			x	x	
Energy storage				x				x	x				x
Intelligent resource management	x		x			x	x			x			x
Operation and control		x				x	x	x	x	x			
Renewable energy sources		x		x					x	x			
Total	**5**	**3**	**4**	**4**	**5**	**3**	**4**	**4**	**4**	**4**	**2**	**3**	**4**

As shown in Table 1, most of the publications focus on DR and distributed generation, while four papers include energy storage. Most of the papers are dedicated to end-user aspects, namely consumer comfort, and to electricity markets and resource aggregation.

In the work presented in [1], Faria et al. presented a methodology to schedule distributed energy resources in order to minimize operation costs. While the focus is given to DR programs, distributed generation units were also addressed. The consumers and the generation units were aggregated using clustering methods in order to establish remuneration groups.

A multi-agent coordination scheme was proposed by Castillo-Cagigal et al. in [2] with a deep focus on contexts of high penetration of renewables, mainly photovoltaic generation. Swarm intelligence and coupled oscillators were combined in the innovative methodology using the SwarmGrid algorithm. The objective was to increase the stability of and reduce stress on the electrical grid, allowing for an increase in renewables penetration.

Gao et al. [3] proposed a bidding mechanism that enables a virtual power plant (VPP) to internally find the resources needed for each context, considering uncertainties. A multi-agent and game theory-based approach was used in a double-layer nested dynamic game bidding model. The game played between the VPP and the internal resources was played with the goal of maximum self-interest.

From a market perspective, Minniti et al. [4] provided insights into the most important aspects that can enable flexibility trading. The current implementation of aggregators in Europe is discussed, pointing out that there is a long way to go before the desired situation was achieved.

Song [5] et al. presented a multi-disciplinary study of distributed resources in China. A comprehensive benefit evaluation index system was developed, and the benefits are evaluated through a case analysis.

Smart buildings are addressed in [6]: Casado-Vara et al. proposed the use of continuous-time Markov chains and a cooperative control algorithm in a novel model-based predictive hybrid control algorithm. The maintenance and reliability of an internet of things network was improved by optimal sensor positioning, and by replacing the sensors that the algorithm predicts will not properly work in the future.

With focus on microgrids, demand side management was managed using a multi-objective approach proposed by Singh and Jha in [7]. Usability indices (peak shaving for example) were determined in order to analyze the contribution of demand side management to the optimal operation of the microgrid.

Moon et al. [8] proposed a method for real-time active power control in a microgrid with energy storage systems and diesel generators operating in island mode. Distributed energy resources were managed in order to ensure an active power balance while keeping the frequency and voltage within determined levels.

In [8], the energy storage unit was addressed as a resource for diesel generators that contribute to maintaining an active power balance in real-time. In [9], Galván et al. proposed a method to optimally schedule the energy storage unit in a microgrid based on priorities. The battery's state of charge was properly considered, and a particle swarm approach is used to find the optimal solution.

A control framework for the strategic operation of wider distribution networks was proposed by Kotsalos et al. in [10]. The proposed approach addressed a large set of distributed energy resource types, namely photovoltaics, energy storage, electric vehicles, and controllable loads. The objective was to improve the integration of renewable resources while keeping within the bounds of the voltage and the load in the secondary substation.

In the review presented in [11], Šujanová et al. discussed several methods for the evaluation of the comfort levels in buildings. The paper highlighted the need for a human-centric design of the built environment, where the efficiency of technology can be measured only if it is successfully implemented and used by a building's occupants.

In [12], Gazafroudi et al. proposed an agent-based approach for energy trading at the local level. The proposed iterative approach allowed for interaction between end-users and the distribution network operator. The developed algorithm is also adequate for energy resource aggregators.

Finally, in [13], optimal energy resource management at the residential house level was obtained using particle swarm optimization, as proposed by Faia et al. The energy storage unit was managed by considering the available photovoltaic generation, the energy cost, and the use of flexible loads.

Author Contributions: Investigation, P.F. and Z.V.; Writing—original draft, P.F.; Writing—review & editing, Z.V. All authors have read and agreed to the published version of the manuscript.

Funding: The present work was done in the scope of CEECIND/02887/2017 and UID/EEA/00760/2019 funded by FEDER Funds through COMPETE Program.

Conflicts of Interest: The authors declare no conflicts of interest.

References

1. Faria, P.; Spínola, J.; Vale, Z. Distributed Energy Resources Scheduling and Aggregation in the Context of Demand Response Programs. *Energies* **2018**, *11*, 1987. [CrossRef]

2. Castillo-Cagigal, M.; Matallanas, E.; Caamaño-Martín, E.; Martín, Á. SwarmGrid: Demand-Side Management with Distributed Energy Resources Based on Multifrequency Agent Coordination. *Energies* **2018**, *11*, 2476. [CrossRef]

3. Gao, Y.; Zhou, X.; Ren, J.; Wang, X.; Li, D. Double Layer Dynamic Game Bidding Mechanism Based on Multi-Agent Technology for Virtual Power Plant and Internal Distributed Energy Resource. *Energies* **2018**, *11*, 3072. [CrossRef]

4. Minniti, S.; Haque, N.; Nguyen, P.; Pemen, G. Local Markets for Flexibility Trading: Key Stages and Enablers. *Energies* **2018**, *11*, 3074. [CrossRef]

5. Song, X.; Shu, M.; Wei, Y.; Liu, J. A Study on the Multi-Agent Based Comprehensive Benefits Simulation Analysis and Synergistic Optimization Strategy of Distributed Energy in China. *Energies* **2018**, *11*, 3260. [CrossRef]

6. Casado-Vara, R.; Vale, Z.; Prieto, J.; Corchado, J. Fault-Tolerant Temperature Control Algorithm for IoT Networks in Smart Buildings. *Energies* **2018**, *11*, 3430. [CrossRef]

7. Singh, M.; Jha, R. Object-Oriented Usability Indices for Multi-Objective Demand Side Management Using Teaching-Learning Based Optimization. *Energies* **2019**, *12*, 370. [CrossRef]

8. Moon, H.; Kim, Y.; Chang, J.; Moon, S. Decentralised Active Power Control Strategy for Real-Time Power Balance in an Isolated Microgrid with an Energy Storage System and Diesel Generators. *Energies* **2019**, *12*, 511. [CrossRef]

9. Galván, L.; Navarro, J.; Galván, E.; Carrasco, J.; Alcántara, A. Optimal Scheduling of Energy Storage Using A New Priority-Based Smart Grid Control Method. *Energies* **2019**, *12*, 579. [CrossRef]

10. Kotsalos, K.; Miranda, I.; Silva, N.; Leite, H. A Horizon Optimization Control Framework for the Coordinated Operation of Multiple Distributed Energy Resources in Low Voltage Distribution Networks. *Energies* **2019**, *12*, 1182. [CrossRef]

11. Šujanová, P.; Rychtáriková, M.; Sotto Mayor, T.; Hyder, A. A Healthy, Energy-Efficient and Comfortable Indoor Environment, a Review. *Energies* **2019**, *12*, 1414. [CrossRef]

12. Shokri Gazafroudi, A.; Prieto, J.; Corchado, J. Virtual Organization Structure for Agent-Based Local Electricity Trading. *Energies* **2019**, *12*, 1521. [CrossRef]

13. Faia, R.; Faria, P.; Vale, Z.; Spinola, J. Demand Response Optimization Using Particle Swarm Algorithm Considering Optimum Battery Energy Storage Schedule in a Residential House. *Energies* **2019**, *12*, 1645. [CrossRef]

14. Faria, P. Distributed Energy Resources Management. *Energies* **2019**, *12*, 550. [CrossRef]

energies

MDPI

Article

Distributed Energy Resources Scheduling and Aggregation in the Context of Demand Response Programs

Pedro Faria [1,2,*], João Spínola [1,2] and Zita Vale [2]

[1] GECAD—Research Group on Intelligent Engineering and Computing for Advanced Innovation and Development, 4200-072 Porto, Portugal; jafps@isep.ipp.pt

[2] IPP—Polytechnic Institute of Porto, Rua DR. Antonio Bernardino de Almeida, 431, 4200-072 Porto, Portugal; zav@isep.ipp.pt

* Correspondence: pnf@isep.ipp.pt; Tel.: +351-228-340-511; Fax: +351-228-321-159

Received: 10 May 2018; Accepted: 30 July 2018; Published: 31 July 2018

Abstract: Distributed energy resources can contribute to an improved operation of power systems, improving economic and technical efficiency. However, aggregation of resources is needed to make these resources profitable. The present paper proposes a methodology for distributed resources management by a Virtual Power Player (VPP), addressing the resources scheduling, aggregation and remuneration based on the aggregation made. The aggregation is made using *K*-means algorithm. The innovative aspect motivating the present paper relies on the remuneration definition considering multiple scenarios of operation, by performing a multi-observation clustering. Resources aggregation and remuneration profiles are obtained for 2592 operation scenarios, considering 548 distributed generators, 20,310 consumers, and 10 suppliers.

Keywords: clustering; demand response; distributed generation; smart grids

---

1. Introduction

The end-consumer's demand reduction or shed, due to technical or economic issues, in response to price signals or incentives, is commonly defined as Demand Response (DR) [1]. Price signals can be given using Real-Time pricing (RTP) programs, as in [2]. DR resources can be used together with Distributed Generation (DG) successfully, at distribution network levels, in order to contribute to the implementation of the Smart Grid (SG) [3,4]. The high potential of DR is discussed in [5–7] for different countries and implementation scenarios, and DG in [8,9].

Despite recent efforts and actual achievements concerning the implement of DR in small-size resources, many DR programs are focused on large-size resources [10]. The full integration of DR programs in the energy markets should be performed using the aggregation of small-size consumers and producers that are usually connected to distribution networks [11]. It is therefore needed to find the most appropriate way to aggregate and remunerate such small-size resources for their participation in electricity markets.

The aggregation of distributed resources can be done under several approaches, which can consider the economic models and entities that perform the aggregation [12,13]. For the specific case of DR resources aggregation [14], one can find the Curtailment Service Provider (CSP), which acts in several current DR implementations [15]. With a wider resource types integration, one can refer to a Virtual Power Player (VPP). A VPP is an entity responsible for the small-size resources management at the level of distribution networks. It performs the resources scheduling, in order to provide the required means to supply the demand and to enable the participation in the electricity markets. It is also responsible for the remuneration of the DG and DR resources [16].

Given the importance of the resources aggregation, adequate methods and tools should be used. Since the aggregation will define the groups of resources and result in the remuneration for each single resource, such task is very critical in the way that it will represent an incentive for the consumers and producers in order to actively participate in the optimal management of the VPP area. Clustering algorithms are used to ensure that the groups are formed considering common characteristics or patterns.

The clustering algorithms can be classified into several types, as hierarchical, partition, fuzzy and many others [17]. The one used in this paper is *K*-means, a partition clustering algorithm. At each iteration, the algorithm computes the distance between objects (based on each observation-value of objects for a given situation) and the center of each of the groups. The objects change amongst groups based upon the calculated distance, i.e. an object is assigned to a group if the distance to it is the lowest when comparing to all other groups [18,19].

The remuneration of the resources is usually done equally for the resources of the same type [20]. It can be performed individually as shown in [21], in which the payment would be proportional to each individual contribution. In [22], the author proposes a unit commitment mechanism to reduce the cost of using wind generation considering the load and spinning reserve implemented at a certain time in the network. The aggregation and remuneration of the resources is, in most situations, largely related to the resources scheduling, namely in the VPP operation context.

In [23], an evolutionary algorithm is used to perform the scheduling and to obtain the Pareto-front optimal solutions. Reference [24] discusses the resources scheduling and aggregation. In [25], several DR programs are presented, as well as onsite generation (OG) and energy storage (ES). The authors refer to the aggregator as an entity able to aggregate the consumers' reduction capacity and make them profitable, using OG and ES for load balance. These works discuss and comment on results about resources scheduling, aggregation and remuneration, without extensive analyses and proposal of remuneration methodologies for small-size energy resources. Some of the authors in the present paper showed preliminary results in distributed resources management considering their scheduling, aggregation and remuneration [26]. However, many aspects are detailed and explored further in the present paper, explained detailed in Section 2.

The present paper proposes a methodology to support the VPP, performing the resource's schedule (optimization algorithms), aggregation (using *K*-means), and remuneration. After the scheduling, the aggregation and remuneration of distributed resources is made considering *K*-means algorithm and maximum price per group—all the resources in a given group are paid considering the same tariff, namely, the maximum value of all the resources belonging to the same group, respectively.

Previous research in this field addresses the problem by analyzing distinct operation scenarios one by one; the methodology proposed in this paper advances in considering the whole set of operation scenarios at the same time. Additionally, it combines incentive-based and price-based demand response programs. The proposed methodology is flexible enough in order to be used by several types of aggregators, whether they do or do not have both the consumption and generation resources.

The *K*-means algorithm allows the consideration of several operation scenarios, determining the groups to be formed on a schedule shape basis. The maximum price per group allows a fair remuneration to less-efficient resources, at the same time that incentives for the ones that are efficient to continue participating.

After this introductory section, Section 2 explains the proposed methodology, including a detailed explanation of its contributions. Then, Section 3 presents the resource scheduling formulation. The case study is presented in Section 4, and Section 5 includes the obtained results. Finally, Section 6 presents the main conclusions of the work.

2. Proposed Methodology

The proposed methodology aims to support the Virtual Power Player (VPP), in optimally scheduling, aggregating and remunerating the resources. The VPP is considered to be an external independent entity and thus, with no interference in technical aspects of the network where it acts. Moreover, it is assumed that the resources are in a contract with this entity, and therefore must respond to DR events when so requested.

2.1. VPP Approach

The VPP is able to optimally use generation and demand response according to distinct types, such as external suppliers, Distributed Generation (DG), Demand Response incentive-based programs (Incentive Demand Response (IDR))—in which the consumers are remunerated proportionally to their contribution in reducing load; and Demand Response price-based programs (Real-Time Pricing (RTP))—in which the consumers are modelled by their elasticity, making changes to their load in response to price signals. The final objective is to support the VPP in the definition of tariffs for DR and DG resources.

The methodology comprehends four different phases, according to Figure 1: input data definition, scheduling, aggregation and remuneration tariffs definition. The four phases run independently and subsequently.

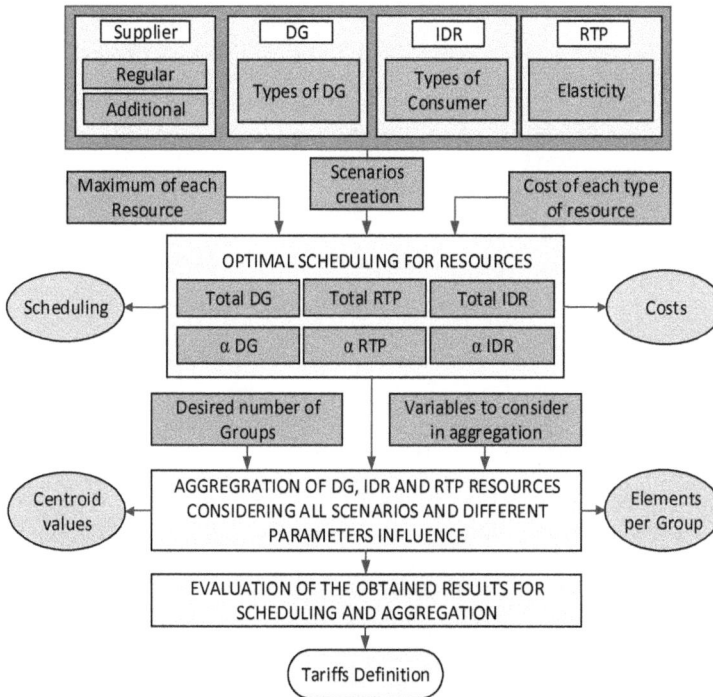

Figure 1. Diagram of the proposed methodology.

In the first phase, a set of data is provided as input to the optimization block which runs the second phase. The data includes all of the resources characteristics, including the maximum capacity for DG units, and for the suppliers, and the reduction capacity for the consumers. Also, prices for each resource and loads elasticity are provided.

The methodology is able to cope with the different time scales of the resources operation. For example, if we are dealing with a scenario of 5 min, where the data is available for each second, it is calculated the average values for each period of 5 min. If the available data belongs to 15 min, the same value has been considered for all 3 of the five-minute periods.

In the second phase, the optimal scheduling of resources is performed for each scenario. The scenarios with unfeasible solution in the optimization are discarded so it is not provided to the aggregation phase. The scheduling is performed using MATLAB (2014A, MathWorks, Natick, MA, USA). Also in this phase, several parameters are diversified with the purpose of creating distinct operation scenarios. Those parameters are: regular supplier and distributed generators cost, the maximum reduction allowed for IDR and RTP consumers programs, the total initial load and, finally, the DG, IDR and RTP reduction contributions. As it will be detailed in Section 3, in Equations (15)–(17), it is possible for the VPP to choose an amount of energy obtained from each type of resource, relative to the total scheduling obtained. The scheduling is discussed and analyzed in particular in Section 5.

2.2. Aggregation of Resources

After the scheduling, the aggregation of the resources is performed (third phase) using a partition clustering algorithm; *K*-means algorithm. This algorithm allows us to define an entry matrix that contains the variables (objects) as rows and different scenarios (values) on the columns enabling the clustering considering all of the scenarios. This allows a better standardized group formation and not concentrated as hierarchical clustering algorithms. The algorithm has as outputs the centroid values of each group and also the group identification for each resource. The centroid value represents the point in the group for which the distance to the rest of the elements in that group is minimum. As for the distance minimization function, several can be considered. In this case, the one used is the squared Euclidean distance, modeled as in Equation (1) [27]

$$d(x, c) = (x - c)(x - c)\prime \tag{1}$$

where x is the sample point and c, the centroid value.

The aggregation is made separately for each type of resource (DG, IDR and RTP), and considers all of the implemented scenarios. The aggregation using *K*-means allows for considering several observations of the variables. In this way, each scenario is considered as an observation of the variables related to each resource.

The resources are clustered according to all scenarios, being obtained the centroid profile, which is the representative profile of all the resources in a group. This centroid profile contains the centroid single values for each scenario. In this phase, the *K* parameter is part of *K*-means algorithm in order to set the number of tariff or remuneration groups that will be assigned to the consumers or producers. It is possible to define different *K* ranges. $K = 1$ means that all the resources will be in one group, so we start with $K = 2$. For the highest *K*, we have the reference of the number of consumer types. The maximum *K* is limited in the sense that it is not possible in practice to implement a large number of tariffs. In the limit, we could have one tariff for each single consumer/producer. The *K* parameter has no influence on the scheduling, since it is made a priori.

2.3. Remuneration

After the aggregation, the remuneration process is addressed using the maximum price inside each group-fourth phase. The remuneration is obtained considering the highest energy price in each group and applying it to all the resources in that group. The consumers and producers with lower initial prices will be remunerated at a higher price after the computation of the proposed methodology since the highest price within a group will be applied to all the resources inside the group.

2.4. Tariffs Comparison

With the results obtained as described in the previous sub-sections, it is possible for the VPP to compare the costs for each number of defined clusters, and the costs with the payment by type of resource. The payment by type can have two distinct approaches which will be compared with the payment for each group in different numbers of defined clusters. In the first one, considering the initial price that corresponds to the price established for all the resources of such type, each resource will be paid for the scheduled power, assuming that this schedule will be respected. This approach can somehow be not so realistic in some scenarios, in the sense that all the resources of the same tariff should be scheduled without discrimination. In the second approach, all the resources of the same type will be paid at the maximum capacity available, as it is made for the resources in different clusters.

The present methodology aims at minimizing operation costs, at optimally scheduling resources, and at defining aggregation and remuneration.

The innovative aspects of the proposed methodology, when comparing to the previous works, including in [26], are related with: the consideration of several distinct scheduling scenarios; inclusion of a mathematical formulation that considers the DR programs possibilities; exhaustive discussion regarding the use of clustering algorithms on resource's aggregation; and finally, the proposal of remuneration models that consider the aggregation of resources according to their use in multiple scenarios. In this way, it is possible to provide the VPP with optimal and practicable solutions to the management of distributed resources, addressing their remuneration. The authors have published work related to the present paper in [27–29]. In [27], it is shown a similar work of this paper; however, the focus of the present paper is given to the consideration of multi-parameters in the clustering process, made possible by the use of *K*-means algorithm instead of the hierarchical clustering. This allows the analysis of the evolution of the group's centroid across the different scenarios evaluated.

3. Scheduling Formulation

In this section, the resources scheduling is presented.

The objective function of the optimization model is as presented in Equation (2), which targets the minimization of operation costs for the VPP. The optimal scheduling of generation and demand response programs use is obtained. In the implemented objective function, multiple distributed energy resources, as distributed generators and consumers are considered. The connection between the network managed by the VPP and other systems is implemented by considering external suppliers, which can also be useful for the management of the VPP's network balance. It is important to stress that the external suppliers are divided into regular and additional ones, so the most expensive ones are only activated when really needed. According to the price elasticity of demand, the price signal to be given to each consumer and the expected consumption reduction are obtained.

The resources are scheduled for each scenario context, concerning the restrictions modeled by Equations (3)–(17). DR programs are divided into IDR and RTP programs.

The optimization problem class is quadratic programming, due to the variable multiplication in the objective function, referring to the use of real-time pricing demand response initiatives.

$$
\begin{aligned}
\text{Min } OC = &\sum_{p=1}^{P} P_{\text{DG}(p)} \times C_{\text{DG}(p)} + P_{\text{NSP}} \times C_{\text{NSP}} \\
&+ \sum_{s=1}^{S} \left[\begin{array}{c} P_{\text{Supplier}(s)}^{\text{reg}} \times C_{\text{Supplier}(s)}^{\text{reg}} \\ + P_{\text{Supplier}(s)}^{\text{add}} \times C_{\text{Supplier}(s)}^{\text{add}} \end{array} \right] \\
&+ \sum_{c=1}^{C} \left[\begin{array}{c} P_{\text{IDR}(c)} \times C_{\text{IDR}(c)} \\ -\left(P_{\text{RTP}(c)}^{\text{Initial}} - P_{\text{RTP}(c)}^{\text{Reduct}}\right) \times \left(C_{\text{RTP}(c)}^{\text{Initial}} + C_{\text{RTP}(c)}^{\text{Increase}}\right) \end{array} \right]
\end{aligned}
\tag{2}
$$

The balance equation is presented in Equation (3). In this equation, the balance between production and demand must be accomplished, i.e., the initial consumption deducted from the reduction verified by IDR and RTP consumers must be equal to the production sum of the DG with suppliers. The equation also includes the non-supplied power (NSP), which refers to the amount of lost load in case the generation is not sufficient to supply the demand.

$$\sum_{c=1}^{C} \left[P_{\text{Load}(c)}^{\text{Initial}} - P_{\text{IDR}(c)} - P_{\text{RTP}(c)}^{\text{Reduct}} \right] = \sum_{p=1}^{P} P_{\text{DG}(p)}$$
$$+ \sum_{s=1}^{S} \left[P_{\text{Supplier}(s)}^{\text{reg}} + P_{\text{Supplier}(s)}^{\text{add}} \right] + P_{\text{NSP}} \tag{3}$$

Equations (4)–(7) are related to the technical limitations of the external suppliers regarding their generation output. Equations (4) and (6) consider the individual limitation of each external supplier, while Equation (5) and (7) limit the total amount bought from external suppliers, for regular and additional type suppliers.

$$P_{\text{Supplier}(s)}^{\text{reg}} \leq P_{\text{Supplier}(s)}^{\text{reg Max}}, \ \forall s \in \{1, \dots, S\} \tag{4}$$

$$\sum_{s=1}^{S} P_{\text{Supplier}(s)}^{\text{reg}} \leq P_{\text{Supplier}}^{\text{reg Total}} \tag{5}$$

$$P_{\text{Supplier}(s)}^{\text{add}} \leq P_{\text{Supplier}(s)}^{\text{add Max}}, \ \forall s \in \{1, \dots, S\} \tag{6}$$

$$\sum_{s=1}^{S} P_{\text{Supplier}(s)}^{\text{add}} \leq P_{\text{Supplier}}^{\text{add Total}} \tag{7}$$

In a similar way to the previous Equations (4)–(7), for distributed generators the same energy limitations are applied considering the generator's output, whether in individual by Equation (8) or total by Equation (9).

$$P_{\text{DG}(p)} \leq P_{\text{DG}(p)}^{\text{Max}}, \ \forall p \in \{1, \dots, P\} \tag{8}$$

$$\sum_{p=1}^{P} P_{\text{DG}(p)} \leq P_{\text{DG}}^{\text{Total Max}} \tag{9}$$

Concerning the participation of consumers in DR events, the constraints modelled are: maximum consumption reduction for each consumer in IDR program, Equation (10); the maximum consumption reduction of the sum between IDR and RTP programs, Equation (11); the maximum cost increase in RTP programs, Equation (12); maximum consumption reduction for each consumer in RTP programs, Equation (13); consumer's elasticity considering scheduled power and price, especially important in RTP programs, Equation (14) [30]. In the RTP demand response program, the consumer reacts to a given rise in the energy price, with a consumption reduction. This rise in the energy price and consumption reduction of the consumer is limited by Equations (12) and (13), respectively, ensuring that the differences between initial and final energy price/consumption are within a given range of values. This reaction feature of the consumer to changes in the energy price is defined as elasticity, and is modelled by Equation (14). It considers that the consumer has a given elasticity value that represents the responsive capacity of the consumer's current operation condition to price changes.

$$P_{\text{IDR}(c)} \leq P_{\text{IDR}(c)}^{\text{Max}}, \ \forall c \in \{1, \dots, C\} \tag{10}$$

$$P_{\text{IDR}(c)} + P_{\text{RTP}(c)}^{\text{Reduct}} \leq P_{\text{RTPDR}(c)}^{\text{Max}}, \ \forall c \in \{1, \dots, C\} \tag{11}$$

$$C_{\text{RTP}(c)}^{\text{Increase}} \leq C_{\text{RTP}(c)}^{\text{Increase Max}}, \ \forall c \in \{1, \dots, C\} \tag{12}$$

$$P_{\text{RTP}(c)}^{\text{Reduct}} \leq P_{\text{RTP}(c)}^{\text{Reduct Max}}, \ \forall c \in \{1, \dots, C\} \tag{13}$$

$$\varepsilon_{(c)} = \frac{P_{\text{RTP}(c)}^{\text{Reduct}} \times C_{\text{RTP}(c)}^{\text{Initial}}}{P_{\text{Load}(c)}^{\text{Initial}} \times C_{\text{RTP}(c)}^{\text{Increase}}}, \ \forall c \in \{1, \dots, C\} \tag{14}$$

In Equations (15)–(17), a usage limitation constraint is implemented to enable control over the contribution of distributed generation and demand response programs. This provides the VPP with an additional tool to manage the resources, considering its operation context and/or other constraints. These three α parameters take values between 0 and 1. These parameters are modelled as presented in Equations (15)—DG, (16)—IDR, and (17)—RTP. In this way, as an example, for α_{DG} equal to 0.6, it will result in a contribution of DG resources to supply the demand in that specific scenario lower or equal to 60%.

$$\frac{\sum\limits_{p=1}^{P} P_{DG(p)}}{\sum\limits_{s=1}^{S} \left[\begin{array}{c} P_{Supplier(s)}^{reg} \\ +P_{Supplier(s)}^{add} \end{array} \right] + \sum\limits_{p=1}^{P} P_{DG(p)} + \sum\limits_{c=1}^{C} \left[\begin{array}{c} P_{IDR(c)} \\ +P_{RTP(c)}^{Reduct} \end{array} \right] + P_{NSP}} \leq \alpha_{DG} \qquad (15)$$

$$\frac{\sum\limits_{c=1}^{C} P_{IDR(c)}}{\sum\limits_{s=1}^{S} \left[\begin{array}{c} P_{Supplier(s)}^{reg} \\ +P_{Supplier(s)}^{add} \end{array} \right] + \sum\limits_{p=1}^{P} P_{DG(p)} + \sum\limits_{c=1}^{C} \left[\begin{array}{c} P_{IDR(c)} \\ +P_{RTP(c)}^{Reduct} \end{array} \right] + P_{NSP}} \leq \alpha_{IDR} \qquad (16)$$

$$\frac{\sum\limits_{c=1}^{C} P_{RTP(c)}^{Reduct}}{\sum\limits_{s=1}^{S} \left[\begin{array}{c} P_{Supplier(s)}^{reg} \\ +P_{Supplier(s)}^{add} \end{array} \right] + \sum\limits_{p=1}^{P} P_{DG(p)} + \sum\limits_{c=1}^{C} \left[\begin{array}{c} P_{IDR(c)} \\ +P_{RTP(c)}^{Reduct} \end{array} \right] + P_{NSP}} \geq \alpha_{RTP} \qquad (17)$$

4. Case Study

The proposed methodology is applied to a case study concerning a real 30 kV distribution network, supplied by a high-voltage substation (60/30 kV) with 90 MVA of maximum power capacity. There are 937 buses, 20,310 consumers, and 548 distributed generators of several types. The consumers are classified into five distinct types: Domestic (DM), Small Commerce (SC), Medium Commerce (MC), Large Commerce (LC) and Industrial (ID). The data concerning consumers have been obtained by measurements, while for the DG units it has been specified according to DG implementation studies [31].

The data concerning each of the resources included in this case study initially had different time scales. In order to address this issue, for the data given for each second, the average values for each period of 5 minutes have been calculated. For the data given for each 15 min, the same value has been considered for all 3 of the five-minute periods.

The total power demand for the considered network, without any parameters variation, is 62,630 kW. On the side of distributed generation, the resources are classified into seven different types: Wind, Photovoltaic (PV), Co-generation (Combined Heat and Power (CHP)), Biomass, Waste-to-energy (Municipal Solid Waste (MSW)), Fuel cell and Small Hydro.

The VPP is responsible for specifying the important characteristics of each resource, in order to obtain proper schedule, aggregation and remuneration scenarios. Due to space limitations, in the present paper, only limited information is presented; further details can be found in [31].

Several scenarios based on the variation of the parameters of both producers and consumers have been defined according to [27]. However, the choice of these parameters reflects the outcoming results, making this procedure very important. The parameter variation must consider reliable and feasible scenarios, thus its values when assigned need to be coherent with the resources characteristics and scenario conditions.

The way that parameters are combined in order to build the 2592 scenarios is explained in Table 1. In the last column of this table, the set of parameters values is presented for the selected scenario, which affect the scheduling.

Table 1. Scenarios Definition.

Parameter	Parameter Variation			# Scenarios	Selected Scenario
	Min	Step	Max		
Regular supplier cost	0.8	0.2	1.2	3	1
Distributed generators cost	1	0.2	1.2	2	1
Total initial load	1.2	−0.2	0.8	3	1
Maximum reduction for IDR	1	0.2	1.2	2	1.2
Maximum reduction for RTP	1	0.2	1.2	2	1.2
α_{DG}	0	0.15	0.3	3	0.3
α_{IDR}	0	0.075	0.15	3	0.15
α_{RTP}	0	0.1	0.3	4	0.3
# Scenarios				2592	

The input parameters are the regular supplier and distributed generators cost, the maximum reduction allowed for IDR and RTP consumers programs, the total initial load and, finally, the alpha parameters described before.

The result matrix, obtained from the scheduling optimization, consists of 21,515 lines (variables) per 2592 columns (scenarios); therefore, considerable computational means are required to handle this amount of data.

5. Results

In this section, the results obtained from the optimization, aggregation and remuneration are presented for a selected scenario in Table 1, last column. Section 5.1 presents an extensive analysis of the selected scenario defined by the parameters, showing the main contributions of each type of resource and technology. Section 5.2 concerns the aggregation of resources.

5.1. Selected Scenario—Resource Schedule

It is important to note that the influence of alpha parameters can be clearly seen in the results. Figure 2a,b demonstrate the optimized power scheduling for distributed producers and real-time-pricing consumers, respectively. Figure 2c presents the production totals and respective percentage of total schedule, to make the analysis simpler and quicker to perform. Due to space limitations and consistent presentation purposes, only 100 resources are presented for each type. Figure 2a presents the scheduling for the first 100 distributed producers, being that some of which are not generating. The black line represents the maximum energy capacity available in that specific moment. Looking at Figure 2c, we can see that the scheduling fulfilled the implemented constraints, namely the alpha parameters, as demonstrated by total DG, IDR and RTP. The RTP consumers, in this scenario, have contributed 30% of the total peak demand (Figure 2c), using in its majority, industrial type consumers, for applying power reductions.

In Figure 2b, the black line represents the maximum power reduction for each RTP consumer, and the grey area is the actual reduced power for each of the RTP consumers that was obtained from the optimization. The authors have discussed the cost function to be applied to the resources payment. In fact, it can impact the scheduling of the resources and consequently the clustering and remuneration. However, due to the small size of the DG units, these usually are paid as fixed price and we want to make an easy remuneration methodology to the resources owners.

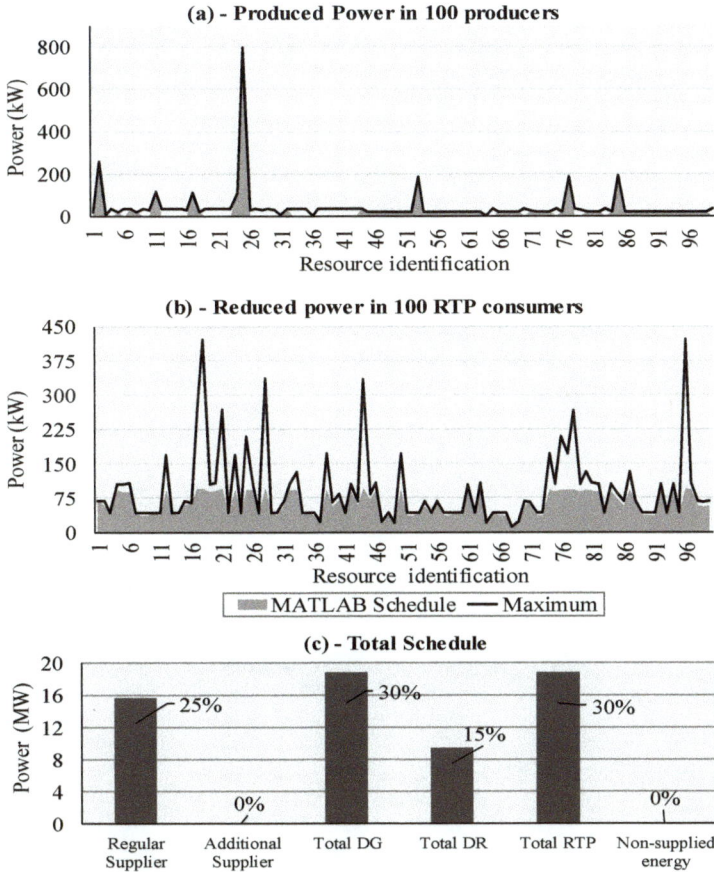

Figure 2. Selected scenario results: (**a**) produced power in 100 producers; (**b**) reduced power in 100 Real-Time pricing (RTP) consumers; (**c**) total schedule.

In this scenario, only 25% of the total power delivered was obtained from external suppliers. Figure 3 presents the comparison between consumption and reduction considering 200 different buses. One can see that some buses do not have any production or consumption allocated to them (for example, 92 to 112). Also, Figure 3 shows that most of the buses export excess energy to other locations and do not just consume it on site. For example, one can see that bus number 148 has a significantly larger consumption when comparing with other buses. However, the production level in bus 148 is too low for it to be able to meet the local demand; therefore, others that have excess energy can supply the remaining energy to bus 148 (ex. 36). The objective function value is 2297.1 m.u., where the minus sign symbolizes that in the total operation management, the resources outcome was profitable to the VPP. Table 2 shows the results for the selected scenario, corresponding to the power schedule in each type of distributed generation and consumer.

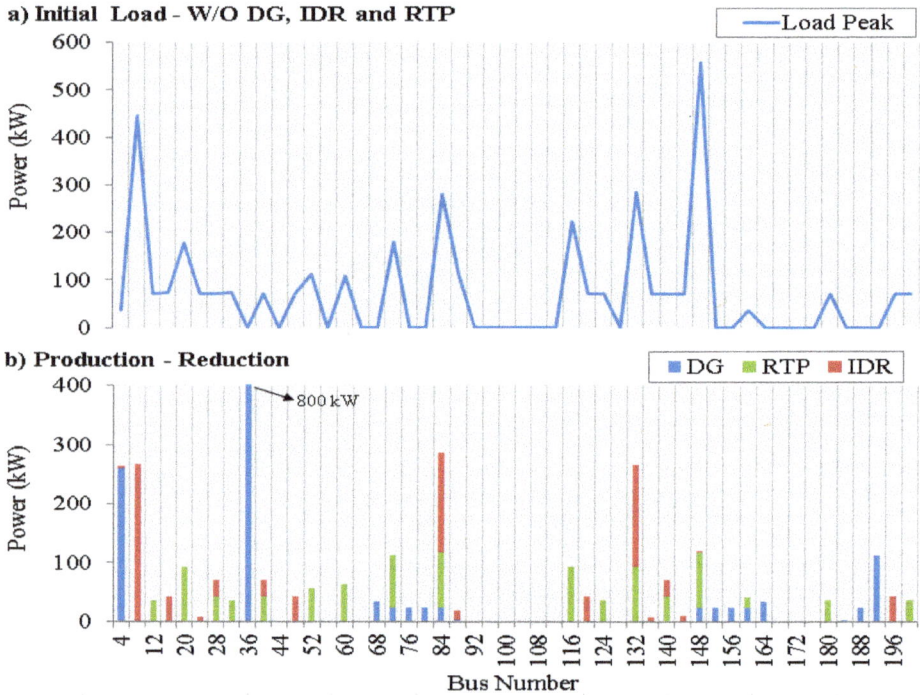

Figure 3. Production versus Consumption: (**a**) initial load—W/O Distributed Generation (DG), Incentive Demand Response (IDR) and RTP; (**b**) production-reduction.

Table 2. Selected Scenario Results.

Resource	Resources Type	Percentage (%)	Power (kW)	Maximum Capacity (kW)	Price (per Type) (m.u./kWh)	Total (kW)
DG	Wind	31.2	5866.09	5866.09	0.071	18,789.13 (25,388.79) 74.0%
	PV	2.5	461.62	7061.28	0.150	
	Biomass	15.0	2826.58	2826.58	0.086	
	Fuel cell	13.1	2457.60	2457.60	0.028	
	MSW	0.3	53.10	53.10	0.056	
	Hydro	1.1	214.05	214.05	0.042	
	CHP	36.8	6910.10	6910.10	0.001	
IDR	DM	14.0	1316.51	5621.61	0.20	9394.56 (17,164.46) 54.7%
	SC	51.0	4789.99	4789.99	0.16	
	MC	35.0	3288.07	6752.85	0.19	
	LC	0.0	0.00	0.00	0.18	
	ID	0.0	0.00	0.00	0.14	
RTP	MC	26.7	5025.70	6752.85	0.20	18,789.13 (27,166.63) 69.2%
	LC	24.3	4568.56	6528.29	0.19	
	ID	48.9	9194.87	13,885.49	0.15	

As demonstrated in Table 2, the distributed generators, IDR and RTP consumers were very used by the scheduling, coming up to more than 50% of their total capacity (DG-74%, IDR-54.7% and RTP-69.2%) supplying a total of 46,972.8 kW, approximately, 46.9 MW of power. The remaining power necessary for the satisfaction of demand is made by the external suppliers, 15,657.6 kW or 15.7 MW.

5.2. All Scenarios—Aggregation and Remuneration

The aggregation was made using the *K*-means function of MATLAB, and also, considering separated clusters, i.e., the resources were clustered considering the same type of resource to group (DG, IDR and RTP). This function allows for obtaining the centroid value for each group and considering 648 different scenarios, there is the possibility of analysing the variation of the centroid values along all these scenarios. The remuneration was performed considering the indices obtained from the aggregation and with the power scheduled in the selected scenario. Figure 4 presents the total power in each of the groups formed for a total number of clusters equal to 4, 6 and 8. Also, performing a quick analysis, one can see that the power per group decreases, as the total number of groups rises.

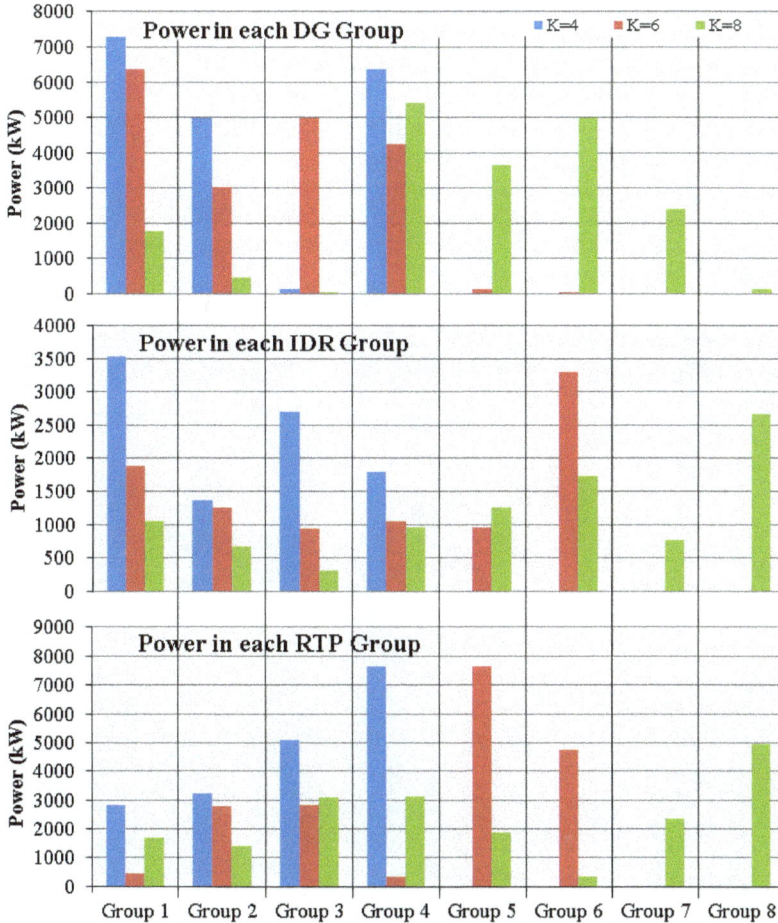

Figure 4. Power in each group.

The centroid values, obtained from the *K*-means algorithm, allows us to estimate a median value of generated/reduced power for each type of resource, making it easier to automatically attribute a group to a new resource. Figure 5 presents the centroid power values for each type of resource. The data series displayed with solid lines are belonging to the left axis and the dashed lines are belonging to the right axis.

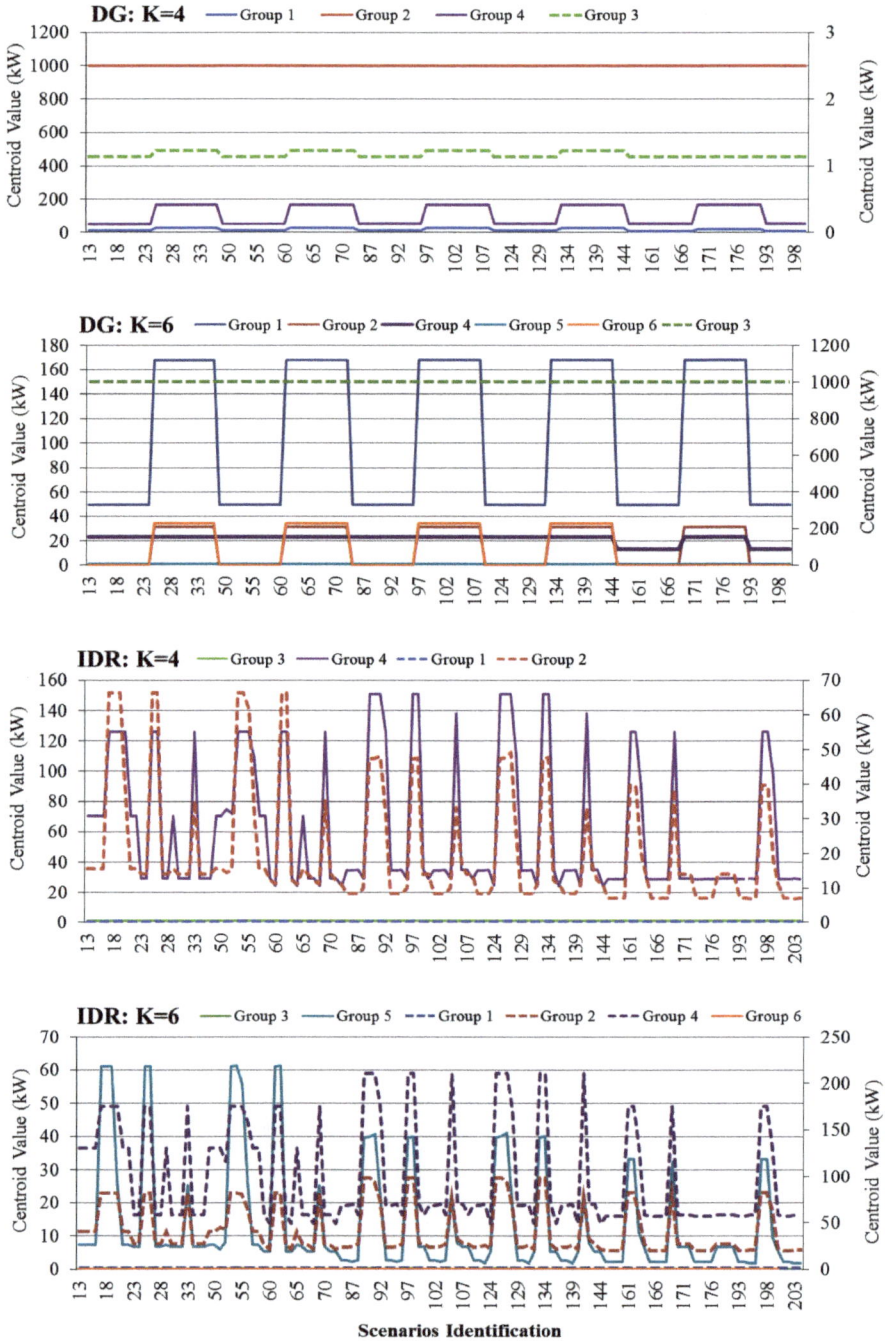

Figure 5. Centroid values of DG and IDR for a total number of clusters of $K = 4$ and $K = 6$.

Through these values, it is possible to estimate a power behaviour in each of the groups, being that group characterized by that centroid power level. This enables the VPP a quick assignment of new resources to the already existing groups. For example, for DG with $K = 4$, there is one group that is composed by large producers (1000 kW, Group 2), another that clusters medium resources between 50 and 200 kW of power (Group 4). Finally, the two remaining groups, 1 (5–30 kW) and 3 (~1 kW), consist of medium-and small-size producers, respectively, that are clustered into the same groups for enabling their power negotiation in the energy markets. In this way, due to the rather different scales of values to be presented, if only one *y*-axis is considered, it would not be possible to see, for example, the curve of group 3; that is why it appears in a secondary *y*-axis on the right. The same evaluation can be performed for the other representations of the centroid value, being $K = 6$. After the aggregation is made, the next step will be to remunerate each distributed producer and consumer in demand response programs.

The tariffs definition was made considering the maximum price of the respective group. Table 3 defines the total remuneration to be done in each group and each cluster scenario ($K = 4$, $K = 6$ and $K = 8$). Given that all the resources in a group will be paid at the highest price, most of these will have an increased remuneration price. For example, consumer #101 has an initial price (before aggregation) of 0.1546 m.u./kWh. After aggregation, it was assigned to group 3, in $K = 4$ clustering scenario, obtaining a final remuneration price of 0.1991 m.u./kWh.

Table 3. Remuneration Results.

Type	# Group	Number of Clusters								
		1	**2**	**3**	**4**	**5**	**6**	**7**	**8**	**Total**
	$K = 4$	1053.4	5.2	19.0	611.3	-	-	-	-	1689
DG	$K = 6$	611.3	426.2	5.2	292.2	19.0	2.9	-	-	1357
	$K = 8$	1.8	65.3	0.0	517.9	254.2	5.2	161.8	19.0	1025
Single DG by type		-	-	-	-	-	-	-	-	950
All DGs by type		-	-	-	-	-	-	-	-	1806
	$K = 4$	740.7	271.3	537.4	344.4	-	-	-	-	1894
IDR	$K = 6$	377.2	241.9	198.4	199.9	190.0	656.7	-	-	1864
	$K = 8$	199.9	142.1	60.0	190.0	241.9	291.5	152.1	447.7	1725
Single IDR by type		-	-	-	-	-	-	-	-	1633
All IDRs by type		-	-	-	-	-	-	-	-	2645

The focus of this method is to find the optimal point between a suitable group power level to be negotiated in the energy markets and a minimization of the costs associated with the energy resources.

In Table 3, in line with the tariffs comparison described in Section 2.4, the total costs for the VPP are presented for the different number of groups under comparison and for the remuneration by type of resource, as in Table 2. According to the initial price defined for each resource of a certain type, the costs are compared for the remuneration of each single resource as scheduled in the scenario, and for the remuneration of all the resources of that type according to the available capacity. In the remuneration of all resources by type, the VPP is paying for the available capacity, which is an approach that can be seen as a fair solution for the participation of resources since, in the event that the VPP is not using the developed methodology, it would accommodate all the available generation from DG and schedule all the DR resources enrolled in the same program.

Using the proposed methodology, considering 8 groups, tariffs are provided resulting in higher costs than simple payment by DG/IDR resource type as single resources, i.e. considering that only the scheduled resources are remunerated. In fact, such higher costs can be advantageous for the VPP since, with the implemented tariff, the actual response from DR/DG resources is more reliable as the resources with same characteristics will be included in the same group and paid at the highest price.

Comparing now the same $K = 8$ costs with the remuneration of all resources available, we can say that if the VPP uses the proposed methodology, the costs will be lower than the ones achieved in a scenario where all the available resources are paid according to the availability. In fact, this approach can lead to an excess of generation in the VPP area, which can be used for selling in the market or integrating in storage means. This comparison shows the interest of the proposed methodology as a whole.

It is important to note that this discussion on the cost comparison is being done for only a selected scenario; it should be replicated by the VPP for each implemented scenario. According to Section 2, the VPP must start by defining the set of scenarios as input for the methodology. This should consider the variability of the available resources by including multiple scenarios. In the aggregation phase, the groups of resources are defined by the clustering method taking into account the complete set of scenarios specified by the VPP. However, the discussion/definition of the better tariff scheme must be done individually for each scenario so the resources scheduled for each group/tariff take into account the obtained optimal scheduling of the resources but that does not necessarily mean that the proposed tariff scheme is always more advantageous than the remuneration by type. The developed methodology provides the means for the VPP to make decisions on tariffs definition, which can be difficult to make in some scenarios.

6. Conclusions

The methodology presented in this paper fits the subject of the remuneration tariffs definition for demand response and distributed generation. The specific focus of the method in this paper is given to an aggregator, a Virtual Power Player that needs to group and remunerate the resources for the operation of its own area or for participation in electricity markets.

The previously published works in the literature address the remuneration of resources by grouping them by consumer's types or by generator primary source. More recent works include the definition of clusters of resources for a single scenario. Moreover, even in the cases that several scenarios are analyzed, they are analyzed one by one.

The methodology developed in this paper goes forward by performing the aggregation of resources according to a complete set of scenarios. In this way, the obtained groups consider attributes behavior along those operation scenarios. With this methodology, an aggregator having a set of resources, in a large or reduced number, is able to obtain the most advantageous tariffs in a planning phase. Additionally, it combines incentive-based and price-based demand response programs. Finally, by all the consumers in a group paying the same price, the highest initial price, one can expect a more accurate response of the resources, integrating their scheduling, aggregation, and remuneration in a single methodology. This methodology is more advantageous than the ones addressing the operation scenarios one by one.

It has been found that the developed methodology provides the VPP with a remuneration scheme that allows to reduce the payment to the consumers providing DR and DG units when comparing to a basis situation in which all the resources of a certain type would be paid according to its availability, disregarding the optimal scheduling of the resources available in each scenario or set of scenarios. Also, it makes it possible for the VPP to perform the definition of tariffs in different time frames. This aspect is very important in the sense that it is very flexible. Also, it allows the VPP to define a large set of scenarios representing the operation conditions that can occur in the VPP operation context.

The developed methodology will be improved as future work regarding several important aspects. Due to space limitation and the need to define a focused paper subject, such aspects have not been addressed in the present paper. The related future work will include: addressing the network capacity in what concerns cables capacity, power losses, and bus voltages; integrating other resources as electric vehicles and storage means; and addressing the availability and controllability of wind and sun-based power plants.

Author Contributions: Conceptualization, P.F. and Z.V.; Methodology, P.F.; Software, J.S.; Validation, P.F.; Formal Analysis, P.F.; Data Curation, J.S.; Writing—Original Draft Preparation, J.S.; Writing–Review and Editing, P.F.

Energies **2018**, *11*, 1987

Funding: This work has received funding from the following projects: SIMOCE Project (ANI | P2020); and from FEDER Funds through the COMPETE program and from National Funds through FCT under the project UID/EEA/00760/2013. This work was also supported by the European Union's Horizon 2020 Research and Innovation Programme under the Marie Sklodowska-Curie Grant Agreement 641794–DREAM-GO Project.

Conflicts of Interest: The authors declare no conflict of interest.

Nomenclature

Variables

$C^{Increase}{}_{RTP}$	Electricity cost increase for RTP
OC	Operation costs
$P^{add}{}_{Supplier}$	Scheduled power for an additional supplier
P_{DG}	Scheduled power for a DG unit
P_{IDR}	Scheduled power reduction for IDR resources
P_{NSP}	Non-supplied power
$P^{Reduct}{}_{RTP}$	Consumption reduction for RTP
$P^{reg}{}_{Supplier}$	Scheduled power for a regular supplier

Parameters

α_{DG}	Maximum allowed DG contribution
α_{IDR}	Maximum allowed IDR contribution
α_{RTP}	Minimum aimed RTP contribution
ε	Price elasticity of demand
$C^{add}{}_{Supplier}$	Additional supplier cost
C_{DG}	Distributed generation cost
C_{IDR}	Incentive based Demand Response cost
$C^{Increase\,Max}{}_{RTP}$	Maximum cost increase in an RTP resource
$C^{Initial}{}_{RTP}$	Initial electricity cost for RTP resources
C_{NSP}	Non supplied power cost
$C^{reg}{}_{Supplier}$	Regular supplier cost
$P^{add\,Max}{}_{Supplier}$	Maximum power from additional suppliers
$P^{add\,Total}{}_{Supplier}$	Maximum total power of additional suppliers
$P^{Initial}{}_{Load}$	Initial consumption of the consumers
$P^{Max}{}_{DG}$	Maximum power schedule in a DG resource
$P^{Max}{}_{IDR}$	Maximum power schedule in a IDR resource
$P^{Max}{}_{RTPDR}$	Maximum total power schedule in a consumer participating in both IDR and RTP
$P^{Min}{}_{DG}$	Minimum power schedule in a DG resource
$P^{Reduct\,Max}{}_{RTP}$	Maximum power schedule in a RTP resource
$P^{reg\,Max}{}_{Supplier}$	Maximum power from a regular supplier
$P^{reg\,Total}{}_{Supplier}$	Maximum allowed total power from all the regular suppliers
$P^{Total\,Max}{}_{DG}$	Maximum allowed total power from all the DG units

Indexes

C	Maximum number of consumers c
P	Maximum number of producers p
S	Maximum number of suppliers s
K	Number of Clusters i
N	Number of Objects j

References

1. MacCormack, J.; Zareipour, H.; Rosehart, W.D. Long-Term Market Equilibrium Model with Strategic, Competitive, and Inflexible Generation. *IEEE Trans. Power Syst.* **2012**, *27*, 2291–2292. [CrossRef]
2. Faria, P.; Vale, Z. Demand response in electrical energy supply: An optimal real time pricing approach. *Energy* **2011**, *36*, 5374–5384. [CrossRef]

3. SEPA. Association for Demand Response & Smart Grid. Available online: http://www.demandresponsesmartgrid.org/ (accessed on 1 May 2018).

4. Behboodi, S.; Chassin, D.P.; Crawford, C.; Djilali, N. Renewable resources portfolio optimization in the presence of demand response. *Appl. Energy* **2016**, *162*, 139–148. [CrossRef]

5. Wang, G.; Zhang, Q.; Li, H.; McLellan, B.C.; Chen, S.; Li, Y.; Tian, Y. Study on the promotion impact of demand response on distributed PV penetration by using non-cooperative game theoretical analysis. *Appl. Energy* **2017**, *185*, 1869–1878. [CrossRef]

6. Siano, P.; Sarno, D. Assessing the benefits of residential demand response in a real time distribution energy market. *Appl. Energy* **2016**, *161*, 533–551. [CrossRef]

7. D'hulst, R.; Labeeuw, W.; Beusen, B.; Claessens, S.; Deconinck, G.; Vanthournout, K. Demand response flexibility and flexibility potential of residential smart appliances: Experiences from large pilot test in Belgium. *Appl. Energy* **2015**, *155*, 79–90. [CrossRef]

8. Balkhair, K.S.; Rahman, K.U. Sustainable and economical small-scale and low-head hydropower generation: A promising alternative potential solution for energy generation at local and regional scale. *Appl. Energy* **2017**, *188*, 378–391. [CrossRef]

9. Gonzalez-Salazar, M.A.; Venturini, M.; Poganietz, W.-R.; Finkenrath, M.; Kirsten, T.; Acevedo, H.; Spina, P.R. Development of a technology roadmap for bioenergy exploitation including biofuels, waste-to-energy and power generation & CHP. *Appl. Energy* **2016**, *180*, 338–352.

10. Jessica, S. Explicit Demand Response in Europe—Mapping the Markets 2017. Available online: http://www.smarten.eu/wp-content/uploads/2017/04/SEDC-Explicit-Demand-Response-in-Europe-Mapping-the-Markets-2017.pdf (accessed on 10 April 2017).

11. Karangelos, E.; Bouffard, F. Towards Full Integration of Demand-Side Resources in Joint Forward Energy/Reserve Electricity Markets. *IEEE Trans. Power Syst.* **2012**, *27*, 280–289. [CrossRef]

12. Hernandez, L.; Baladron, C.; Aguiar, J.M.; Carro, B.; Sanchez-Esguevillas, A.; Lloret, J.; Chinarro, D.; Gomez-Sanz, J.J.; Cook, D. A multi-agent system architecture for smart grid management and forecasting of energy demand in virtual power plants. *IEEE Commun. Mag.* **2013**, *51*, 106–113. [CrossRef]

13. Palensky, P.; Dietrich, D. Demand Side Management: Demand Response, Intelligent Energy Systems, and Smart Loads. *IEEE Trans. Ind. Inform.* **2011**, *7*, 381–388. [CrossRef]

14. Chassin, D.P.; Rondeau, D. Aggregate modeling of fast-acting demand response and control under real-time pricing. *Appl. Energy* **2016**, *181*, 288–298. [CrossRef]

15. Heffner, G.C.; Goldman, C.A.; Moezzi, M.M. Innovative approaches to verifying demand response of water heater load control. *IEEE Trans. Power Deliv.* **2006**, *21*, 388–397. [CrossRef]

16. Morais, H.; Pinto, T.; Vale, Z.; Praca, I. Multilevel Negotiation in Smart Grids for VPP Management of Distributed Resources. *IEEE Intell. Syst.* **2012**, *27*, 8–16. [CrossRef]

17. Xu, R.; Wunsch, D. Survey of clustering algorithms. *IEEE Trans. Neural Netw.* **2005**, *16*, 645–678. [CrossRef] [PubMed]

18. Ramos, S.; Duarte, J.M.M.; Soares, J.; Vale, Z.; Duarte, F.J. Typical load profiles in the smart grid context—A clustering methods comparison. In Proceedings of the 2012 IEEE Power and Energy Society General Meeting, San Diego, CA, USA, 22–26 July 2012; pp. 1–8.

19. Duarte, F.J.; Duarte, J.M.M.; Ramos, S.; Fred, A.; Vale, Z. Daily wind power profiles determination using clustering algorithms. In Proceedings of the 2012 IEEE International Conference on Power System Technology (POWERCON), Auckland, New Zealand, 30 October–2 November 2012; pp. 1–6.

20. Herrero, I.; Rodilla, P.; Batlle, C. Electricity market-clearing prices and investment incentives: The role of pricing rules. *Energy Econ.* **2015**, *47*, 42–51. [CrossRef]

21. Zobaa, A.F.; Bansal, R.C. *Handbook of Renewable Energy Technology*; World Scientific: Singapore, 2011.

22. Gupta, R. Economic implications of non-utility-generated wind energy on power utility. *Comput. Electr. Eng.* **2002**, *28*, 77–89. [CrossRef]

23. Salinas, S.; Li, M.; Li, P. Multi-Objective Optimal Energy Consumption Scheduling in Smart Grids. *IEEE Trans. Smart Grid* **2013**, *4*, 341–348. [CrossRef]

24. Vrba, P.; Mark, V.; Siano, P.; Leitão, P.; Zhabelova, G.; Vyatkin, V.; Strasser, T. A Review of Agent and Service-Oriented Concepts Applied to Intelligent Energy Systems. *IEEE Trans. Ind. Inform.* **2014**, *10*, 1890–1903. [CrossRef]

25. Parvania, M.; Fotuhi-Firuzabad, M.; Shahidehpour, M. ISO's Optimal Strategies for Scheduling the Hourly Demand Response in Day-Ahead Markets. *IEEE Trans. Power Syst.* **2014**, *29*, 2636–2645. [CrossRef]

26. Faria, P.; Vale, Z. Remuneration Structure Definition for Distributed Generation Units and Demand Response Participants Aggregation. In Proceedings of the 2014 IEEE PES General Meeting I Conference & Exposition, National Harbor, MD, USA, 27–31 July 2014; pp. 1–5.

27. Faria, P.; Spínola, J.; Vale, Z. Aggregation and Remuneration of Electricity Consumers and Producers for the Definition of Demand-Response Programs. *IEEE Trans. Ind. Inform.* **2016**, *12*, 952–961. [CrossRef]

28. Spinola, J.; Faria, P.; Vale, Z. Scheduling and aggregation of distributed generators and consumers participating in demand response programs. In Proceedings of the 2015 IEEE Eindhoven PowerTech, Eindhoven, The Netherlands, 29 June–2 July 2015.

29. Spínola, J.; Faria, P.; Vale, Z. Remuneration of distributed generation and demand response resources considering scheduling and aggregation. In Proceedings of the 2015 IEEE Power & Energy Society General Meeting, Denver, CO, USA, 26–30 July 2015; pp. 1–5.

30. Arnold, R.A. *Economics*; Cengage Learning: Boston, MA, USA, 2008.

31. Faria, P.; Soares, J.; Vale, Z.; Morais, H.; Sousa, T. Modified Particle Swarm Optimization Applied to Integrated Demand Response and DG Resources Scheduling. *IEEE Trans. Smart Grid* **2013**, *4*, 606–616. [CrossRef]

energies

MDPI

Article

SwarmGrid: Demand-Side Management with Distributed Energy Resources Based on Multifrequency Agent Coordination

Manuel Castillo-Cagigal [1], Eduardo Matallanas [1], Estefanía Caamaño-Martín [2] and Álvaro Gutiérrez Martín [1,*]

[1] E.T.S. Ingenieros de Telecomunicación, Universidad Politécnica de Madrid, Av. Complutense 30, 28040 Madrid, Spain; manuel.castillo@upm.es (M.C.-C.); eduardo.matallanas@upm.es (E.M.)
[2] Instituto de Energía Solar, E.T.S. Ingenieros de Telecomunicación, Universidad Politécnica de Madrid, Av. Complutense 30, 28040 Madrid, Spain; estefan@ies-def.upm.es
* Correspondence: a.gutierrez@upm.es; Tel.: +34-91-067-19-00

Received: 3 August 2018; Accepted: 16 September 2018; Published: 18 September 2018

Abstract: This paper focuses on a multi-agent coordination for demand-side management in electrical grids with high penetration rates of distributed generation, in particular photovoltaic generation. This coordination is done by the use of swarm intelligence and coupled oscillators, proposing a novel methodology, which is implemented by the so-call SwarmGrid algorithm. SwarmGrid seeks to smooth the aggregated consumption by considering distributed and local generation by the development of a self-organized algorithm based on multifrequency agent coordination. The objective of this algorithm is to increase stability and reduce stress of the electrical grid by the aggregated consumption smoothing based on a frequency domain approach. The algorithm allows not only improvements in the electrical grid, but also increases the penetration of distributed and renewable sources. Contrary to other approaches, this objective is achieved anonymously without the need for information exchange between the users; it only takes into account the aggregated consumption of the whole grid.

Keywords: demand-side management; multi-agent system; distributed coordination; distributed energy resources; swarm intelligence

---

1. Introduction

Historically, electricity power systems have been designed by following a vertical integration scheme: large power generators supply energy to multiple consumers through a hierarchical transport and distribution network. Nowadays, Distributed Energy Resources (DERs) are changing this situation. They consist of a wide range of local generators and storage systems, which are geographically dispersed, generally close to consumption centers and locally managed [1]. They bring a new conception of electric power systems, with the local user becoming not only a consumer, but also a generator.

Distributed Generation (DG) is a generation structure where small generators are spread over the grid, usually on the consumer side. Distributed Generation (DG) has been growing over the last two decades, arousing a major interest among electric power system planners and operators, energy policy makers and regulators, as well as developers and consumers [2].

On the other hand, the management of the consumption on electrical grids is receiving increasing attention by research and industry with Demand-Side Management (DSM) [3,4]. One of the main objectives of DSM is to smooth the aggregated consumption by reducing the difference between the maximum and minimum consumption power [5]. This smoothing requires the coordination of thousands or even millions of elements spread over a certain area. The coordination of these elements, to properly use the available resource, becomes a complex task, which is addressed by the smart grids and must be tackled from the multi-agent and distributed system point of view [6–9].

Therefore, DSM in smart grids enhances its capabilities regarding the classical electrical grids. The convergence of information and communications technologies (ICTs) with power system engineering allows new levels of automation. DSM in this context is usually formulated as optimization problems, which may be solved by various approaches [10–17]. These approaches are also moving towards a new paradigm where the consumer is the center of the grid and goes from being a passive element that only consumes to an active element involved in the management [18].

The interaction between agents is a key factor in this coordination, which can become highly complex depending on both internal and external factors and is affected by different elements of the environment. Previous works have studied multi-agent coordination with the presence of noise or disturbances [19]. However, the environment may affect the coordination process, not only as noise or disturbance, but the nature of the coordination process itself. In such a case, the coordination between agents may be considered as an adaptive process to the environment where they are. This adaptive process may be designed so that the environment meets certain predefined characteristics or goals.

In this paper, the environment is an electrical grid, and SwarmGrid focuses on the problem of smoothing the aggregated consumption in an environment where Distributed Generation (DG) is part of the system. SwarmGrid takes advantage of one main feature of the consumption in the electrical grids: its periodicity. The consumption profile almost repeats every day, week or season. SwarmGrid focuses on the coordination of distributed multi-agent systems from the frequency domain perspective, based on previous works on multifrequency coupled oscillators [20]. This work develops an algorithm that is able to coordinate an n-elements system by using the periodicity of the environment.

The remainder of this paper is as follows. The environment approach and the definition of the multi-agent electrical system are presented in Section 2. The SwarmGrid algorithm is defined in Section 3. Section 4 describes the simulated environment and its generation and consumption parameters. In Section 5, experiments with different penetrations of Distributed Generation (DG) are evaluated. A discussion is presented in Section 6, and the paper is concluded in Section 7.

2. Environment Definition

2.1. Facility

A facility is defined as an electric power system that belongs to a particular consumer. In this paper, it may be composed of two types of elements: generation and consumption. The proposed algorithm will control the consumption in order to modify the power flow between facilities and the grid.

2.1.1. Consumption

Consumption is the most heterogeneous part of an electrical power system. There is a huge variety of devices that consume electricity with different characteristics: nominal power (from an LED lamp to industrial ovens), intermittent consumption (e.g., a fridge pump that activates periodically), peak power (e.g., the use of a drill), etc. For the implementation of Demand-Side Management (DSM) techniques and the deployment of the SwarmGrid algorithm, it is particularly interesting to classify the consumption by means of its controllability. In this way, consumption can be divided into fixed, deferrable and elastic types [21]. The electric vehicle can be also included as elastic consumption that can be positive when it is charging or negative when it is discharging. In this aspect, the SwarmGrid algorithm could be also used for electric vehicle fleet charging management [22].

2.2. Environment

In this paper, the electrical grid is defined as a single node disregarding losses and transmission delays. The electrical grid contains facilities of different users, which contain in turn different electrical loads. The consumption in time of the facility i is represented as the power signal $p_i(t)$. The sum of all facilities in the electrical grid is the aggregated consumption such that $p(t) = \sum_{i=1}^{M} p_i(t)$, where M is the number of facilities in the electrical grid.

Each facility is divided into a controllable and a non-controllable part. The non-controllable ($p_i^{nc}(t)$) part represents all consumptions that cannot be managed, that is a fixed consumption. The controllable part ($p_i^c(t)$) represents consumption that can be managed, deferrable or elastic.

Therefore, the aggregated consumption $p(t)$ of the complete electrical system is defined as the total energy consumed by all facilities, which can be decomposed into a sum of controllable or non-controllable signals, so that:

$$p(t) = p^{nc}(t) + p^c(t) = \sum_{i=1}^{M} p_i^{nc}(t) + \sum_{i=1}^{M} p_i^c(t) \tag{1}$$

2.3. Metrics

Two different metrics will be used to study the behavior of the network, the self-consumption factor and the crest factor.

2.3.1. Self-Consumption Factor

The self-consumption factor (ζ) represents the fraction of the electrical energy consumed by the loads, which is only supplied by the local generation sources [21,23]:

$$\zeta = \frac{E_{PV,L}}{E_L} \tag{2}$$

where $E_{PV,L}$ is the energy directly supplied by the Photovoltaic (PV) generator to the loads and E_L is the total energy consumed by the loads. The range of ζ is $[0, 1]$, because this factor is normalized by the total consumption of the local facility.

2.3.2. Crest Factor

In order to analyze the variability of the aggregated consumption, the crest factor C has been used. It is a measure of a waveform, showing the ratio of peak values to the average value, such that:

$$C = \frac{|x|^{peak}}{x^{rms}}$$

$$x^{rms} = \sqrt{\frac{x_1^2 + x_2^2 + \cdots + x_N^2}{N}} \tag{3}$$

where N is the number of samples taken from the aggregated consumption, $|x|^{peak}$ is the absolute value of the maximum peak and x^{rms} is the root mean square. The crest factor makes reference to a concrete time interval in which the signals are evaluated together with a concrete sample period. This factor is in the range $[1, \infty)$, one being the optimal value where the aggregated consumption is completely flat. In this paper, the time interval is denoted with a subscript, for example C_{year}, C_{month}, C_{week} and C_{day}.

3. SwarmGrid

Two specific objectives can be tackled when trying to schedule loads in the electric system: to maximize the use of local generation or to maximize the aggregated consumption smoothing. The first one is a local objective, taking into account local generation and consumption without the need for any information or interest in the environment. This objective maximizes the self-consumption factor. On the other hand, the second objective is a global objective, which takes into account the aggregated consumption to reduce the difference between peaks and valleys, being related with the crest factor.

Unfortunately, both objectives can be very antagonistic. The proposed SwarmGrid algorithm is designed to meet a compromise between both objectives by scheduling deferrable loads in every facility of the system. Therefore, SwarmGrid schedules loads in local facilities, which affect the global shape of the aggregated consumption by means of a self-organized coordination.

3.1. Modeling

In the SwarmGrid algorithm, the environment is modeled from the signal processing point of view dividing the consumption into its periodic components. In this way, the consumption smoothing problem can be seen as a frequency filtering where the controllable consumption filter is the non-controllable part. When extended to multiple agents in the system, this process is performed in a distributed way. Therefore, the controllable part of every facility can be represented as a sum of sinusoidals, such that:

$$p_i^c(t) = \sum_{j=1}^{N} A_j^c sin(\omega_j^c t + \phi_j^c(t)) \tag{4}$$

where A_j^c is the amplitude, ω_j^c the frequency and ϕ_j^c the phase of each component of each of the N signals in which $p_i^c(t)$ is decomposed.

For the sake of privacy, facilities do not have a direct communication system, and they can only coordinate by observing the aggregated signal $p(t)$. Therefore, the shape of $p_i^c(t)$ must be controlled in a distributed way such that A_i^c and ϕ_i^c are modified to smooth the aggregated consumption. This modeling is exhaustively described in [20]. This approach takes advantage of the periodic behavior of the consumption allowing load schedules without time limit and in a stable way.

3.2. Load Scheduling

In all facilities, users request to activate some loads at specific times. If these loads are fixed, they must be instantaneously activated. However, if they are deferrable or elastic, they may be scheduled in terms of time and amplitude. In the SwarmGrid algorithm, the scheduling consists of the assignment of an activation time $t_{i,j}^{act}$ for each deferrable load controlled, where the subscript i denotes that the deferrable load belongs to facility i and the subscript j is an identifier of the deferrable load. When the user requires performing a deferrable load, he/she must indicate which load must be performed together with its running range $[t_{i,j}^{beg}, t_{i,j}^{end}]$. The algorithm is responsible for scheduling the task deciding $t_{i,j}^{act}$, satisfying that this time instant is within the running range. The consumption of these deferrable loads is expressed by the following equation:

$$p_{i,j}^{def}(t) = \begin{cases} P & t \in [t_{i,j}^{act}, t_{i,j}^{act} + \tau] \\ 0 & t \notin [t_{i,j}^{act}, t_{i,j}^{act} + \tau] \end{cases} \tag{5}$$

As previously mentioned, the scheduling consists of the assignment of $t_{i,j}^{act}$ for each incoming deferrable load. This assignment is performed by using a local consumption pattern, denoted by $f_i(t)$. This pattern defines the objective shape of the local consumption. It means that the deferrable load should be scheduled such that the local consumption has the same shape as $f_i(t)$. In order to achieve this objective, the local consumption pattern is defined as a probability density function

$pdf(t_{i,j}^{act})$. Therefore, the activation time is considered as a random variable, which takes on a value with a certain probability described by $pdf(t_{i,j}^{act})$. The following equation connects $f_i(t)$ with $pdf(t_{i,j}^{act})$:

$$pdf(t_{i,j}^{act}) = \frac{1}{K} f_i(t_{i,j}^{act}) \quad \text{for} \quad t_{i,j}^{act} \in [t_{i,j}^{beg}, t_{i,j}^{end}]$$

$$K = \int_{t_{i,j}^{beg}}^{t_{i,j}^{end}} f_i(t)dt \tag{6}$$

where K is a normalizing factor, which is required for $pdf(t_{i,j}^{act})$, being a probability density function.

However, the local consumption patterns have two objectives on the SwarmGrid algorithm: to smooth the aggregated consumption and to increase the self-consumption. In order to achieve these objectives, $f_i(t)$ has been divided into two parts. The first part is a function based on a multi-agent coordination algorithm, Multi-Frequency Coupled Oscillators (MuFCO) [20]. This function is able to smooth the aggregated consumption regardless of the number of elements that participate. The second part is the Photovoltaic (PV) generation forecast function $f_i^{PV}(t)$. This function is able to increase the self-consumption of the local facility. Both parts are combined in the following equation:

$$f_i(t) = \beta \cdot f_i^{PV}(t) + (1 - \beta) \cdot f_i^{MuFCO}(t) \tag{7}$$

where $\beta \in [0, 1]$ is a parameter that regulates the importance of the previous two functions in $f_i(t)$. This means that if $\beta \to 0$, the local consumption pattern tends to $f_i^{MuFCO}(t)$ and the scheduling tends to smooth the aggregated consumption. On the other hand, if $\beta \to 1$, the local consumption pattern tends to $f_i^{PV}(t)$ and the scheduling tends to increase the self-consumption.

3.3. Scheduling with MuFCO

If the SwarmGrid algorithm is configured with $\beta = 1$, it only uses $f_i^{MuFCO}(t)$ to generate the local consumption pattern. The Multi-Frequency Coupled Oscillators (MuFCO) algorithm is in charge of generating consumption patterns to smooth the aggregated consumption where n individuals participate. n has no upper theoretical limit, getting a more sinusoidal form of the signal as the number of individuals increases [20]. This smoothing process is summarized in the reduction of the amplitude of $p(t)$ through the control of amplitudes A_j^c and phases ϕ_j^c of every facility.

Thus, a facility is considered as an individual oscillator, which is shifted in magnitude, such that:

$$f_i^{MuFCO}(t) = \frac{1 + x_i(t)}{2} = \frac{1 + sin(\omega_i t + \phi_i)}{2} \tag{8}$$

Therefore, the sum of all facilities has the same properties as the collective of oscillators generated by the Multi-Frequency Coupled Oscillators (MuFCO) algorithm. The facilities observe the aggregated consumption signal and generate sinusoidal functions such that they adapt to the non-controllable consumption and smooth the aggregated consumption.

The sinusoidal function for a local facility i is directly used as a probability density function when $\beta = 1$, such that:

$$pdf(t_{i,j}^{act}) = \frac{1}{K^{MuFCO}} f_i^{MuFCO}(t_{i,j}^{act}) \quad \text{for} \quad t_{i,j}^{act} \in [t_{i,j}^{beg}, t_{i,j}^{end}]$$

$$K^{MuFCO} = \int_{t_{i,j}^{beg}}^{t_{i,j}^{end}} f_i^{MuFCO}(t)dt \tag{9}$$

3.4. Scheduling with PV Forecast

In order for the SwarmGrid algorithm to improve the self-consumption of the locally-generated electricity, a local energy generation forecast should be included in the scheduling process [24].

In this section, the SwarmGrid algorithm is configured with $\beta = 0$; thus, it only uses $f_i^{PV}(t)$ to generate the local consumption pattern. $f_i^{PV}(t)$ is the normalized Photovoltaic (PV) generation forecast for the facility i. This function is normalized by the maximum generation of the Photovoltaic (PV) generator. This means that $f_i^{PV}(t)$ is in the range $[0,1]$. The normalized Photovoltaic (PV) generation forecast for a local facility i is directly used as a probability density function when $\beta = 0$, such that:

$$pdf(t_{i,j}^{act}) = \frac{1}{C} f_i^{PV}(t_{i,j}^{act}) \quad \text{for} \quad t_{i,j}^{act} \in [t_{i,j}^{beg}, t_{i,j}^{end}]$$

$$C = \int_{t_{i,j}^{beg}}^{t_{i,j}^{end}} f_i^{PV}(t)dt \tag{10}$$

3.5. Real-Time Execution

The SwarmGrid algorithm has been designed to run in real-time. It is locally executed when a new deferrable load is required by the user, and this event can occur at any time. Thus, the SwarmGrid algorithm is an asynchronous algorithm. When a new deferrable load is required, this algorithm is executed. The algorithm begins by obtaining the information of the deferrable load to be scheduled. This information comes directly from the user. Once the user has indicated the required information, the SwarmGrid algorithm calculates $f_i^{MuFCO}(t)$ from the Multi-Frequency Coupled Oscillators (MuFCO) algorithm. This means that the SwarmGrid algorithm asks for ω_i and ϕ_i for the Multi-Frequency Coupled Oscillators (MuFCO) algorithm and calculates the sinusoidal function; see Equation (9). $f_i^{PV}(t)$ is also calculated through Equation (10). The local consumption pattern $f_i(t)$ is calculated by using Equation (7). Once $f_i(t)$ is obtained, $pdf(t_{i,j}^{act})$ can be calculated in the range $[t_{i,j}^{beg}, t_{i,j}^{end}]$ by using Equation (6). Finally, $t_{i,j}^{act}$ is calculated for this deferrable load. The value of $t_{i,j}^{act}$ is taken on from $pdf(t_{i,j}^{act})$ as the execution of a random variable.

4. Simulator Definitions

4.1. GridSim Simulation Framework

All examples of the SG algorithm operation have been performed in the GridSim simulation framework (https://github.com/Robolabo/gridSim). GridSim is a simulator developed to analyze the power balances on a virtual electrical grid. The grid is composed of lines. A line represents a set of nodes with the same characteristics. The nodes are complex elements connected to the lines, which can be equipped with different types of consumption, DERs and control systems. These nodes can represent a single device to a fully-equipped facility. In addition, a base consumption function can be added to the grid. It represents a non-controllable consumption, which is added to the aggregated consumption of the simulated grid.

All experiments have been done through simulations of an electrical grid divided into 600 nodes. Each node is a facility that is equipped with a Photovoltaic (PV) generator, a storage system and local consumption. The time step of simulation is 1 min. In this analysis, a virtual user is used in order to generate the consumption profile. This user belongs to a concrete facility; this means that if there are 600 nodes, there are 600 virtual users.

4.2. Virtual Users

The consumption of the local facilities (nodes) is created by virtual users in the GridSim simulator. In this analysis, the virtual users only create deferrable loads. Thus, the user of each facility requires a number of deferrable loads during the execution of any simulation. This requirement is an asynchronous event, which represents a new deferrable load creation in the local facility. It could happen at any time step of the simulation. Every new deferrable load is created at a concrete time instant denoted by $t_{i,j}^{NDL}$. If no controller schedules the load, it is activated when it is created such that $t_{i,j}^{act} = t_{i,j}^{NDL}$. Figure 1 shows an example of this procedure for two facilities where one facility is equipped with the SwarmGrid algorithm and the other has no controller.

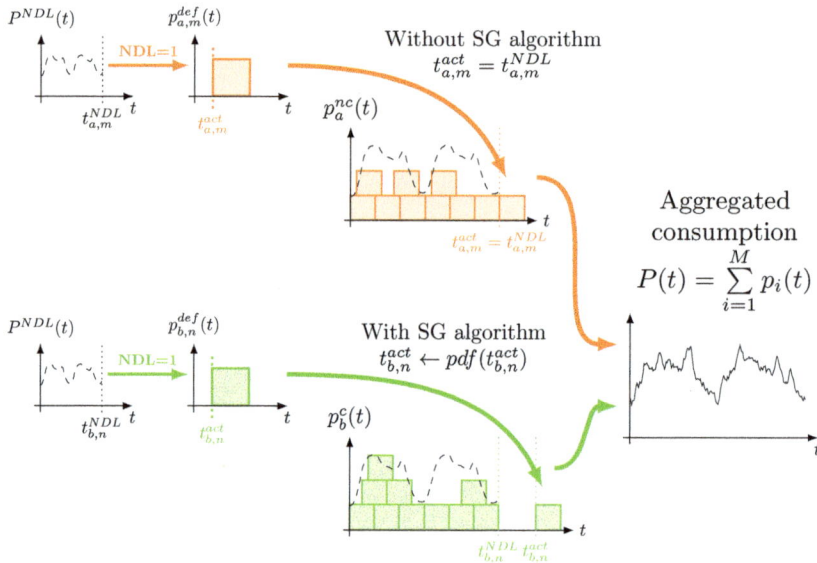

Figure 1. Conceptual example of the deferrable consumption created by the virtual users. SG, SwarmGrid.

Virtual users create new deferrable loads randomly. There is a probability density function that defines the probability that a new deferrable load is created at a given time instant. This function is denoted by $P^{NDL}(t)$. Thus, the creation of a new deferrable load is a binary random variable NDL. Every time step of the simulation, it is checked if there is a new deferrable load $P^{NDL}(t) \rightarrow NDL = 1$. For the analysis presented in this paper, $P^{NDL}(t)$ is the real aggregated consumption of peninsular Spain during 2017. This implies that the consumption of the virtual users has a similar shape as this aggregated consumption. SwarmGrid controllers modify the activation time of the deferrable loads; thus, it is satisfied that $t_{i,j}^{act} \neq t_{i,j}^{NDL}$.

Moreover, deferrable loads are modeled as energy packets. These packets consume a constant nominal power P during a certain time interval τ; see Figure 1. To perform the following analysis, the nominal power is $P = 50$ MW, and duration is $\tau = 60$ min. In addition, it is considered that the deferrable loads have a running range of one day (1440 min): $t_{i,j}^{beg} = t_{i,j}^{NDL}$ and $t_{i,j}^{end} = t_{i,j}^{NDL} + 1440$. Each user creates around 12,200 deferrable loads per year. Thus, a local facility consumes around 410 GWh per year. The sum of all local facilities has a yearly energy consumption of around 246 TWh, equal to the Spanish yearly consumption. In a real electrical grid, the facilities do not consume as much electrical energy as the simulated facilities. These energy amounts have been chosen because the grid has been divided into 600 facilities to reduce the computing load. Simulating an electrical grid with thousands or millions of facilities causes a very high computation time. On the other hand, 600 facilities are enough to observe the coordination and smoothing capacity of the SG algorithm.

4.3. Local Generation

Local, specifically PV, generation should be included in the facilities in order to use the local controllers. In the simulated experiment, the Photovoltaic (PV) generation is considered as a negative consumption [25]. If a facility has a generation excess of 1 kW, it is consuming -1 kW. This consideration allows to introduce the Photovoltaic (PV) generation in the aggregated consumption and observe its effects. As a final aspect to be appreciated, the aggregated consumption is always positive. This means that if the Photovoltaic (PV) generation is greater than the whole consumption of the grid, the aggregated consumption will be zero instead of taking negative values.

The generation profiles from six different cities of Spain have been used: Cáceres, Ciudad Real, Logroño, Madrid, Santiago and Soria. The calculation of the Photovoltaic (PV) generation profiles has been done by using the real irradiance data from these cities during 2017. This irradiance has been translated to AC power by using a model of the Photovoltaic (PV) generators and weather forecasting of an energy self-sufficient solar house located in Madrid [24]. There are 100 virtual facilities (nodes) deployed in each city. The combination of these cities allows to consider different climate regions. This heterogeneous generation is representative of the national generation.

The maximum nominal Photovoltaic (PV) power generation in each facility is 240 MW, such that the total power generation of all facilities is 144 GW. This generation power corresponds to a yearly energy generation of around 246 TWh, which is the same energy amount as the yearly peninsular energy consumption of Spain.

5. Simulation Experiments

In this section, the SwarmGrid algorithm's operation is analyzed. This analysis focuses on validating the SwarmGrid algorithm and studying the improvement that it brings to the electrical grid. All analyses presented in this Section are based on the 2017 consumption data of the Spanish grid. This section analyzes two different scenarios:

- The grid without DG: in this scenario, there are no Distributed Energy Resources (DERs), and the SwarmGrid algorithm just manages the consumption of the electrical grid to smooth the aggregated consumption. The SwarmGrid algorithm schedules the deferrable loads only by using the sinusoidal patterns generated by Multi-Frequency Coupled Oscillators (MuFCO); thus, $\beta = 0$ for this scenario.
- The grid with DG: in this scenario, local Photovoltaic (PV) generators are included in the local facilities as Distributed Generation (DG) without electrical storage.

5.1. The Grid without DG

In this scenario, the facilities are purely consumers without local generation. This is the common scenario if the SwarmGrid algorithm is deployed in buildings without Distributed Energy Resources (DERs) where the classical Demand-Side Management (DSM) mechanisms are performed. Part of this consumption is controlled by the SwarmGrid algorithm, and the other part is directly set by the user. The percentage of consumption controlled by the algorithm is denoted by ρ^{ctr}. For example, if $\rho^{ctr} = 50\%$ and the yearly consumption is 246 TWh, this means that the SwarmGrid algorithm controls 123 TWh of deferrable loads along the year. Notice that $\beta = 0$ for all experiments performed in this scenario because there is no local generation.

A campaign of experiments has been performed to study how the percentage of consumption controlled by the SwarmGrid algorithm affects the crest factors of the aggregated consumption. For each ρ^{ctr} value, 100 experiments have been performed with different seeds of the random number generator. An experiment consists of the simulation of the previously explained electrical grid during one year and a half (788,400 min) with a concrete ρ^{ctr} value and a concrete seed. The first half of the year is used to adapt the SwarmGrid algorithm to the grid; after that, the crest factors are calculated for the remainder of the year.

Figure 2 shows the development of the crest factors for different percentages of consumption controlled by the SwarmGrid algorithm. For each ρ^{ctr} value, the mean and the maximum and minimum values have been calculated from the 100 different seeds. Figure 2 is divided into four graphs where each graph plots the statistical analysis for each type of crest factor. The common trend is that the greater the amount of consumption controlled by the algorithm, the lower the crest factors are. In addition, the dispersion of the experiments is higher the greater is the range of time covered by the crest factor. For instance, the difference between *max* and *min* is much lower for the daily crest factor average than for the yearly crest factor average. In all graphs, the crest factors achieve a plateau value when the SwarmGrid algorithm controls over 50% of the whole consumption. The following conclusions come from these results:

- Regardless of ρ^{ctr}, the algorithm reduces the electrical grid variability: in general, the higher ρ^{ctr}, the lower the crest factors. This implies that the smoothing objective is met by the SwarmGrid algorithm. This relationship is satisfied until the crest factors achieve the plateau value. When this value is achieved, the algorithm self-organizes, and the crest factors may not be reduced.
- The algorithm self-organizes: For $\rho^{ctr} < 50\%$ values, the controllable part of the grid's consumption is lower than the non-controllable one. The SwarmGrid algorithm adapts to the non-controllable consumption, reducing the variability of the grid. On the other hand, when $\rho^{ctr} > 50\%$, the SwarmGrid algorithm should also self-organize. This means that the controllable consumption must coordinate with itself to achieve a neutral effect. For example, if $\rho^{ctr} = 60\%$, there are 240 facilities without the SwarmGrid algorithm; thus, 240 facilities with the SwarmGrid (SG) algorithm are required at least to fully smooth their consumption. The other 120 facilities with the SwarmGrid algorithm should coordinate among each other to provoke a constant aggregated consumption. This feature is verified because the crest factors do not increase for $\rho^{ctr} > 50\%$. This implies that the controlled consumption is able to coordinate to reduce the crest factors for $\rho^{ctr} < 50\%$ values and to maintain them for $\rho^{ctr} > 50\%$.

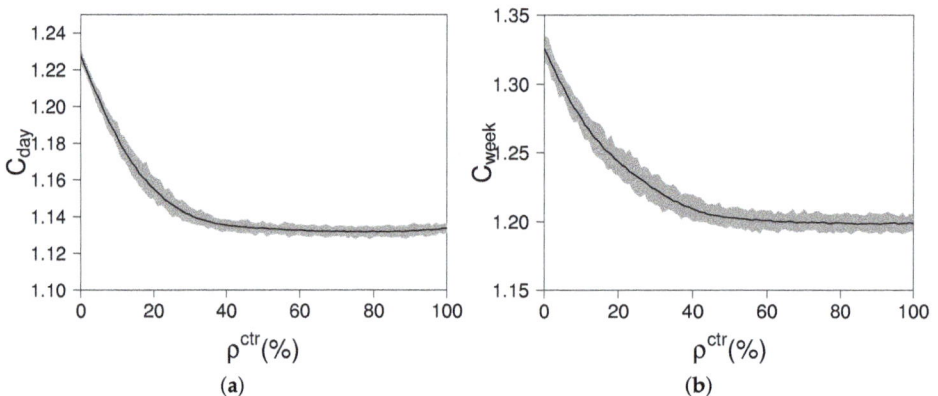

Figure 2. *Cont.*

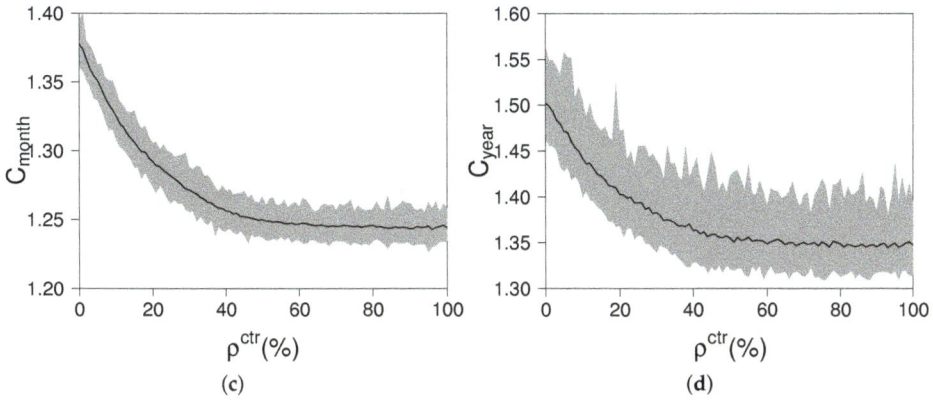

Figure 2. Development of the crest factors for different percentages of consumption controlled by the SwarmGrid algorithm. The solid line represents the mean of the crest factors for the 100 simulated seeds. The shaded area is between the maximum and the minimum value obtained from the 100 simulated seeds: (a) $\bar{C}_{day}$; (b) $\bar{C}_{week}$; (c) $\bar{C}_{month}$ and (d) C_{year}.

5.2. The Grid with DG

The presence of Distributed Generation (DG) in the electrical grid may increase the variations in the aggregated consumption. In this scenario, the effects of the Photovoltaic (PV) generation together with the SwarmGrid algorithm are analyzed. The generation profiles from six different cities of Spain have been used. There are 100 virtual facilities (nodes) for each city. The combination of these cities allows different climate regions to be considered. The maximum nominal Photovoltaic (PV) power generation is 240 MWp in each facility, such that the sum of the generation of all facilities is 144 GWp. This generation power corresponds to a yearly energy generation of around 246 TWh, which is the same energy amount as the yearly energy consumption of Spain. In this section, the effects of the Photovoltaic (PV) penetration in the electrical grid have also been studied by using the Photovoltaic (PV) penetration factor ρ^{PV}.

The SwarmGrid algorithm modifies the aggregated consumption shape and the self-consumption of the local facilities depending on the β parameter when there is Photovoltaic (PV) generation. A campaign of experiments has been performed to study how ρ^{PV} and β affect the crest factors of the aggregated consumption and the self-consumption. In this campaign, $\rho^{ctr} = 100\%$ for all experiments so that the effect of the β parameter could be better observed. Different combinations of ρ^{PV} and β have been studied. For each combination of these parameters, 30 experiments have been performed with different seeds of the random number generator. An experiment consists of the simulation of the previously explained electrical grid during one year and a half (788,400 min) with a concrete combination of ρ^{PV} and β and a concrete seed. The first half of the year is used to adapt the SwarmGrid algorithm to the grid, and after that, the crest factors are calculated for the remainder of the year. The crest factors and the self-consumption of the local facilities are calculated for each experiment. The SwarmGrid algorithm operates with the tuned parameters to reduce the daily variability obtained in [20]: $W = 16$, $T_{smp} = 90$ min, $K = -0.03$ and $P_{switch} = 0.02$.

Figure 3 shows the development of the crest factors for different combinations of ρ^{PV} and β with $\rho^{ctr} = 100\%$. In general, the lower the β parameter is, the lower the electrical grid variability. When β takes values close to one, the variability of the grid increases exponentially. The greater the β parameter is, the greater the importance that is given to the Photovoltaic (PV) generation forecast. Analyzing the effects of the forecast error is not the objective of this paper; for this reason, the ideal forecast model has been used. The Photovoltaic (PV) generation forecast of a day is exactly the generation of that day. In these situations, all facilities schedule the deferrable loads following a similar pattern (the Photovoltaic (PV) forecast) without synchronization. This implies that days with low Photovoltaic (PV) generation or forecast errors cause a great mismatch between consumption and generation. These results suggest that the SwarmGrid algorithm works better for $\beta = 0$ from the grid point of view.

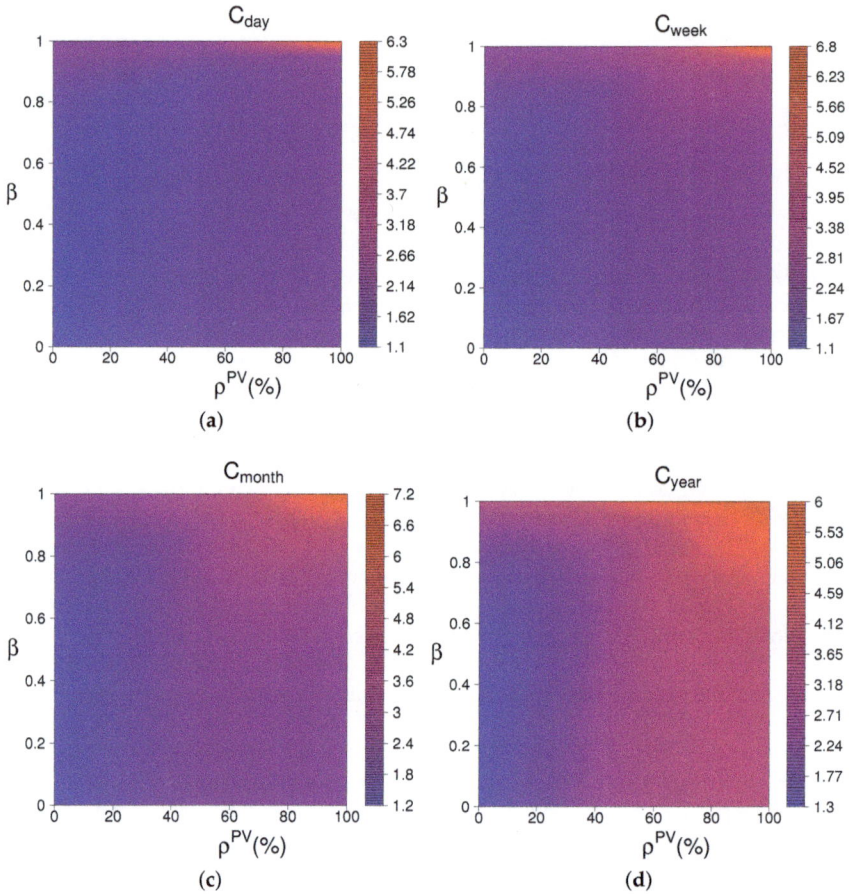

Figure 3. Heat map representing the development of the crest factors for different combinations of ρ^{PV} and β with $\rho^{ctr} = 100\%$: (a) $\bar{C}_{day}$; (b) $\bar{C}_{week}$; (c) $\bar{C}_{month}$ and (d) C_{year}.

During the remainder of this section, β is zero to focus on the smoothing of the aggregated consumption.

The effect of the percentage of consumption controlled by the SwarmGrid algorithm on the aggregated consumption depends on the Photovoltaic (PV) penetration. A campaign of experiments has been performed to study how ρ^{PV} and ρ^{ctr} affect the crest factors of the aggregated consumption and the self-consumption. In this campaign, β is zero because of the results of the previous analysis. Different combinations of ρ^{PV} and ρ^{ctr} have been studied. For each combination of these parameters, 30 experiments have been performed with different seeds of the random number generator. An experiment consists of the simulation of the previously explained electrical grid during one year and a half (788,400 min) with a concrete combination of ρ^{PV} and ρ^{ctr} and a concrete seed. The first half of the year is used to adapt the SwarmGrid algorithm to the grid, and after that, the crest factors are calculated for the remainder of the year. The crest factors and the self-consumption of the local facilities are calculated for each experiment.

Figure 4 shows the development of the crest factors for different combinations of ρ^{PV} and ρ^{ctr} with $\beta = 0$. The effect of the SwarmGrid algorithm on these factors changes depending on the Photovoltaic (PV) penetration. For low values of ρ^{PV}, the higher ρ^{ctr}, the smoother the aggregated consumption. The SwarmGrid algorithm is able to synchronize with the electrical grid and to reduce the crest factors. On the other hand, for high values of ρ^{PV}, the increase of ρ^{ctr} intensifies the variability introduced by the Photovoltaic (PV) generation.

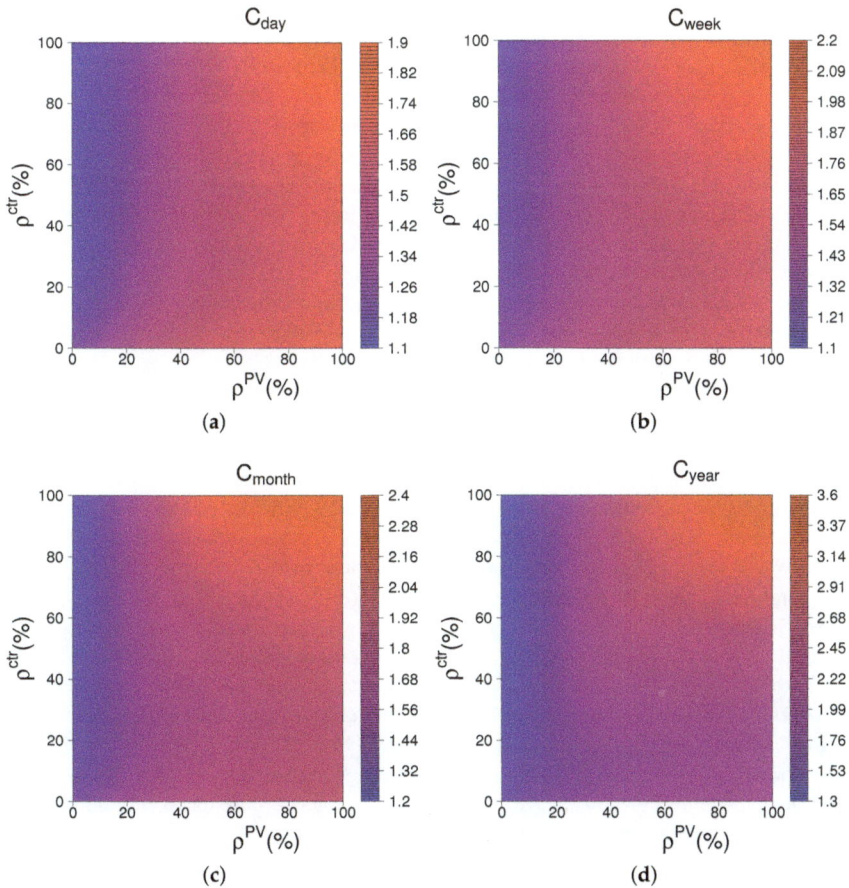

Figure 4. Heat map representing the development of the crest factors for different combinations of ρ^{PV} and ρ^{ctr} with $\beta = 0$: (a) $\bar{C}_{day}$; (b) $\bar{C}_{week}$; (c) $\bar{C}_{month}$ and (d) C_{year}.

Figure 4 shows the development of the self-consumption for different combinations of ρ^{PV} and ρ^{ctr} with $\beta = 0$. As in previous analysis, the self-consumption increases with ρ^{PV} because of the Photovoltaic (PV) resource availability. Moreover, ρ^{ctr} enhances the self-consumption, as well. This implies that the SwarmGrid algorithm synchronizes with the Photovoltaic (PV) generation because it is included in the aggregated consumption signal.

6. Discussion

The SwarmGrid algorithm has a number of valuable features for Demand-Side Management (DSM) that it borrows from swarm intelligence and coupled oscillators algorithms: self-organization, adaptability, low information exchange, local conditions and anonymous communication:

- Self-organization: An electrical grid with facilities equipped with the SwarmGrid algorithm is able to self-organize. All facilities schedule the deferrable loads following a common goal. Without an explicit information exchange and without the presence of a central agent, the consumption of these facilities organizes over time. This feature could be observed in Section 5.1, when the amount of controllable consumption exceeds the amount of non-controllable consumption. This feature differs from centralized and distributed algorithms based on other techniques. The centralized algorithms cannot self-organize by definition, and the loss of the central agent causes a complete operation failure. Other distributed techniques typically perform DSM through a negotiation between agents. These processes requires direct communication between agents, causing operation failures when this communication is lost.

- Adaptability: Facilities are able to adapt to the aggregated consumption of the grid. A non-controllable consumption was introduced in the simulated electrical grid. The SwarmGrid algorithm schedules the deferrable loads adapting to the non-controllable consumption and smoothing the aggregated consumption. All DSM algorithms require certain adaptability; however, the main difference of SwarmGrid in relation to other techniques is that this adaptability is performed for each agent by dividing this feature into thousands or millions of elements. This feature is also implemented by other distributed algorithms, being an advantage in comparison with centralized approaches.

- Low information exchange: The SwarmGrid algorithm requires a very low information exchange. The facilities only require the aggregated consumption signal, which should be provided by the electrical grid operator. In addition, the sample period of this signal implies a low computing load compared to the current electronics: $T^{smp} = 90$ min in the examples shown. Swarm intelligence-based algorithms are based on relative simple calculations for each agent, but they achieve complex behaviors when working as a swarm. This feature is acquired by SwarmGrid , which implies lower computing requirements in relation to other DSM algorithms.

- Local conditions: The SwarmGrid algorithm takes into account local conditions. It schedules the deferrable loads by considering the grid and the local Photovoltaic (PV) generation forecast. In addition, this algorithm could be designed to satisfy other local conditions because its distributed in nature. This feature is a general advantage of the distributed approaches regards the centralized ones. Each agent participates in the DSM process, but at the same time, it can improve its local performance.

- Anonymous communication: The SwarmGrid algorithm does not require communicating the local state to other elements of the grid. This is an advantage in comparison with other centralized and distributed algorithms. In centralized algorithms, the local state must be communicated to the central agent. Moreover, other distributed approaches typically require information exchange between agents. SwarmGrid ensures that local information is not transmitted, keeping the system as simple as possible and improving the security of the user.

The smoothing of the aggregated consumption by the SwarmGrid algorithm could be observed in different scenarios in Section 5. In general, the SwarmGrid algorithm reduces the crest factors of the electrical grid proportionally to the amount of controlled energy. This issue is easily observed in the analysis of the SwarmGrid operation without DG technologies. In this case, the SwarmGrid algorithm schedules a certain percentage of energy of the electrical grid. The results conclude that for an amount of energy controlled by the SwarmGrid algorithm lower than 50%, the higher the percentage of controlled energy, the lower the crest factors. For percentages higher than 50%, the crest factors achieve a plateau value. This is because the SwarmGrid algorithm adapts to the non-controllable consumption and schedules the remaining controllable consumption so that the variability of the aggregated consumption does not increase. This is indicative of the self-organization capacity of the algorithm. Two examples have been proposed to show how the SwarmGrid algorithm affects the shape of the aggregated consumption and its difference between peaks and valleys. Thanks to the use of the algorithm, the maximum yearly peak has been reduced from 41.9 GW to 36.1 GW, maintaining the same average consumption. This implies that the size of the electrical grid has been improved because the average consumption is closer to the minimum peak: less electrical infrastructure is required, which is used more often on average. In addition, the average daily difference between peaks and valleys has been reduced from 12.2 GW to 4.7 GW. This implies that the daily use of the electrical grid has also been improved: the difference between peaks and valleys is much lower, and thus, the grid does not have to respond to such abrupt power variations, increasing its stability.

Generally, the presence of large amounts of Photovoltaic (PV) generation makes the smoothing process of the aggregated consumption difficult. The cause of this malfunction is typically the same: all facilities schedule the deferrable loads to run during the same time period without considering other facilities. This causes a consumption peak at this period, which may be increased because of a lack of generation or a bad forecast. On the other hand, the local self-consumption is enhanced when $\beta \to 1$. Thus, a compromise between aggregated consumption smoothing and self-consumption should be found. This compromise depends on the specific requirements where the SwarmGrid algorithm is implemented.

7. Conclusions

In this paper, Demand-Side Management (DSM) has been tackled from the electrical grid point of view, but actions are taken on the consumer side. This means that the local consumption is managed to smooth the aggregated consumption of the grid, where every user takes only into account its local conditions. A Demand-Side Management (DSM) algorithm, SwarmGrid, has been developed to address this issue. The proposed algorithm is developed from a distributed point of view; thus, the Demand-Side Management (DSM) is not performed by a central agent, but each consumer is actively involved in its implementation. Direct communication between users or facilities has also been avoided. Thus, the coordination is performed with the sole information of the aggregated consumption of the grid.

The proposed SwarmGrid (SwarmGrid) algorithm uses the frequency consumption patterns generated by the Multi-Frequency Coupled Oscillators (MuFCO) algorithm and takes into account the local generation. It generates a pattern by combining the aggregated consumption and Photovoltaic (PV) generation forecasts to schedule the deferrable loads. In general, the SwarmGrid algorithm is able to smooth the aggregated consumption in every scenario. The improvement caused by the SwarmGrid algorithm is more relevant for low percentages of controllable power. It achieves a plateau value after the controllable energy of the whole electrical grid reaches 50%. This means that the deferrable power controlled by the SwarmGrid algorithm is able to adapt to the non-controllable power. Thanks to the use of this algorithm, the maximum yearly peak and the daily difference between peaks and valleys are reduced.

Some future developments of the SwarmGrid algorithm are discussed below. The SwarmGrid algorithm has been designed to schedule deferrable loads, but it is not the only type of load. Elastic loads, whose consumption can be controlled directly, could be also included. For example, the power of an electric pump can be modified to shape a sinusoidal pattern. Several loads are a mix between deferrable and elastic consumption. For example, the power of HVAC systems may be modified, but only in a number of discrete values in a certain range, and it usually has time constraints such as minimum operation time. In general, many devices have a complex operation that could be introduced into SwarmGrid . By combining these developments, the control of one of the major future electricity consumption challenges can be addressed from the local perspective: the Electric Vehicle (EV). The charge of large fleets of EVs could be addressed through SwarmGrid implementing the previously mentioned developments. Another research line is related to the structure of the grid. Actually, the grid has different zones with different generation and consumption capacities such that the produced energy does not have to travel long distances. This could be linked with another new concept about electrical grid design: the microgrids. The SwarmGrid algorithm could be use to coordinate the consumption of different microgrids.

The SwarmGrid algorithm tackles the aggregated consumption by smoothing it from a technical point of view. This issue can be also addressed from other perspectives, for example from the economical point of view. The electricity market can also be considered as a signal. It varies continuously during the day. The electricity market is not only affected by the demand, but by the resources' availability and the financial market, as well. The SwarmGrid algorithm can be used to adapt to this market by considering other variables that affect the electricity price. The required variables to achieve this new objective should be studied, as well as how they could be included in the algorithm.

Author Contributions: Conceptualization, Á.G.M. Data curation, M.C.-C. Formal analysis, M.C.-C. Funding acquisition, Á.G.M. Investigation, M.C.-C., E.M. and E.C.-M Methodology, M.C.-C., E.C.-M and Á.G.M. Project administration, Á.G.M. Resources, Á.G.M. Software, M.C.-C. and E.M. Supervision, E.C.-M and Á.G.M. Validation, Á.G.M. Visualization, E.M. Writing, original draft, M.C.-C. Writing, review editing, E.C.-M and Á.G.M.

Funding: This work was partially supported by the "DEMS: Sistema Distribuido de Gestión de Energía en Redes Eléctricas Inteligentes", funded by the Programa Estatal de Investigación Desarrollo e Innovación orientada a los retos de la sociedad of the Spanish Ministerio de Economía y Competitividad (TEC2015-66126-R).

Conflicts of Interest: The authors declare no conflict of interest.

References

1. Rahimi, F.; Ipakchi, A. Demand Response as a Market Resource Under the Smart Grid Paradigm. *IEEE Trans. Smart Grid* **2010**, *1*, 82–88. [CrossRef]
2. Lopes, J.A.P.; Hatziargyriou, N.; Mutale, J.; Djapic, P.; Jenkins, N. Integrating distributed generation into electric power systems: A review of drivers, challenges and opportunities. *Electr. Power Syst. Res.* **2007**, *77*, 1189–1203. [CrossRef]
3. Palensky, P.; Dietrich, D. Demand Side Management: Demand Response, Intelligent Energy Systems, and Smart Loads. *IEEE Trans. Ind. Inform.* **2011**, *7*, 381–388. [CrossRef]
4. Tran, N.H.; Ren, S.; Han, Z.; Huh, E.N.; Hong, C.S. Reward-to-Reduce: An Incentive Mechanism for Economic Demand Response of Colocation Data Centers. *IEEE J. Sel. Areas Commun. Spec. Issue Green Commun.* **2016**, *34*, 3438–3450. [CrossRef]
5. Strbac, G. Demand side management: Benefits and challenges. *Energy Policy* **2008**, *36*, 4419–4426. [CrossRef]
6. Mohsenian-Rad, A.H.; Wong, V.W.S.; Jatskevich, J.; Schober, R.; Leon-Garcia, A. Autonomous Demand-Side Management Based on Game-Theoretic Energy Consumption Scheduling for the Future Smart Grid. *IEEE Trans. Smart Grid* **2010**, *1*, 320–331. [CrossRef]
7. Wang, Y.; Saad, W.; Mandayam, N.B.; Poor, H.V. Load Shifting in the Smart Grid: To Participate or Not? *IEEE Trans. Smart Grid* **2016**, *7*, 2604–2614. [CrossRef]
8. Asrari, A.; Lotfifard, S.; Ansari, M. Reconfiguration of Smart Distribution Systems With Time Varying Loads Using Parallel Computing. *IEEE Trans. Smart Grid* **2016**, *7*, 2713–2723. [CrossRef]
9. Antoniadou-Plytaria, K.E.; Kouveliotis-Lysikatos, I.N.; Georgilakis, P.S.; Hatziargyriou, N.D. Distributed and Decentralized Voltage Control of Smart Distribution Networks: Models, Methods, and Future Research. *IEEE Trans. Smart Grid* **2017**, *8*, 2999–3008. [CrossRef]
10. Deng, R.; Yang, Z.; Chow, M.Y.; Chen, J. A Survey on Demand Response in Smart Grids: Mathematical Models and Approaches. *IEEE Trans. Ind. Inform.* **2015**, *11*, 570–582. [CrossRef]
11. Rahman, M.; Oo, A. Distributed multi-agent based coordinated power management and control strategy for microgrids with distributed energy resources. *Energy Convers. Manag.* **2017**, *139*, 20–32. [CrossRef]
12. Vu, D.H.; Muttaqi, K.M.; Agalgaonkar, A.P.; Bouzerdoum, A. Short-Term Electricity Demand Forecasting using Autoregressive Based Time Varying Model Incorporating Representative Data Adjustment. *Appl. Energy* **2017**, *205*, 790–801. [CrossRef]

13. Vahid-Pakdel, M.; Nojavan, S.; Mohammadi-ivatloo, B.; Zare, K. Stochastic optimization of energy hub operation with consideration of thermal energy market and demand response. *Energy Convers. Manag.* **2017**, *145*, 117–128. [CrossRef]

14. Andoni, M.; Robu, V.; Fruh, W.G.; Flynn, D. Game-Theoretic Modeling of Curtailment Rules and Network Investments with Distributed Generation. *Appl. Energy* **2017**, *201*, 174–187. [CrossRef]

15. Yang, X.; Zhang, Y.; Zhao, B.; Huang, F.; Chen, Y.; Ren, S. Optimal energy flow control strategy for a residential energy local network combined with demand-side management and real-time pricing. *Energy Build.* **2017**, *150*, 177–188. [CrossRef]

16. Ahmad, T.; Chen, H.; Guo, Y.; Wang, J. A comprehensive overview on the data driven and large scale based approaches for forecasting of building energy demand: A review. *Energy Build.* **2018**, *165*, 301–320. [CrossRef]

17. Nikmehr, N.; Najafi-Ravadanegh, S.; Khodaei, A. Probabilistic optimal scheduling of networked microgrids considering time-based demand response programs under uncertainty. *Appl. Energy* **2017**, *198*, 267–279. [CrossRef]

18. Saad, W.; Glass, A.L.; Mandayam, N.B.; Poor, H.V. Toward a Consumer-Centric Grid: A Behavioral Perspective. *Proc. IEEE* **2016**, *104*, 865–882. [CrossRef]

19. Cao, Y.; Yu, W.; Ren, W.; Chen, G. An Overview of Recent Progress in the Study of Distributed Multi-Agent Coordination. *IEEE Trans. Ind. Inform.* **2013**, *9*, 427–438. [CrossRef]

20. Castillo-Cagigal, M.; Matallanas, E.; Monasterio-Huelin, F.; Caamaño-Martínn, E.; Gutiérrez, A. Multifrequency-Coupled Oscillators for Distributed Multiagent Coordination. *IEEE Trans. Ind. Inform.* **2016**, *12*, 941–951. [CrossRef]

21. Castillo-Cagigal, M.; Gutiérrez, A.; Monasterio-Huelin, F.; Caamaño-Martín, E.; Masa-Bote, D.; Jiménez-Leube, J. A semi-distributed electric demand-side management system with PV generation for self-consumption enhancement. *Energy Convers. Manag.* **2011**, *52*, 2659–2666. [CrossRef]

22. Hu, J.; Morais, H.; Sousa, T.; Lind, M. Electric vehicle fleet management in smart grids: A review of services, optimization and control aspects. *Renew. Sustain. Energy Rev.* **2016**, *56*, 1207–1226. [CrossRef]

23. Castillo-Cagigal, M.; Caamaño-Martín, E.; Matallanas, E.; Masa-Bote, D.; Gutiérrez, A.; Monasterio-Huelin, F.; Jiménez-Leube, J. PV self-consumption optimization with storage and Active DSM for the residential sector. *Sol. Energy* **2011**, *85*, 2338–2348. [CrossRef]

24. Masa-Bote, D.; Castillo-Cagigal, M.; Matallanas, E.; Caamaño-Martín, E.; Gutiérrez, A.; Monasterio-Huelin, F.; Jiménez-Leube, J. Improving photovoltaics grid integration through short time forecasting and self-consumption. *Appl. Energy* **2014**, *125*, 103–113. [CrossRef]

25. Calpa, M.; Castillo-Cagigal, M.; Matallanas, E.; Caamaño-Martín, E.; Gutiérrez, A. Effects of Large-scale PV Self-consumption on the Aggregated Consumption. In Proceedings of the 6th International Conference on Sustainable Energy Information Technology (SEIT-2016), Madrid, Spain, 23–26 May 2016; Elsevier: New York, NY, USA, 2016; pp. 816–823.

![energies logo] *energies*

MDPI

Article

Double Layer Dynamic Game Bidding Mechanism Based on Multi-Agent Technology for Virtual Power Plant and Internal Distributed Energy Resource

Yajing Gao [1,*], Xiaojie Zhou [2,*], Jiafeng Ren [2], Xiuna Wang [1] and Dongwei Li [1]

[1] China Electric Power Enterprise Association Power Construction Technology and Economic Consultation Center, Beijing 100000, China; wangxiuna@cec.org.cn (X.W.); lidongwei@cec.org.cn (D.L.)

[2] State Key Laboratory of Alternate Electrical Power System with Renewable Energy Sources (North China Electric Power University), Baoding 071003, China; renjf1028@163.com

* Correspondence: gaoyajing@cec.org.cn (Y.G.); 15631292227@163.com (X.Z.)

Received: 27 September 2018; Accepted: 29 October 2018; Published: 7 November 2018

Abstract: As renewable energies become the main direction of global energy development in the future, Virtual Power Plant (VPP) becomes a regional multi-energy aggregation model for large-scale integration of distributed generation into the power grid. It also provides an important way for distributed energy resources (DER) to participate in electricity market transactions. Firstly, the basic concept of VPP is outlined, and various uncertainties within VPP are modeled. Secondly, using multi-agent technology and Stackelberg dynamic game theory, a double-layer nested dynamic game bidding model including VPP and its internal DERs is designed. The lower layer is a bidding game for VPP internal market including DER. VPP is the leader and each DER is a subagent that acts as a follower to maximize its profit. Each subagent uses the particle swarm algorithm (PSA) to determine the optimal offer coefficient, and VPP carries out internal market clearing with the minimum variance of unit profit according to the quoting results. Then, the subagents renew the game to update the bidding strategy based on the outcomes of the external and internal markets. The upper layer is the external market bidding game. The trading center (TC) is the leader and VPP is the agent and the follower. The game is played with the goal of maximum self-interest. The agent uses genetic algorithms to determine the optimal bid strategy, and the TC carries out market clearance with the goal of maximizing social benefits according to the quotation results. Each agent renews the game to update the bidding strategy based on the clearing result and the reporting of the subagents. The dynamic game is repeated until the optimal equilibrium solution is obtained. Finally, the effectiveness of the model is verified by taking the IEEE30-bus system as an example.

Keywords: virtual power plant; distributed energy resources; multi-agent technology; bidding strategy; stackelberg dynamic game

---

1. Introduction

With the continuous increase of power demand and the increasingly severe problems of global energy shortages and environmental pollution, distributed generations (DG) have been adopted by more and more countries due to their reliability, economy, flexibility and environmental protection.

Nowadays, the global power industry is rapidly transforming, and the power system should be based on market operations. However, because of the characteristics of DG, such as small capacity and being random, intermittent and volatile, it is not feasible for them to join power market operations alone. At present, most studies use the concept of microgrid (MG) as the grid connection form of DG [1]. MG can well coordinate the technical contradiction between large power grids and DGs, and has certain energy management functions. However, MG takes the local application of DGs and users as the main control target, and is subject to geographical restrictions. There are also some limitations on the effective use of multi-regional and large-scale DGs and the economies of scale in the power market [2]. The concept of Virtual Power Plant (VPP)-provides new ideas for solving these problems.

The term "virtual power plant" is derived from Dr. Shimon Awerbuch's book "Virtual Public Facilities: Description, Technology, and Competitiveness of Emerging Industries" [3]. VPP does not change the way each DG is connected to the grid. Instead, it uses advanced technologies such as metrology and communications to aggregate DGs and energy storage systems. Different types of distributed energy resources, such as controllable loads and electric vehicles (EV), achieve coordinated and optimized operation through higher-level software architecture, which is more conducive to the rational allocation and utilization of resources. This method can aggregate DERs without transforming the power grid, and provides stable power transmission to the power grid, becoming an effective method for DERs to join the power market, reducing the risk of their independent operation in the market and the impact of grid-connected DER on the power grid [4–7].

The existing research of VPP focuses on two aspects: one is optimizing the internal resources of a single VPP, and the other is participating in overall power system dispatch. In terms of VPP internal resource allocation, Sun S. et al. [8] studied the VPP optimization scheduling model with EV, using direct, hierarchical and distributed methods to control the charging and discharging of EV. Zhao H. et al. [9] constructed an internal VPP scheduling model with interruptible load, and determined the VPP optimization scheduling model with uncertain wind power through scene analysis method. Liu Y.Y. et al. [10] established a bidding model for VPP and considered uncertainty to solve the VPP optimal scheduling problem under multi-market mode. Zamani A.G. et al. [11] considered the energy storage system and demand response, established the VPP optimization scheduling model, and conducted the related researches on the VPP bidding problem. Zang H.X. et al. [12] divided the VPP into incentive demand-responsive VPP and price demand-responsive VPP according to the internal demand response type, and established a day-ahead dispatch model considering the demand response. Guo H.X. et al. [13] constructed a bidding model for VPP. However, the VPP studied was only used as a price receiver and could not affect the outcome of system clearing. Its essence is still the internal optimization problem of a virtual power plant.

In the context of VPP's participation in competition in the electricity market, Rahimiyan M. and Baringo L. [14] considered the electricity market under the hybrid model. A VPP that includes wind farms, pumped storage power plants, and gas turbines participates in the operation of the medium-term contract market, daily market, and balance market. Mnatsakanyan A. and Kennedy S.W. [15] proposed a bidding model in which VPP participates in the energy market and ancillary service markets, and determined the bidding power of VPP in each market. Liu J.N. et al. [16] adopted the point estimation method to deal with the uncertainty of electricity price and new energy generation, and proposed a bidding strategy of VPP in the current electricity market. Song W. et al. [17] considered the power supply-demand balance and safety constraints of VPP, and put forward the joint bidding model of VPP in the power energy and hot standby market. Yang J.J. et al. [18] considered the uncertainty of the EVs' number and wind generations' power, and established a robust optimization model for VPP to participate in the day-ahead energy market and regulate market bidding. Peik-Herfeh M. et al. [19] reviewed the participation of VPP in the electricity market, and compared and analyzed the similarities and differences between VPP and MG. Fang Y.Q. et al. [20] used game theory to study the participation of multiples VPPs in the electricity market. VPPs mainly include distributed generation, energy storage, and time-of-use electricity price demand responses. Mashhour E. and Moghaddas-Tafreshi S.M. [21,22] studied the VPP participating in the main energy and ancillary service markets theoretically, and established a non-equilibrium model that considered the demand response and the VPP safety constraints. Zhou Y.Z. et al. [23] introduced agency theory and used energy storage of EV to reduce the impact of randomness of wind power on the system. VPP participated in the electricity market as a whole and responded to market price information. Yuan G.L. et al. [24] proposed a multi-agent power market simulation mechanism using game theory and machine learning methods, and used scene analysis methods to analyze the market behavior of VPPs.

Wang Y. et al. [25] took maximizing the profit of day-ahead market bidding and balancing market reward as goals. A commercial VPP model that considered interruptible loads, small wind farms, pumped storage power stations, and gas turbines was established.

However, in the operation of VPP, different types of DER are most likely to belong to different owners of property rights. Conventionally, it is no longer appropriate to consider VPP as a whole to participate in bidding. Therefore, it is necessary to consider the independence of internal entities and study bidding issues involving multiple types of DER from multiple entities. To guide VPP to occupy a more favorable position in the power market competition, and maximize its internal DER interests, this paper establishes a double-layer nested Stackelberg dynamic game model based on multi-agent technology. In Section 2, based on analyzing the main structure of VPP, a probabilistic output model for various uncertainties within VPP is established. In Section 3, multi-agent technology and Stackelberg dynamic game theory are used to design the underlying VPP internal market bidding mechanism, and establish the dynamic game model of each DER subagent. In combination with the real-time clearing results of the lower layer games, a multi-agent day-before-day power market dynamic game model with VPP is designed. Through repeated games, until the upper and lower levels jointly clear out, and obtain the optimal bid strategy of VPPs and its internal DERs. In Section 4, an IEEE 9 system is used as an example to demonstrate the validity of the proposed model and method. Section 5 contains some conclusions about the research.

2. Uncertain Factors Modeling in VPP

The VPP studied in this paper contains wind turbine (WT), photovoltaic (PV), energy storage (ES) devices, demand side resources (DSR), and diesel generators (DE). The power of DE is controllable, and the internal uncertainty of VPP mainly comes from the prediction error and random fluctuation of each DER, ES device and various loads.

2.1. Error Correction Based Fixed Load and DER Output Prediction

When forecasting the output of fixed load and DERs (PV and WT), this paper uses modified wavelet neural network (WNN) prediction method [26] which adds additional adaptive dynamic programming correction links. The idea of "prediction–correction" is introduced, and actual measurement data are used to update the WNN parameters to improve the prediction accuracy. The prediction principle is shown in Figure 1.

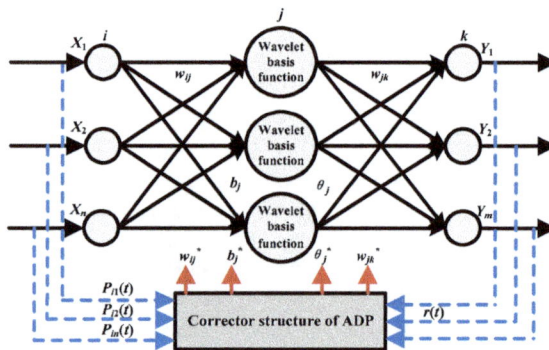

Figure 1. VPP model diagram.

In the VPP model diagram, x_1, x_2, ..., x_n are input parameter sequences; Y_1, Y_2, ..., Y_m are output parameter sequences; w_{ij} is the weight of the input layer to the hidden layer; w_{jk} is the weight value from the hidden layer to the output layer; b_j is the wavelet basis function scaling factor; θ is the threshold; $P_{i1}(t)$, $P_{i2}(t)$, ..., $P_{in}(t)$ are the measured data during the sampling period; w_{ij}^*, w_{jk}^*, b_j^* and

θ_j^* are updated parameter of the optimized structure using adaptive dynamic programming; and $r(t)$ is the cost function.

The specific process for forecasting using the WNN model is as follows:

Step 1: Pre-process data; cull or correcti bad data in various types of data; and normalizing them.

Step 2: Determine the number of nodes n, l and m of the WNN input layer, the hidden layer, and the output layer according to the original data (where the number of hidden layer nodes adopts an empirical value, that is, the default $l = 2n - 1$), and determine the maximum iteration (number of times N_{max} and iteration accuracy e_0).

Step 3: Randomly initialize the weights of the WNN (input layer to implicit layer weight wij and implicit layer to output layer weight w_{jk}) and wavelet basis function related parameters (scaling factor a_j and translation factor b_j), that is, w_{ij} = randn (n,l), w_{jk} = randn (l,m), a_j = randn ($1,l$) and b_j = randn ($1,l$).

Step 4: Initialize the learning rates $\eta 1$ and $\eta 2$ of w_{ij} and w_{jk} (both defaults to 0.01) and the learning rates $\eta 3$ and $\eta 4$ of a_j and b_j (both default to 0.001).

Step 5: Input the pre-processed data obtained in Step 1 into the WNN. Meanwhile, input the measured data of the sampling period and the obtained neural network output sequence into the ADP correction optimization structure. Obtain network weights w^*_{ij} and w^*_{jk} with strong fitting ability for the original data and wavelet parameters a^*_j and b^*_j.

Step 6: Input the data used for the prediction into the trained WNN network, obtain the predicted value by prediction, calculate the error, and derive the prediction result and analyze it.

2.1.1. Fixed Load Forecast

According to the statistical theory of probability, the normal distribution (ND) has good properties and can be used to approximate many probability distributions. Therefore, the ND is used to estimate the error caused by the fixed load forecast.

2.1.2. WT Forecast

For the error produced by WT output forecasting, the actual forecasting error of WT shows a large kurtosis and skewness, and simply using the normal distribution description will produce a larger error. Wang et al. [27] pointed out that the wind speed prediction error was the main cause of the WT forecasting error, and the different WT forecasting values corresponding to the WT forecast error had different probability distributions. The article also gives the applicable wind speed interval (m/s) for different distributions: the applicable range of the "0 error" distribution is [0, 2.1], [12.6, ∞]; the applicable range of the exponential distribution is (2.1, 5.1]; the ND applies to (5.1, 9.8); and the extreme value distribution applies to [9.8, 12.6). This paper draws on the idea of the literature and uses the error distribution of the corresponding wind speed to establish different probability density functions to estimate the error produced by WT output prediction.

2.1.3. PV Forecast

For the error generated by PV output prediction, simply using the ND description will also produce a larger error. Yadav et al. [28] pointed out that the output level of photovoltaic power generation is closely related to the weather conditions, and the distribution of forecast error under different weather types is different. This paper draws on the idea of the literature, and uses the idea of different weather types to establish different probability density distribution functions to estimate the error generated by PV output prediction.

2.2. Demand Response Modeling

This paper considers transfer load (TL) and incentive-based interruptible load (IL) based on time-sharing price (TP) mechanism.

2.2.1. Transfer Load

The usage of TL is affected by the real-time electricity price, and a considerable proportion of it will fluctuate with the electricity price with a certain probability. The proportion of TL defined by the tariff is the load transfer rate α_{TL} [29].

The relationship between TL usage and electricity price $C(t)$ at time t is shown in Equation (1).

$$P_{TL}(t) = \alpha_{TL}(t)P'_{TL}(t)k + (1 - \alpha_{TL}(t))P'_{TL}(t) \tag{1}$$

$$k = \left(\frac{C(t)}{C_{TL}}\right)^{\beta} \tag{2}$$

where $P_{TL}(t)$ is the total amount of TL after transfer at time t; $\alpha_{TL}(t)$ is the load transfer rate of TL at time t; $P'_{TL}(t)$ is the total amount of TL that does not transfer at time t; k is the influence coefficient of electricity price to TL; C_{TL} is the threshold electricity price of TL transfer and transfer; and β is the demand price elasticity coefficient.

Assuming that the total electricity demand of users within one day is constant, the controlled demand that is reduced in one period will be delayed to other periods. Therefore, the sum of changes in TL usage within one day should be zero, as shown in Equation (3).

$$\sum_{t=1}^{24} \Delta P_{TL}(t) = \sum_{t=1}^{24} (P_{TL}(t) - P'_{TL}(t)) = \sum_{t=1}^{24} \alpha_{TL}(t)P'_{TL}(t)(k-1) = 0 \tag{3}$$

Considering TL access, $P'_{TL}(t) \neq 0$, thus obtaining Equation (4).

$$\sum_{t=1}^{24} \alpha_{TL}(t)(k-1) = 0 \tag{4}$$

In addition, β is usually less than -1, so, when $C(t) > C_{TL}$, that is, TL is in roll-out state, $k < 1$. When $C(t) < C_{TL}$, TL is in transition state, $k > 1$. In view of this, using the NDs of two different parameters to carry out probabilistic modeling of the load transfer rate α_{TL} at the time of TL roll-out and TL transition, Equation (4) can be modified as Equation (5).

$$\sum_{t \in T_I} \alpha_{TL,I}(t)(k-1) = \sum_{t \in T_O} \alpha_{TL,O}(t)(1-k) \tag{5}$$

where T_I and T_O are the collections of TL transfer in and out in a day, respectively; and $\alpha_{TL,I}$ and $\alpha_{TL,O}$ are the load transfer rates when TL is transferred in and out, respectively.

Since the α_{TL} is sampled according to its probability density function, when the number of samples is large, the sample mean will be close to the expectation of its probability density function. Thus, Equation (5) can be further reduced to:

$$\gamma = \frac{\mu_{TL,I}}{\mu_{TL,O}} \approx \frac{t_O \sum\limits_{t \in T_O}(1-k)}{t_I \sum\limits_{t \in T_I}(k-1)} \tag{6}$$

where γ is the ratio coefficient between the expected $\mu_{TL,I}$ and $\mu_{TL,O}$ of the distribution $\alpha_{TL,I}$ and $\alpha_{TL,O}$; and t_I and t_O are the number of times included in T_I and T_O, respectively.

When the next day electricity price is known, the threshold price set value, γ, is constant.

2.2.2. Interruptible Load

IL users can sign contracts with power companies to allow them to interrupt power supply under different specific conditions. In this paper, the power-off condition is characterized by the power-off electricity price threshold. Before interrupting the power supply to the IL, the prediction method is the same as the fixed load, and the prediction error is fitted by using the normal distribution. Define S_{IL} as the 0–1 state variable of IL, and characterize the actual use of IL [29], as shown in Equation (7).

$$S_{IL}(t) = \begin{cases} 0, & C(t) \geq C_{IL} \\ 1, & C(t) < C_{IL} \end{cases} \tag{7}$$

where $S_{IL}(t)$ is a state variable at time t; $C(t)$ is a real-time electricity price at time t; and C_{IL} is a power-off valve price.

2.3. Energy Storage Unit Modeling

The ES unit can be charged and discharged within its allowable range. To extend its service life, it is generally not allowed to overcharge. The basic model of ES is as follows:

$$\begin{cases} SOC(t) = SOC(t-1) + \Delta SOC_{ch}(t), t \in T_{ch} \\ SOC(t) = SOC(t-1) + \Delta SOC_{dis}(t), t \in T_{dis} \end{cases} \tag{8}$$

$$\begin{cases} SOC_{ch}(t) = P_{ch}(t)\eta_{ch}\Delta T S_{es}(t)/S_{ES}, t \in T_{ch} \\ SOC_{dis}(t) = P_{dis}(t)\eta_{dis}\Delta T S_{es}(t)/S_{ES}, t \in T_{dis} \end{cases} \tag{9}$$

$$\begin{cases} S_{es} = 1, \lambda_{es} < \lambda_0 \\ S_{es} = 0, \lambda_{es} < \lambda_0 \end{cases} \tag{10}$$

where $SOC(t)$ and $SOC(t-1)$ are the charge states of the ES device at t and $t-1$, respectively; $\Delta SOC_{ch}(t)$ and $\Delta SOC_{dis}(t)$ are the charging and discharging capacities of ES at time t; $P_{ch}(t)$ and $P_{dis}(t)$ are charging and discharging powers of ES at time t; η_{ch} and η_{dis} are charging and discharging efficiency of ES,; S_{ES} is rated capacity of ES; $S_{es}(t)$ is the charging and discharging state parameter of ES at time t; $\lambda_{es}(t)$ is the charging and discharging action parameter of ES at time t; λ_0 is the charging and discharging action parameter threshold of ES; and T_{ch} and T_{dis} are the charging and discharging time periods of ES, where they are equivalent to the peak time of the electricity price T_f and the time interval of the electricity price valley T_g, respectively.

The charging and discharging rules of the ES are closely related to the price of electricity. The relationship between the charging and discharging power at time t and the price $C(t)$ is shown in Equation (11).

$$\begin{cases} P_{ch}(t) = a_{ch}C(t) + b_{ch}, t \in T_{ch} \\ P_{dis}(t) = a_{dis}C(t) + b_{dis}, t \in T_{dis} \end{cases} \tag{11}$$

where a_{ch}, b_{ch}, a_{dis} and b_{dis} are the relation coefficients of the charging price and the charging and discharging power at the charging time and the discharging time, respectively.

The ES element's initial charge state is SOC (1) = 0, and the charge state returns to zero at the end of each day, that is, SOC (24) = 0. Thus, all ESs will be completely discharged when the price of electricity peaks at the end of the day. If there is an ES with a non-zero amount of electricity, when the end time of the discharge period is the length of time required for its current discharge, it is forcibly discharged to restore the initial state of charge.

3. Double Layer Bidding Model Design Based on Stackelberg Dynamic Game

3.1. Framework Design of Double Layer Bidding Model

Agents in this paper refer to VPPs and traditional thermal power plants. Subagents refer to various investment entities within VPPs. Multiple VPPs and traditional thermal power plants participate in the day-ahead market bidding of the power system, and each DER only participates in the VPP internal day-ahead market auctions to which it belongs. The market is divided into 24 periods, and only one period is selected when building a model. The software used in the model solving process is "Matlab 2014a".

The Stackelberg dynamic game bidding model established in this paper can be divided into two layers, where one layer nests a game mode. The lower layer is VPP internal day-ahead market bidding stage. Each DER within the VPP acts as a subagent to obtain forecasted output values based on historical data. The Monte Carlo method is used to generate the quote scenario set and determine the quoted coefficient, and the bid strategy is obtained and submitted to the VPP data center. The VPP data center predicts the load demand based on historical data, and carries out unified internal market clearing. Before the bidding deadline, each subagent may make new bidding strategies.

At the upper layer, VPP acts as an agent to participate in the day-ahead market bidding of the power system with other agents. At this point, the output of VPP is the total power of the internal market in the lower layer, and the unit cost is the clear price of the internal market. Each agent simulated the game through the genetic algorithm (GA) to determine the bidding strategy and reported it to the TC for unified market clearing. Before the bidding deadline, each agent can update the bidding strategy to obtain the best result. The specific model framework is shown in Figure 2.

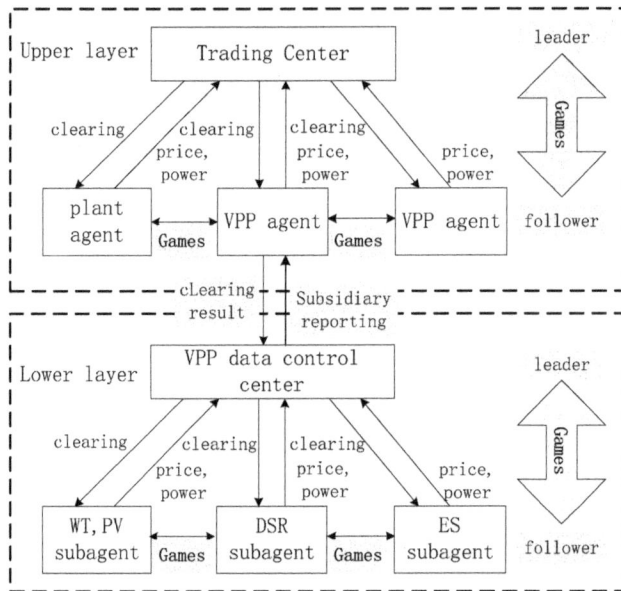

Figure 2. Framework diagram of a double-layer dynamic game model based on multi-agent technology.

3.2. Design of VPP Internal Bidding Model Based on Dynamic Game

After each subagent in VPP conducts the first round of game, it reports to the VPP data control center the availability and quotation for each period. The VPP data control center carries out the unified internal market clearing, that is, the selection of the power generation sequence from low to high electricity prices until the supply and demand balance is satisfied. After determining the clear price and the clearing power quantity, the information is fed back to each subagent. Each subagent formulates a new bidding strategy based on the feedback information, so that the game is repeated until an optimal equilibrium strategy is obtained.

3.2.1. Bidding Strategy of Subagent in VPP Internal Market

This section examines the bid strategies for a single subagent i. Assume that each subagent uses a linear quoting function [30], as shown in Equation (12).

$$P_i(q_i) = \alpha_i + \beta_i q_i \tag{12}$$

where α_i and β_i are quote parameters of the subagent i and q_i is the inflow and outflow power of the subagent i. The specified energy outflow is positive, and the inflow is negative.

In addition to considering operational costs and operational constraints, subagent i must also consider possible quotations from other subagents when quoting. Under normal circumstances, subagent i does not know the price coefficient of the competitor j, and can only perform the probability estimation by using historical information. Under the power market clearing rule, the price coefficient of the subagent j conforms to the two-dimensional normal distribution [31,32], and its probability density function is shown as in Equation (13).

$$(\alpha_{ij}, \beta_{ij}) \sim N\left(\begin{bmatrix} \mu_{\alpha,ij} \\ \mu_{\beta,ij} \end{bmatrix}, \begin{bmatrix} \sigma_{\alpha,ij}^2 & \rho_{ij}\sigma_{\alpha,ij}\sigma_{\beta,ij} \\ \rho_{ij}\sigma_{\alpha,ij}\sigma_{\beta,ij} & \sigma_{\beta,ij}^2 \end{bmatrix} \right) \tag{13}$$

where the subscript ij represents subagent i's estimate of subagent j; $\sigma_{\alpha,ij}$ and $\sigma_{\beta,ij}$ are the standard deviation estimates of α and β respectively; $\mu_{\alpha,ij}$ and $\mu_{\beta,ij}$ are mean estimates of α and β, respectively; and ρ_{ij} is the correlation coefficient between α and β. When subagent i predicts the power of subagent j is negative, ρ_{ij} is -1. As a supplier of electric energy, to increase the number of successful bids, one factor will be increased while the other coefficient will be decreased. Similarly, it can be obtained that when j needs to purchase electricity from the outside, the ρ_{ij} is 1.

Subagent i can generate subagent j quoting scenario set through Monte Carlo simulation to determine its own optimal bidding strategy. The quota coefficient scene generation step is as follows:

(1) Step 1: Assuming $M = \begin{bmatrix} \sigma_{\alpha,ij}^2 & \rho_{ij}\sigma_{\alpha,ij}\sigma_{\beta,ij} \\ \rho_{ij}\sigma_{\alpha,ij}\sigma_{\beta,ij} & \sigma_{\beta,ij}^2 \end{bmatrix}$, the lower triangular matrix A is generated such that $M = AA^T$.

(2) Step 2: Producing mutually independent two-dimensional standard normal distribution random vectors $\lambda = [\lambda_\alpha, \lambda_\beta]^T$, where $\lambda_\alpha \sim N(0, 1)$, $\lambda_\beta \sim N(0, 1)$.

(3) Step 3: $[\alpha_{ij}, \beta_{ij}]^T = [\mu_{\alpha,ij}, \mu_{\beta,ij}]^T + A\lambda$.

Each subagent in VPP will pursue its own maximization of profit in the bidding process. The expected profit function of subagent i is as follows:

$$\max E_i = (R_{in} - C_i) \cdot q_i^R \tag{14}$$

where R_{in} indicates the price of the VPP internal market at a certain time; q_i^R represents the outgoing power of the agent i at the same time; and C_i is the cost of the subagent i's power generation.

3.2.2. VPP Internal Day-Ahead Market Clearing

During the competition in the VPP internal market, efficient subagents have more advantages because high-performance subagents in the process of re-quotes can sell electricity at a relatively low price and earn profits. Subagents with lower efficiency over time will be eliminated from the internal market. In a small VPP system, this bidding model threatens the balance between supply and demand

and standby demand, and it is not conducive to long-term stable operation of the system. This paper adopts the uniform clearing price (UCP) for settlement. When the VPP internal market clears, the goal of pursuing the average of the unit power profit is to ensure that all players are as profitable as possible. Therefore, the clearing result is generated when the variance of the unit's profit is the smallest, that is, the bidding strategy that guarantees the safe and stable operation of the VPP internal system in the bidding process is easier to win the bid [33].

Introducing the unit power balance objective of subagent i as follows:

$$\min V = \sum_{i=1}^{m} \left(\overline{V} - V_i\right)^2 \tag{15}$$

$$\overline{V} = \frac{1}{m}(V_1 + V_2 + \cdots + V_i + \cdots + V_m) \tag{16}$$

$$V_i = \frac{\sum\limits_{i=1}^{m} \left[(P_i - C_i)^2 \cdot q_i\right]}{\sum\limits_{i=1}^{m} q_i} \tag{17}$$

where the subscript m is the total number of subagents in VPP and V is the subagent's profit variance of per-unit electricity.

This paper ignores the cost of power generation for WT and PV. Other subagents' costs are shown in Equations (18)–(21).

$$F^{DE} = \sum_{i \in m} \left(a q_i^{DE^2} + b q_i^{DE} + c\right) \tag{18}$$

$$F^{DS} = C^C + C^{IL} \tag{19}$$

$$C^C = \delta \cdot q^C, C^{IL} = \gamma \cdot q^{IL} \tag{20}$$

$$F^{ES} = \frac{1}{2} e \cdot \left| q^{ES} \right| \tag{21}$$

where F^{DE}, F^{DS} and F^{ES} are power generation costs of DEs, demand-side response costs, and energy storage operating costs, respectively; q_i^{DE} is the power generated by $_i^{DE}$ (ith diesel generator in VPP) at a certain moment; a, b, and c are its consumption factors; C^C and C^{IL} are the interruption capacity costs and interruption cost of VPP at a certain moment; δ and γ are the interrupted capacity prices and interrupted charge prices of VPP at a certain moment; q^C and q^{IL} represent the interrupt capacity and interrupted power provided by VPP at a certain time; q^{ES} is the charge or discharge power of the ES device at a certain time, which can be positive or negative; and e is the energy storage cost, which is determined by the energy storage efficiency.

The clearing process includes the following constraints:

(1) Power balance constraints

$$\begin{aligned} Q^{VPP} &= Q_G^{VPP} - Q_L^{VPP} + Q_{IL}^{VPP} \\ Q_G^{VPP} &= q^{DE} + q^{WT} + q^{PV} + q^{ES} \end{aligned} \tag{22}$$

where Q^{VPP} indicates the overall external power of VPP at a certain moment. When it is positive, it means that VPP generates electricity to the outside. When it is negative, it means that VPP purchases electricity; Q_G^{VPP} indicates the total power generated by VPP at a certain moment; and q^{DE}, q^{WT}, and q^{PV}, respectively, represent the power generation of all DEs, all WTs, and all PVs at a time.

(2) Power constraints for WT and PV

$$q_{imin}^{WT} \leq q_i^{WT} \leq q_{imax}^{WT}$$
$$q_{imin}^{PV} \leq q_i^{PV} \leq q_{imax}^{PV} \tag{23}$$

where q_{imax}^{WT}, q_{max}^{WT} are the upper and lower limits of the external output of $_i^{WT}$; and q_{imax}^{PV} and q_{max}^{PV} are the upper and lower limits of the external output of $_i^{PV}$, respectively.

(3) Power constraints for DE

$$q_{imin}^{DE} \leq q_i^{DE} \leq q_{imax}^{DE} \tag{24}$$

$$- \Delta q_i^{down} \leq q_{t,i} - q_{t-1,i} \leq \Delta q_i^{up} \tag{25}$$

where q_{imax}^{DE} and q_{min}^{DE} are the upper and lower limits of $_i^{DE}$ output to a certain moment; and Δq_i^{down} and Δq_i^{up} are the power variation limits of $_i^{DE}$ in pre-unit time, that is, the upper and lower limits of climbing rate.

(4) Capacity constraints for ES devices

$$\sum_{t \in T} q_t^{ES} \cdot \Delta t = 0 \tag{26}$$

where Δt is the time interval and T is the charge and discharge cycle.

(5) Charging and discharging constraints for ES devices

$$E_{min} \leq E_0 - \sum_{i=1}^{t} q_t^{ES} \cdot \Delta t \leq E_{max}, \ t = 1, 2, \cdots, T \tag{27}$$

$$- q_{max}^{ch} \leq q^{ES} \leq q_{max}^{dis} \tag{28}$$

where E_{max} and E_{min} are the upper and lower limits of the ES device, respectively; and q_{max}^{ch} and q_{max}^{dis} are the power limits for ES charging and discharging, respectively.

3.2.3. Model Solving

The solution of this section mainly focuses on the determination of the subagent quotation coefficient and the internal market clearing of VPP. The algorithm flow is shown in Figure 3. The PSA is used to solve the optimal quote coefficient. Subagent i simulates and generates a set of possible bid scenarios based on historical bid data of other subagents, and obtains the maximum revenue of the subagent i under each scenario, with the expectation of maximum profit. The target is updated and the particles selected to determine the final quote coefficient. To solve the issue of VPP internal market clearing, the GA toolbox in "Matlab 2014a" was used to solve the problem.

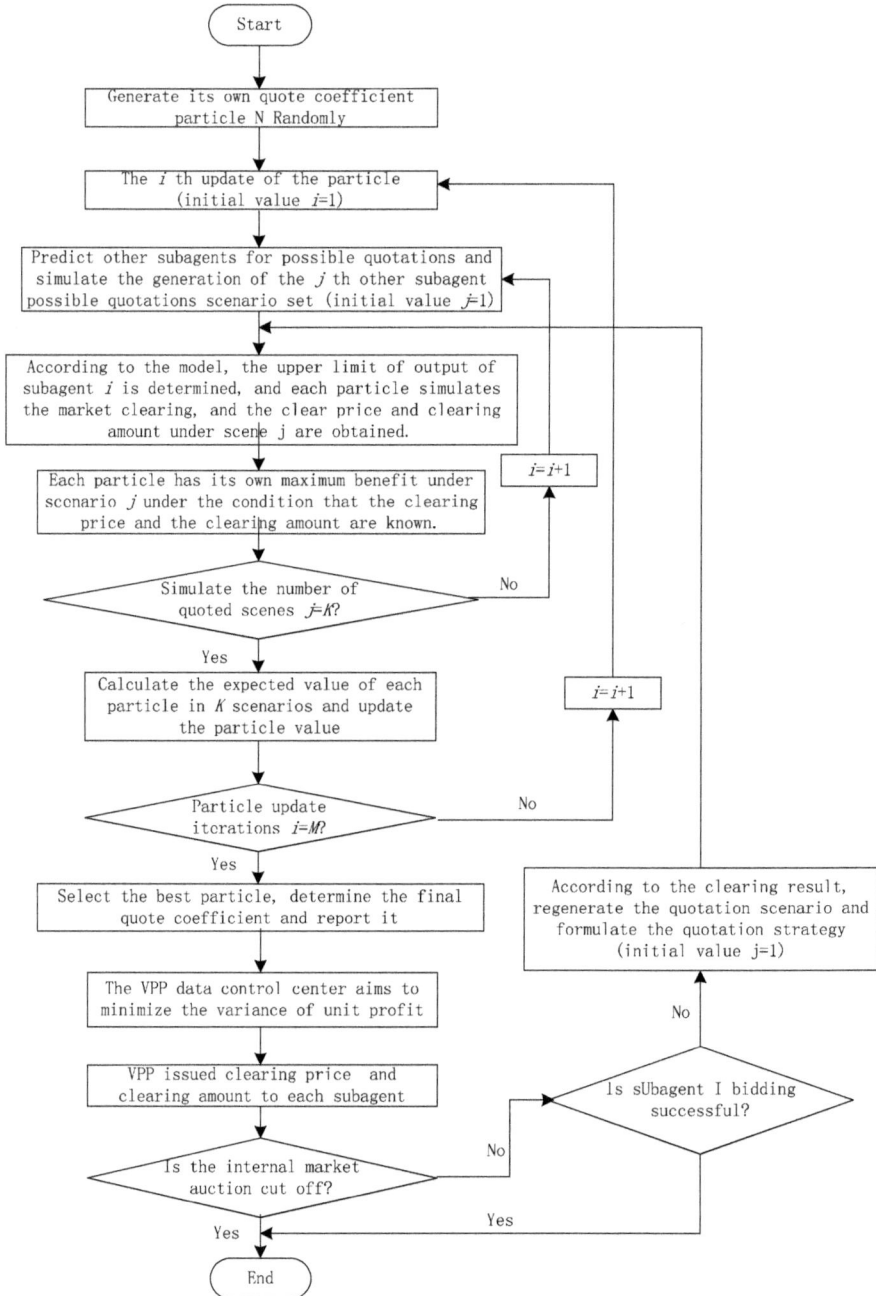

Figure 3. VPP internal market bidding flow chart.

3.3. VPP Dynamic Fame Bidding Model Based on Multi-Agent Technology

3.3.1. Model Establishing

All agents will pursue the maximization of their own interests during the bidding process. For this reason, the function argmax ($\bullet$) is introduced to represent a set of solution sets in the definition domain. Each set of solutions enables the function argmax ($\bullet$) to obtain the maximum value. From this, the Objective Function 1 shown in Equation (28) can be obtained.

$$
\left\{
\begin{array}{l}
P_1^{VPP*} \in \underset{P_1^{VPP}}{\text{argmax}}[(P_1^{VPP} - R_{in}) \cdot Q_1(P_1^{VPP}{}_1, \cdots, P_{n_1}^{VPP*})] \\
\qquad\qquad \vdots \\
P_{n_1}^{VPP*} \in \underset{P_{n_1}^{VPP}}{\text{argmax}}[(P_{n_1}^{VPP} - R_{in}) \cdot Q_{n_1}(P_1^{VPP*}, \cdots, P_{n_1}^{VPP})] \\
P_1^{Gen*} \in \underset{P_1^{Gen}}{\text{argmax}}[(P_1^{Gen} - C_1) \cdot Q_1(P_1^{Gen}, \cdots, P_{n_2}^{Gen*})] \\
\qquad\qquad \vdots \\
P_{n_2}^{Gen*} \in \underset{P_{n_2}^{Gen}}{\text{argmax}}[(P_{n_2}^{Gen} - C_{n_2}) \cdot Q_{n_2}(P_1^{Gen*}, \cdots, P_{n_2}^{Gen})]
\end{array}
\right\}
\tag{29}
$$

where the subscripts n_1 and n_2 represent the number of VPPs and conventional thermal power plants in the day-ahead market; $P_{n_1}^{VPP}$ and $P_{n_2}^{Gen}$ represent the bidding strategies of VPP n_1 and thermal power plant n_2, respectively; $P_{n_1}^{VPP*}$ and $P_{n_2}^{Gen*}$ represent the optimal bidding strategies for VPP n_1 and thermal power plant n_2, respectively; Q_{n_1} and Q_{n_2} refer to the supply of VPP n_1 and thermal power plant n_2, respectively, determined according to the bidding price; ND C_{n_2} is the total cost of external power generated by thermal power plant n_2.

TC uses UCP for settlement, with the goal of maximizing social benefits, and the social benefit is the sum of consumer surplus and producer surplus. The basic diagram is shown in Figure 4. The VPP can be defined as a consumer or producer during the quotation process, but can only be purchased or sold in one direction at a single time. Conventional generation companies generate quotes as producers.

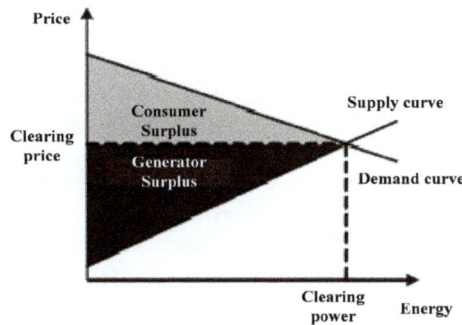

Figure 4. Consumer surplus and producer surplus [34].

The required social benefit is the area between the supply curve and the demand curve, so the Objective Function 2 can be obtained as Equation (30).

$$
\max E = \sum_{m=1}^{N_s} \int_0^{q_{s,m}} (R_{ex} - P_{s,m}(q)) dq + \sum_{n=1}^{N_c} \int_0^{q_{c,n}} (P_{c,n}(q) - R_{ex}) dq
\tag{30}
$$

$$P_{s,m}(q_{s,m}) = R_{ex} \forall m \in N_s \tag{31}$$

$$P_{c,n}(q_{c,n}) = R_{ex} \forall n \in N_c \tag{32}$$

where E is the social benefit at a certain moment; the subscripts s and c represent producers and consumers; N_s and N_c are the number of producers and consumers, respectively; $q_{s,m}$ and $q_{c,n}$ are the amount of production of producers m and consumers n at a certain time; and R_{ex} is clearing price in market at a certain moment.

The constraints during the clearing process include:

(1)   Low and high output constraints

$$Q_{s,m\min} \leq q_{s,m} \leq Q_{s,m\max} \quad \forall m \in N_s \tag{33}$$

$$Q_{c,n\min} \leq q_{c,n} \leq Q_{c,n\max} \quad \forall n \in N_c \tag{34}$$

where $Q_{s,m\max}$ and $Q_{s,m\min}$ are the upper and lower limits of the overall output power of producer m; and $Q_{c,n\max}$ and $Q_{c,n\min}$ are the load's upper and lower limits of consumer n, respectively.

(2)   System Power Balance Constraints

$$\sum_{m=1}^{N_s} q_{s,m} + \sum_{n=1}^{N_c} q_{c,n} = 0 \tag{35}$$

To ensure the feasibility of clearing results and prevent the occurrence of trend overruns, DC currents is used to carry out safety checks on the lines, as shown in Equations (36) and (37).

$$q_l = \sum_{m=1}^{N_s} sf_{m-l} q_{s,m} + \sum_{n=1}^{N_c} sf_{n-l} q_{c,n} \tag{36}$$

$$|q_l| \leq Q_{l\max} \tag{37}$$

where q_l is the power flow of line l at time t; sf_{m-l} and sf_{n-l} are node power transfer factors of node m and n to line l, respectively; and $Q_{l\max}$ is the active power flow upper limit of line l.

3.3.2. Dynamic Game Process

The calculation of this model includes two parts. The first part is the game competition among multi-agents. The second part is the dynamic game between VPP agents and trading centers. GA is used to simulate the process of mutual bidding between multiple agents. For the two objective functions of the bidding process, the hierarchical planning method is used for analysis [35]. The processes of selection, crossover, and mutation in GA are similar to the bidding rules between agents. In the process of selection, the optimal bidding strategy is based on the maximum revenue of each agent (Objective Function 1). That is, in the process of bidding, each agency tends to have a bidding strategy that maximizes the agency's own revenue. In the crossover process, the bid prices between the agents affect each other, and the agents change their own quotes according to the historical quote information of other agents. In the process of variation, in connection with the actual situation, the bids of each agency cannot be kept in good order, and they will suddenly reduce or raise the bid price according to their own situation and information, but the bid price will also be within a certain range. Using GA to simulate the bidding game process is a simulation scheme under the premise of following the actual situation of each agent and the electricity market. After the bidding process is over, the optimal solution is selected according to the maximum social benefit (Objective Function 2) in the population, that is, the optimal bidding scheme. The process is shown in Figure 5.

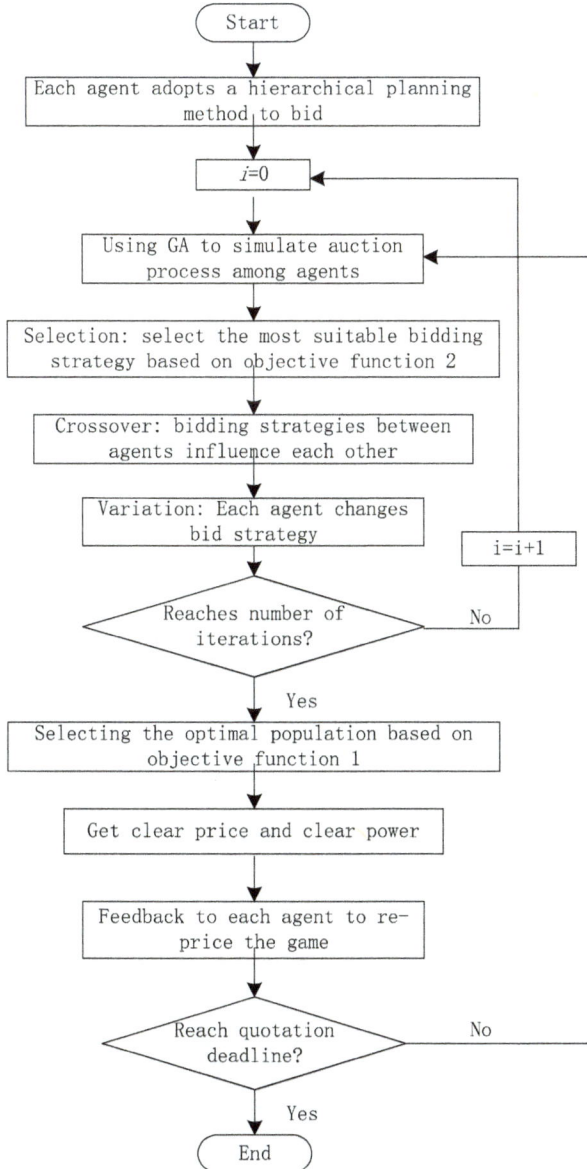

Figure 5. Multi-VPP external electricity market bidding flow chart.

4. Case Study

4.1. Case Description

An IEEE-9 node system with three VPPs was adopted for case verification, as shown in Figure 6. Line parameters can be found in [36]. The VPP Center Controller (VPPCC) is a monitoring and management unit that bears the functions of communication, monitoring, and management with various devices. To reflect the differences between different VPPs, VPP A consists of DE1, WT1, PV1, and ES1. VPP B consists of DE2, WT2, and ES2. VPP C consists of DE3, PV3, and ES3. The specific parameters are shown in Table 1.

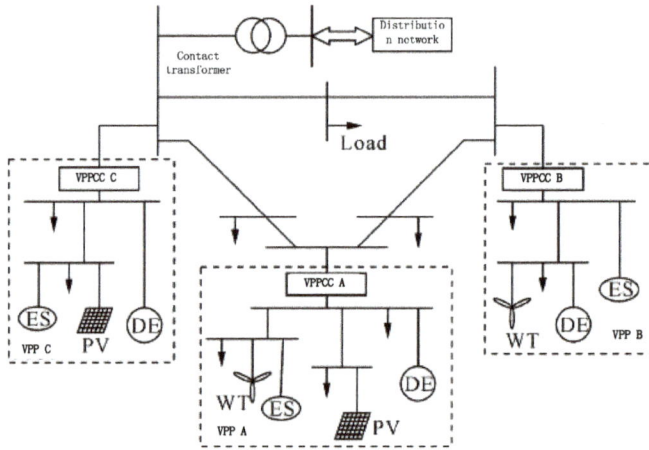

Figure 6. Multi-VPP system based on IEEE-9.

Table 1. DE and ES parameter settings in VPP.

Type	Rated Capacity/kW			Maximum Power/kW			Minimum Power/kW			Cost/(¥/kW)
	VPPA	**VPPB**	**VPPC**	**VPPA**	**VPPB**	**VPPC**	**VPPA**	**VPPB**	**VPPC**	
DE	180	120	180	180	120	180				0.56
ES	100	60	60	65	38	38	−15	−15	−15	0.98

4.2. VPP Internal Bidding Results

The normal distribution model of the load prediction error expects $\mu_L = 0.02$. The PV prediction model expects $\mu_{PV} = 0.07$ (sunny, error is ND). The wind power prediction model expects $\mu_{WG} = 0.05$ (wind speed interval is 5.1–9.8, error is ND). The model of TL load transfer rate expects $\mu_{TL,I} = 0.5$, $\mu_{TL,O} = 0.7554$. The model expects of the ES charge and discharge motion parameters is $\mu_{ES} = 0.8$, and the standard deviation of all models is $\sigma = 0.2$. Using Monte Carlo simulation (MCS) to generate electricity price data, C_{TL} takes the average price of electricity. $B = -1.3$, $\eta_{ch} = \eta_{dis} = 0.85$, $\lambda_0 = 0.8$, $a_{ch} = -0.8327$, $b_{ch} = 0.6255$, $a_{dis} = 1.5895$, $b_{dis} = 0$, $T_f = 9:00–22:00$, $T_g = 1:00–8:00$ and 23:00–24:00. Figure 7 shows the load and DER prediction results for the three VPPs.

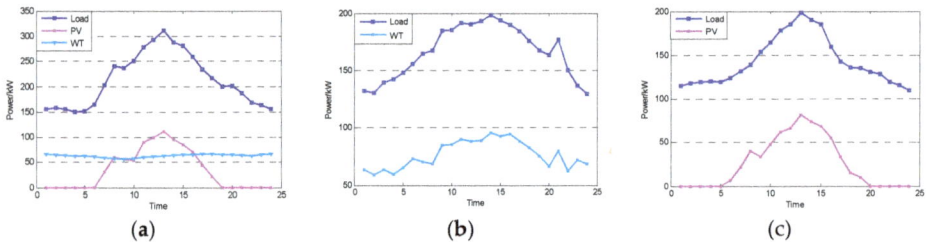

Figure 7. VPP internal load and DER forecast results: (**a**) VPP A; (**b**) VPP B; (**c**) VPP C.

Using Matlab software to program and run the model and algorithm, the dynamic game results of the internal market operation of VPP can be obtained. Taking VPP A as an example, Figure 8 shows the dynamic process of each subagent's bidding strategies in the market as the game times N increases in a certain period. In Figure 8, it can be seen that, after repeated games, the subagents obtain equilibrium solutions after about 200 rounds.

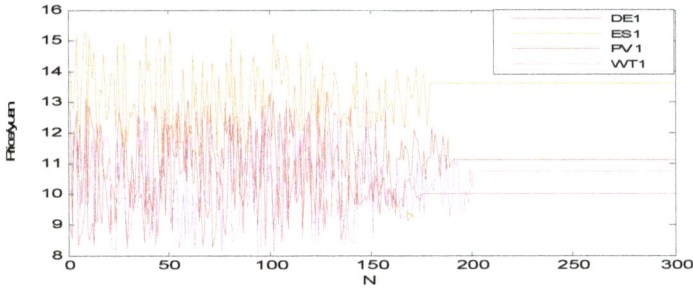

Figure 8. Dynamic game process of quoting strategies for subagents within VPPA.

The process shown in Figure 3 was used to solve the model. Table 2 only lists the quotations, the electricity quantity of acceptance of the bid, and market clearing prices for each subagent within the three VPPs at a given period. Assume that three VPPs participate in the day-ahead market bidding. The clearing electricity quantities obtained at the corresponding time are 310, 193, and 198 kW, respectively. Figure 9 shows the sub-agency quotes for each of the three VPP's internal day markets in 24 periods, while Figure 10 shows the winning bids for each subagent during per period.

Table 2. Subagent quotations, winning bids, and market clearing prices at a certain time within VPP.

		Quotation/¥	Winning Bids/kW	Profit/¥	Clearing Price/¥
VPPA	**DE1**	10.68	135	513	10.68
	ES1	13.60	15	6	
	PV1	11.11	100	200	
	WT1	10.72	60	84	
VPPB	**DE2**	10.88	105	375	10.88
	ES2	13.56	8	14	
	WT2	12.59	80	171	
VPPC	**DE3**	11.01	115	449	10.59
	ES3	13.35	8	10	
	PV2	10.59	75	209	

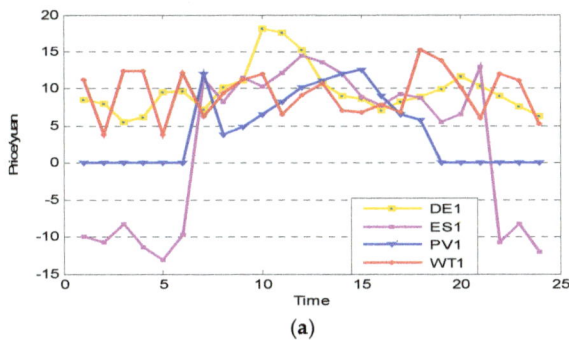

(a)

Figure 9. *Cont.*

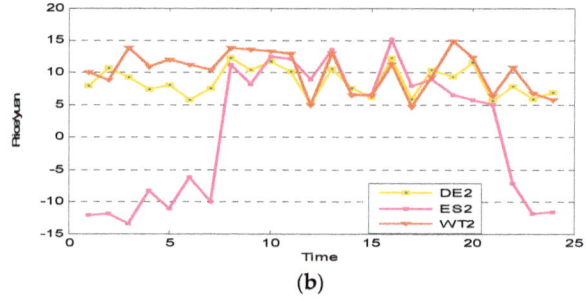

(b)

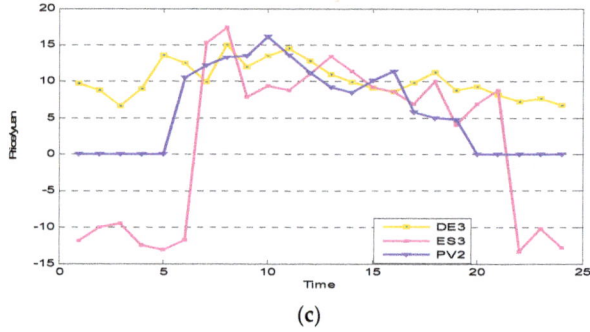

(c)

Figure 9. Price quoted by subagents in the VPP internal market: (**a**) VPP A; (**b**) VPP B; (**c**) VPP C.

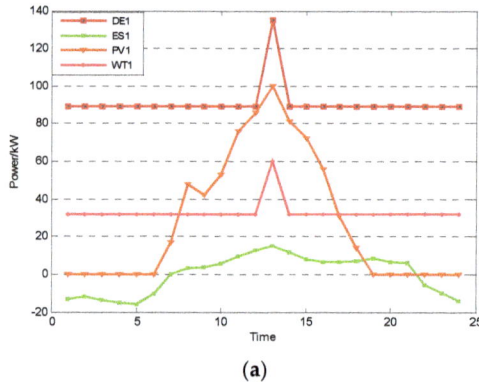

(a)

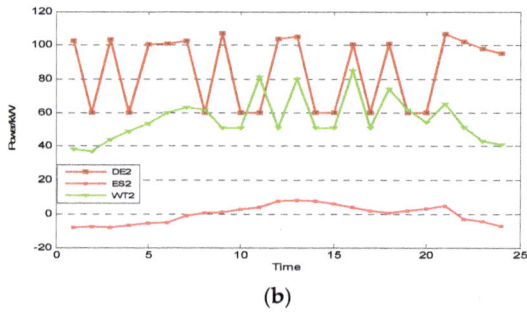

(b)

Figure 10. *Cont.*

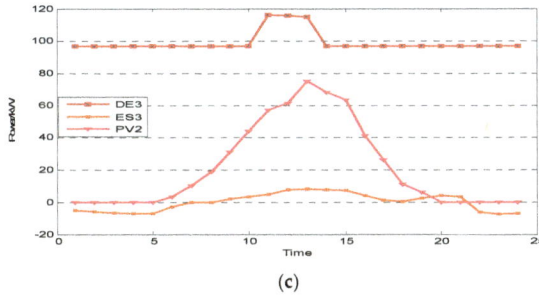

(c)

Figure 10. Electricity quantity of winning bids for each subagent in VPP internal market: (**a**) VPP A; (**b**) VPP B; (**c**) VPP C.

As can be seen from the quotations and market clearances of the three subagents within the VPP, the subagents' winning power curves are significantly different. VPP A has WT and PV at the same time, and it supplies power to the extra when a low load at night and a large surplus of photovoltaic power at noon. VPP B has wind power WT, so contributes surplus power at night. VPP C, which has PV, contributes the remaining power at noon. Renewable energy sources have greater volatility. At noon, the PV generation power is relatively large. ES absorbs the remaining power for charging and discharges the power at peak load. When the ES device absorbs power, it appears to be in the form of power purchase. Therefore, the quote is negative. When the energy is released, the external device shows the form of power sales. The quotation is positive.

According to Figures 9 and 10, analysis shows that, when it is noon (more PV power) or nighttime (more WT power), renewable energy power generation remains surplus, and sales are prioritized. Therefore, the auction prices of various subagents are generally low. Because WT has windy period (22:00–1:00) at night, WT 2 adopts a relatively low bidding strategy. For PV 2 with a large amount of PV residual power, there is no need to adopt such a strategy. There was larger load at noon (11:00–15:00) and PV power could be effectively consumed. At the same time, the overall bidding strategies tend to have a time-shared price, with a higher peak-loading period and a generally lower grain-loading rate.

4.3. Auction Result of VPP in the Day-Ahead Market

The analysis of multi-VPP bidding game model is based on the dynamic game model of subagents of VPP in lower layer. Using the Matlab software to program and run the model and algorithm, the dynamic game results of the day-ahead market operation of the power system with multiple VPPs can be obtained. The number of GA iterations is set to 50, the population size is 100, the crossover probability is set to 0.6, the mutation probability is 0.1, and the variable length is 120. Figure 11 shows the dynamic process of the three VPP agent's bidding strategies in the market as game times N increases.

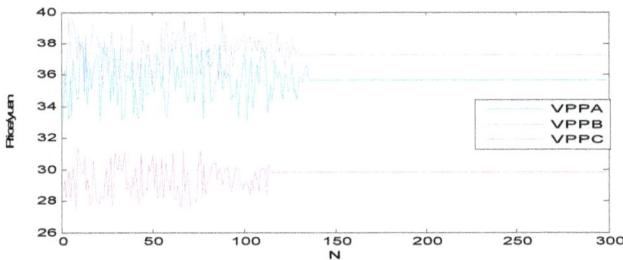

Figure 11. Dynamic game process of agents' quotation strategies.

The process shown in Figure 4 was used to solve the model. Figure 12 shows the quoted prices of three VPP agents in the daily market for 24 periods, and Figure 13 shows the winning bid conditions for each agent in per period.

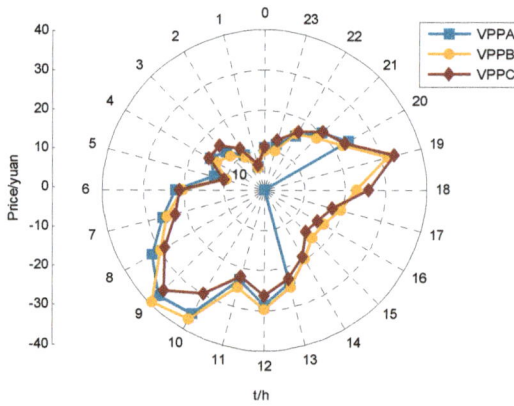

Figure 12. Price quotes for each agent in per period.

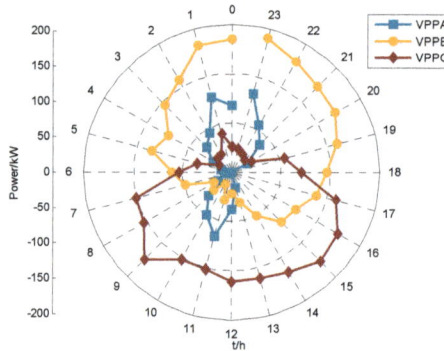

Figure 13. Neutral power of each agent during per period.

In the above figures, there are a few differences between the proxy quotes in each period. During 15:00–20:00, VPP A encounters its own maximum load peak, during which the external supply is 0, thus it does not participate in the market bidding. Considering that VPP B has many wind power sources, it releases more electricity to the outside at night. There are many PV power sources in VPP C, thus there are relatively more efforts in the afternoon. The points in Figure 13 represent the externally available load values of VPP, and the VPP bids from the lowest to the highest order, and the available load is superimposed from the inside to the outside. The points in the figure indicate the value of VPP that can be awarded to the outside during the bidding process. Taking time 13:00 as an example, the blue dots (VPP A), the yellow dots (VPP B), and the red dots (VPP C) are lined up from low to high, which shows that, in the bidding process, VPP A has the lowest price, VPP B has the second, VPP C has the highest price, and the TC prefers to select VPP A. Of the 198 kW demanded load at 13:00, VPP A provides 87.6 kW of total external power, followed by VPP B which provides 43.1 kW of total external power, and the remaining portion is borne by VPP 3.

5. Discussion

With the large-scale grid-connected operation of DG, its characteristics of volatility and intermittent have brought certain negative impact on grid voltage, power quality, and dispatch operation. VPP provides new ideas for the development of new energy by aggregating this type of distributed energy. However, the current research on VPP mainly focuses on the overall external operation mode, and usually ignores the internal distributed energy complementary operation modes and benefits. The research can be divided into two parts. The first part is about the VPP of multi-investment subject. Based on the multi-agent technology and the theory of Stackelberg game, the dynamic game bidding model of VPP internal subagent was constructed. Then, in the second part, referring to the clearing results after each round of internal games, a VPP dynamic game bidding model based on multi-agent technology was established.

6. Conclusions

Through the previous research analysis and argumentation, the main conclusions of this paper can be summarized as follows.

Firstly, based on a brief overview of the basic components of VPP, the uncertainties of photovoltaic power generation, wind power generation, and demand-side response within VPP were modeled separately.

Secondly, regarding DER as a subagent of VPP, a subagent's bidding model in VPP internal market based on dynamic game theory was established. The MCS was used to determine the price coefficient of each subagent, and, after repeated games determining its own optimal price and affecting the power system's day-ahead market clearing, an optimal equilibrium solution was obtained.

Finally, based on the completion of each round of VPP internal market clearing, with reference to the outcome of each round of clearing, a bidding strategy for VPP to participate in the day-ahead market was developed. Based on multi-agent technology, a dynamic competitive game model for electricity market with multiple VPPs was established. VPP and traditional thermal power plants were agents. Based on the methods and strategies that each agency may adopt in the market competition, the dynamic game behavior between VPP and trading market was simulated by GA, and a complete game process of bidding was presented. Verified by examples, the model can provide methods for the efficient use of resources in power systems with multiple VPPs.

Author Contributions: Y.G. carried out the main research tasks and wrote the full manuscript. X.Z. proposed the original idea, analyzed and double-checked the results and the whole manuscript. J.R. contributed to data processing and to writing and summarizing the proposed ideas. X.W. and D.L. provided technical and financial support throughout.

Funding: This work was supported in part by the Nature Science Foundation of China (51607068), the Fundamental Research Funds for the Central Universities (2018MS082), and the Fundamental Research Funds for the Central Universities (2017MS090).

Conflicts of Interest: The authors declare no conflict of interest.

Abbreviations

The following abbreviations are used in this manuscript:

VPP	Virtual power plant
DER	Distributed energy resource
TC	Trading left
PSA	Particle swarm algorithm
WNN	Wavelet neural network
DG	Distributed generation
MG	Micro grid
EV	Electric vehicle
WT	Wind turbine
PV	Photovoltaic

ES	Energy storage
DSR	Demand side resource
DE	Diesel generator
TL	Transfer load
IL	Interruptible load
TP	Time-sharing price
GA	Genetic algorithm
UCP	Uniform clearing price
VPPCC	VPP left Controller
ND	Normal distribution
MCS	Monte Carlo simulation

References

1. Li, P.; Zhang, X.S.; Zhao, B.; Wang, Z.L.; Sun, J.L. Design and mode switching control strategy for multi-microgrid multi-grid point structure microgrid. *Power Syst. Autom.* **2015**, *39*, 172–178.
2. Fang, X.; Yang, Q.; Wang, J.; Yan, W. Coordinated dispatch in multiple cooperative autonomous islanded microgrids. *Appl. Energy* **2016**, *162*, 40–48. [CrossRef]
3. Xue, M.D.; Zhao, B.; Zhang, X.S.; Jiang, Q.Y. Optimized configuration and evaluation of grid-connected microgrid. *Electr. Power Syst. Autom.* **2015**, *39*, 6–13.
4. Kardakos, E.G.; Simoglou, C.K.; Bakirtzis, A.G. Optimal offering strategy of a virtual power plant: A stochastic bi-level approach. *IEEE Trans. Smart Grid* **2016**, *7*, 794–806. [CrossRef]
5. Zhu, J.Q.; Duan, P.; Liu, M.B. Electric power real-time balanced dispatching based on risk and source-net-load bi-level coordination. *Proc. CSEE* **2015**, *35*, 3239–3247.
6. Zhou, Y.Y.; Yang, L. Synergetic scheduling model based on classical scene sets for virtual water power plants. *Power Syst. Technol.* **2015**, *39*, 1855–1859.
7. Niu, W.J.; Li, Y.; Wang, B.B. Demand-responsive virtual power plant modeling considering uncertainties. *Proc. CSEE* **2014**, *34*, 3630–3637.
8. Sun, S.; Yang, Q.; Yan, W. A novel Markov-based temporal-SoC analysis for characterizing PEV charging demand. *IEEE Trans. Ind. Inform.* **2018**, *14*, 156–166. [CrossRef]
9. Zhao, H.; Wu, Q.; Hu, S.; Xu, H.; Rasmussen, C.N. Review of energy storage system for wind power integration support. *Appl. Energy* **2015**, *137*, 545–553. [CrossRef]
10. Liu, Y.Y.; Jiang, C.W.; Tan, S.M.; Hu, J.Z.; Li, Q.S. Optimal scheduling strategy of virtual power plant considering risk-adjusted threshold of capital yield. *Proc. CSEE* **2016**, *36*, 4617–4627.
11. Zamani, A.G.; Zakariazadeh, A.; Jadid, S. Day-ahead resource scheduling of a renewable energy based virtual power plant. *Appl. Energy* **2016**, *169*, 324–340. [CrossRef]
12. Zang, H.X.; Yu, S.; Wei, Z.N.; Sun, G.Q. Two-layer optimal scheduling of virtual power plant considering safety constraints. *Electr. Power Autom. Equip.* **2016**, *36*, 96–102.
13. Guo, H.X.; Bai, H.; Liu, L.; Wang, X.L. Virtual power plant optimization scheduling model under uniform energy market. *J. China Electrotech. Soc.* **2015**, *30*, 136–145.
14. Rahimiyan, M.; Baringo, L. Strategic bidding for a virtual power plant in the day-ahead and real-time markets: A price-taker robust optimization approach. *IEEE Trans. Power Syst.* **2016**, *31*, 2676–2687. [CrossRef]
15. Mnatsakanyan, A.; Kennedy, S.W. A novel demand response model with an application for a virtual power plant. *IEEE Trans. Smart Grid* **2015**, *6*, 230–237. [CrossRef]
16. Liu, J.N.; Li, P.; Yang, D.C. Virtual power plant bidding strategy based on combined wind and solar storage optimization. *Electr. Power Eng. Technol.* **2017**, *36*, 32–37.
17. Song, W.; Wang, J.W.; Zhao, H.B.; Song, X.J.; Li, W. Multi-stage bidding strategy for virtual power plant considering demand-responsive trading market. *Power Syst. Prot. Control* **2017**, *45*, 35–45.
18. Yang, J.J.; Zhao, J.H.; Wen, F.S.; Xue, Y.S.; Li, L.; Lv, H.H. Competitive bidding strategy for virtual power plants with electric vehicles and wind turbines. *Electr. Power Syst. Autom.* **2014**, *38*, 92–102.
19. Peik-Herfeh, M.; Seifi, H.; Sheikh-El-Eslami, M.K. Decision making of a virtual power plant under uncertainties for bidding in a day-ahead market using point estimate method. *Int. J. Electr. Power Energy Syst.* **2013**, *44*, 88–98. [CrossRef]

20. Fang, Y.Q.; Gan, L.; Ai, X.; Fan, S.L.; Cai, Y. Bi-level bidding strategy of virtual power plant based on master-slave game. *Electr. Power Syst. Autom.* **2017**, *41*, 61–69.

21. Mashhour, E.; Moghaddas-Tafreshi, S.M. Bidding strategy of virtual power plant for participating in energy and spinning reserve markets—Part I: Problem formulation. *IEEE Trans. Power Syst.* **2011**, *26*, 949–956. [CrossRef]

22. Mashhour, E.; Moghaddas-Tafreshi, S.M. Bidding strategy of virtual power plant for participating in energy and spinning reserve markets—Part II: Numerical Analysis. *IEEE Trans. Power Syst.* **2011**, *26*, 957–964. [CrossRef]

23. Zhou, Y.Z.; Sun, G.Q.; Huang, W.J.; Xu, Z.; Wei, Z.N.; Chen, S.; Chen, S. Multi-regional virtual power plant integrated energy coordination scheduling optimization model. *Proc. CSEE* **2017**, *37*, 6780–6790.

24. Yuan, G.L.; Chen, S.L.; Liu, Y.; Fang, F. Economical optimal scheduling of virtual power plants based on time-of-use pricing. *Power Syst. Technol.* **2016**, *40*, 826–832.

25. Wang, Y.; Ai, X.; Tan, Z.; Yan, L.; Liu, S. Interactive dispatch modes and bidding strategy of multiple virtual power plants based on demand response and game theory. *IEEE Trans. Smart Grid* **2016**, *7*, 510–519. [CrossRef]

26. Chitsaz, H.; Amjady, N.; Zareipour, H. Wind power forecast using wavelet neural network trained by improved Clonal selection algorithm. *Energy Convers. Manag.* **2015**, *89*, 588–598. [CrossRef]

27. Wang, S.; Zhang, N.; Wu, L.; Wang, Y. Wind speed forecasting based on the hybrid ensemble empirical mode decomposition and GA-BP neural network method. *Renew. Energy* **2016**, *94*, 629–636. [CrossRef]

28. Yadav, A.K.; Chandel, S.S. Solar radiation prediction using Artificial Neural Network techniques: A review. *Renew. Sustain. Energy Rev.* **2014**, *33*, 772–781. [CrossRef]

29. Wei, B.; Suo, Q.; Lin, X.N.; Li, Z.T.; Chen, L.; Deng, K.; Bo, Z.Q.; Huang, J.G.; Muhammad, S.K.; Owolabi, S.A. Daily operation energy control optimization strategy of wind/light/chai/gull islanding microgrid considering transferable load efficiency. *Chin. J. Electr. Eng.* **2018**, *38*, 1045–1053.

30. Ai, X.; Zhou, S.P.; Zhao, Y.Q. Research on an optimal scheduling model with interruptible load based on scenario analysis. *Proc. CSEE* **2014**, *34*, 25–31.

31. Prabavathi, M.; Gnanadass, R. Energy bidding strategies for restructured electricity market. *Int. J. Electr. Power Energy Syst.* **2015**, *64*, 956–966. [CrossRef]

32. Nguyen, D.T.; Le, L.B. Optimal bidding strategy for microgrids considering renewable energy and building thermal dynamics. *IEEE Trans. Smart Grid* **2014**, *5*, 1608–1620. [CrossRef]

33. Wu, S.P.; Hu, Z.C.; Song, Y.H. Optimal configuration method of wind farm energy storage combining stochastic programming and sequential monte carlo simulation. *Power Syst. Technol.* **2018**, *42*, 1055–1062.

34. Zhu, Q.L.; Ma, X.S.; Yuan, J.S. Uncertainty planning method of power generation companies' competitive bidding strategies considering transmission capacity constraints. *J. China Electrotech. Soc.* **2012**, *27*, 216–223.

35. Zhang, L.; Jin, Z.D. Research on hierarchical planning method of distribution network based on particle swarm optimization. *Shaanxi Electr. Power* **2014**, *42*, 44–47.

36. Feng, Y.; Qi, Z.J.; Zhou, Q.; Sun, J.W. Online risk assessment and prevention control considering wind power access. *Electr. Power Autom. Equip.* **2017**, *37*, 61–68.

Review

Local Markets for Flexibility Trading: Key Stages and Enablers

Simone Minniti *, Niyam Haque, Phuong Nguyen and Guus Pemen

Electrical Energy Systems Group, Eindhoven University of Technology, 5600 MB Eindhoven, The Netherlands; a.n.m.m.haque@tue.nl (N.H.); p.nguyen.hong@tue.nl (P.N.); a.j.m.pemen@tue.nl (G.P.)
* Correspondence: s.minniti@tue.nl

Received:16 October 2018; Accepted: 1 November 2018; Published: 8 November 2018

Abstract: The European energy transition is leading to a transformed electricity system, where Distributed Energy Resources (DERs) will play a substantial role. Renewable Energy Sources (RES) will challenge the key operational obligation of real-time balancing and the need for flexibility will consequently increase. The introduction of a local flexibility market (LFM) would allow the trading of flexibility supplied by both producing and consuming units at the distribution level, providing market access to DERs, a support tool for Distribution System Operators (DSOs) and a value stream for energy suppliers. Aggregators and DSOs for different reasons can enhance the valuation of flexible DERs. Several research papers have assumed aggregators fully interacting with the electricity markets and DSOs contracting services with power system actors. These interactions are still not allowed in many European countries. This article aims to analyze the European regulation to identify the most important enablers and pave the way towards the full exploitation of DER flexibility, culminating in the establishment of an LFM. Therefore, three main stages, emerging from the progressive withdrawal of the current regulatory and market barriers, are identified: (1) enabling the aggregator's trading, (2) evolution of the DSO's role, and (3) key-design challenges of an LFM.

Keywords: aggregator; distribution system operator; distributed energy resources; local flexibility market; flexibility service

---

1. Introduction

The currently undergoing transformation of the electrical power system has the integration of renewable energy sources (RES) as one of its main pillars. The traditional, centralized way of producing and managing electrical power is being substituted by a decentralized manner with increased contribution from local resources. This will pose new challenges in keeping the real-time balance between electricity demand and supply while efficiently operating the entire system. RES, in particular, photovoltaics (PV) and wind power, are intermittent, and the difficulty of predicting their power production increases the need for system flexibility [1,2]. Currently, the Transmission System Operator (TSO) is responsible for the system balance, and this can be ensured by procuring reserve from the generation-side resources. While causing an increased need for flexibility, RES production also has dispatch priority and it is slowly pushing the fossil fuel-fired power plants out of the market, thus limiting one of the main sources of flexibility. This is the reason why, nowadays, more and more conventional power-plants are struggling to keep their market share and gain profit [3,4].

Other energy sectors are looking at electrification as the solution to energy transition, with electric vehicles from the transport sector and heat pumps from the heating and cooling sector being the most evident examples. Worldwide, the number of countries that have established renewable-related policies regarding the power sector is way higher than the other two aforementioned energy sectors [5], meaning that the main responsibility is with the power system itself. As a result, the integration of

Distributed Energy Resources in the distribution network (DN) will be at the core of the shift from traditional to smart electricity networks. This includes Electric Vehicles (EVs), Dispatchable Loads (DLs), Energy Storage (ES), as well as Distributed Generation (DG). These four subcategories with related examples are represented in Figure 1.

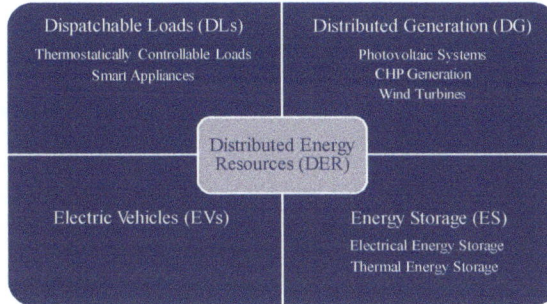

Figure 1. Distributed Energy Resource (DER) subcategories.

Distributed Energy Resources (DERs) are shaping the future of the distribution network (DN), which was not originally designed to host a significant level of local generation, and Distribution System Operators (DSOs) need to reconsider the DN's operations and planning in the presence of distributed generation [6]. The traditional 'fit and forget' approach based on the expected long-term peak load will no longer be suitable, because the installation rate of distributed resources is growing substantially [7]. DERs are raising challenges from the operational point of view by introducing reverse flow, congestion problems, and voltage limit violations. DSOs are used to passively operate, maintain, and upgrade the DN based on a fixed remuneration set yearly by the regulator. The on-going discussion at the European level is focused on the modification of DSO's business model towards the role of an active manager of the grid, since both in the short-term planning and in the real-time phase, it will deal with the dynamics of renewable production and with the increased coincidence factor due to DERs. All these resources are also, in fact, capable of modifying their production or consumption pattern upon request (direct or indirect control) by the owner/energy supplier/network operator. This capability can be exploited to provide a flexibility service to different parties. Therefore, beside increasing technical problems, DERs can also offer a potential solution throughout their inherent controllability. The DSO should be enabled to use the potential benefits deriving from flexibility existing within the distribution network.

The flexibility potential offered by DERs is considerable [8], and it is suitable for tackling location-dependent problems (such as congestions) but also, if properly aggregated, for supporting the traditional flexibility sources at the transmission level. Its exploitation is currently limited for two reasons. Firstly, the flexibility of DERs cannot be used due to a lack of Information and Communication Technology (ICT) infrastructure and regulatory framework; secondly, the majority of the flexibility is disaggregated as it belongs to small and medium-sized customers and consequently it is unable to access the market.

In this context, the aggregator is emerging as a new figure: an energy utility taking care of directly or indirectly managing and operating the flexibility of DERs for small and medium-size prosumers which are otherwise disaggregated. The aggregators can potentially bridge the gap between the electricity markets and small-scale DERs. Consumers transform from passive grid users to active providers of various energy services to the grid, both consuming and producing on-site energy—becoming prosumers. Prosumers will potentially be able to interact in the energy markets and compete directly with traditional energy utilities [9] or indirectly through aggregators.

The European Commission, within the new proposal for a directive on common rules on the internal energy market released in November 2016 (part of the so-called "Clean Energy for

All" package), underlines the importance of distribution-level flexibility and introduces new definitions of related concepts such as 'demand response', 'independent aggregator', 'active customer', and 'dynamic electricity price contracts'. When both wholesale and reserve markets were designed, product requirements and arrangements of access were tailored around the capabilities of traditional generating units. The system operators relied on the supply-side flexibility and there was no interest in procuring flexibility from the demand-side or from distributed resources. In the recent years, European institutions and policymakers have brought attention to this new unexploited flexibility potential alongside the willingness to move the electricity markets closer to the final user, resulting in a growing interest towards local markets [10–12]. In local energy markets, prosumers can (directly or via aggregators) gain more benefit from valuing their resources by offering energy or flexibility services at the local level.

The introduction of a local flexibility market (LFM) with the aim of trading flexibility supplied by both producing and consuming units would provide market access to DERs and would provide a support tool to allow DSOs to handle technical problems. Potentially, energy suppliers also could be enabled to participate in the process of adjusting their market imbalances, but the coexistence of market players and grid operators pursuing flexibility for different purposes is a key discussion point.

This article aims to define a pathway for reaching the appropriate regulatory context in which LFM could be integrated and thus enabling full exploitation of distribution level flexibility from different stakeholders. In particular, we focus on aggregators which are currently not regulated in most of the European countries and on the evolution of DSOs towards a more active role in managing the distribution grid.

The contribution of the paper is threefold:

1. Identification of three main stages to gradually enable the flexibility of DERs for full exploitation, eventually through the establishment of LFM;
2. Identification of the main stakeholders (aggregators and DSOs) of the small-scale flexibility and definitions of market and regulatory barriers to overcome for each of them.
3. Discussion of key challenges of the LFM design including the coupling of grid and market-service and the coordination with overlaying markets.

Section 2 introduces the definition and main characteristics of the flexibility concept focusing on DERs' service provision. Section 3 elaborates on the following three main stages, which will gradually enable the full-exploitation of flexibility at the distribution level, leading to the establishment of the LFM:

*   Enabling the aggregator's trading (Section 4);
*   Evolution of the DSO's role (Section 5);
*   Key design challenges of local flexibility markets (Section 6).

Discussion and conclusions will follow in Section 7, summing up significant findings and individuating future research trends.

2. Flexibility Provision by DERs

RES power output depends on the availability of the primary source. The uncertainty in forecasting, even in the short-term, of these primary sources (e.g., wind, irradiation) is the main obstacle in handling the variability of RES. The resulting technical difficulties increase the need for operational flexibility. TSOs procure this flexibility through two phases: firstly, long-term contracts or market mechanisms in place for the procurement of reserve from generation companies (ancillary services procurement markets) and secondly, market mechanisms for reserve activation (ancillary services activation markets). The regulatory efforts now are focused on re-designing the short-term markets to provide fair market access for all sources of flexibility, including distributed generation and active demand, namely DER flexibility. Many definitions are present in literature for 'flexibility' [13–15].

From a system perspective, it can be defined as the capability of the power system to adapt its production or consumption with respect to sudden changes, expected or not. At the individual level, it is often referred to as the modification of the consumption or injection pattern due to direct or indirect signals. As highlighted in [16], electric flexibility is a heterogeneous commodity with multiple attributes; these main attributes are reported in Figure 2 and are related to the deviation from a baseline, which corresponds to the usual consumption or production pattern.

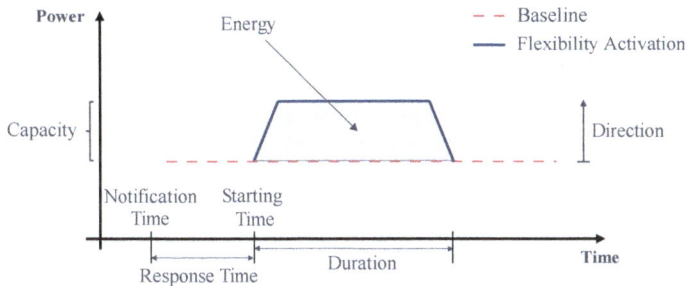

Figure 2. Example of flexibility activation with relevant attributes.

Flexibility cannot be characterized using a single metric. Several dimensions should be taken into account: capacity, duration, ramp rate, direction, energy content, response time, and location [17]. One type of DER might be efficient in one dimension and inefficient in another. For example, PV systems can provide only upward regulation via the curtailment of production, while battery storage can provide both upward and downward regulation. In terms of duration, a residential Thermostatically Controllable Load (TCL) can modify its consumption by up to a few hours; however, the change in power is not significant, because it is bound by the violation of the comfort constraints of the end-user. Extensive categorization of DER regarding different dimensions is conducted in [18]. The relevance of each attribute also depends on the service that the resource has to provide—the location is important if the problem to be solved is location-dependent (congestion, overvoltage), while it is irrelevant if the flexibility is used for adjusting market imbalances. One fundamental difference in the DER flexibility is between residential and industrial types. The exploitation of industrial flexibility for providing services is more convenient—bigger-sized loads imply a smaller amount of resources is required to reach the minimum size for bidding in the market. Furthermore, in the industry sector, the information and communication infrastructure is usually already installed for the purpose of energy management and a smaller investment is required to deploy industrial assets. On the contrary, residential demand flexibility introduces more complications, as follows [19]:

- A layer of uncertainty is added when predicting and procuring the available flexibility from residential customers because these loads have direct relations with the users' behavior and comfort. For example, controlling the set-points of residential heat pumps for space heating and thus changing the power consumption will directly affect the internal temperature of the building;
- Inter-temporal constraints of some resources, such as smart appliances, come in play. The flexibility that is available at time t depends on the flexibility offered and then activated at time $t - 1$; this makes the scheduling process of the flexibility available for a longer time-horizon more challenging;
- The size of the loads and costs to be incurred for a single household make the choice of participating in the market unprofitable.

These drawbacks can be partially overcome by aggregating small resources together. The aggregation of small and medium-sized resources will shortcut the access to the market, with the aggregator acting as an intermediary. Flexibility from distribution level resources can be used for different purposes; specifically, the scope of these flexibility services is threefold [20]:

- Market-oriented services: Balance Responsible Parties (BRPs) can use flexibility to correct unbalanced positions in their portfolios;
- System-oriented services: TSOs can use flexibility for the procurement and activation of reserve for balancing purposes;
- Grid-oriented services: the DSOs can use the DERs' flexibility to solve local problems, such as feeder (or transformer) overloadings.

The exploitation of flexibility from DERs will be fully enabled when the owner (or manager) is acknowledged as a market participant who is allowed to offer services in all organized markets with the only restriction of not jeopardizing the security of the power system. As an example, if the aggregator provides flexibility from distribution level DERs to the TSO, this service provision should not put the safe operation of the distribution network itself at risk. The regulation should consider this facet by harmonizing the provision of multiple flexibility services from DERs existing in the distribution network. In the following sections, we discuss the barriers which, in the current regulatory framework, do not allow the use of small-scale flexibility to interact with the electricity markets.

3. Three Stages towards the Full Exploitation of DERs' Flexibility

As previously said, the aggregator can bridge the gap between DERs and the electricity markets. Thus, the first step towards the use of bottom-up flexibility services is the definition of a regulation concerning the role and responsibilities of the aggregator, allowing him to offer services to available market mechanisms. Market access for aggregated small units remains critical, despite the willingness of the European Union to exploit demand response resources as a flexibility source in the power system. Regulatory changes are needed to allow interactions between the two levels, especially by means of enabling the aggregator's trading. With the aggregator being able to offer its flexibility from the distribution level up to short-term markets, both system-oriented and market-oriented service provision can be realized. This topic is elaborated in Section 4.

A parallel paradigm change will enable DSOs to evolve towards the role of active manager of their networks that is able to procure flexibility within the geographical area where they operate. DERs will then become the main flexibility source to handle local problems, giving DSOs an alternative to avoid or defer grid reinforcement [21]. In the context of a liberalized electricity market, the establishment of a market-based procedure to value DER flexibility is considered to be one of the possible solutions. DSO performs a grid-oriented use of the flexibility, deciding during the planning process whether to call for a market-based procurement or to proceed with grid reinforcement. This market mechanism is intended to be only for the grid-oriented use of flexibility without competition on the buyer side. Enhanced DSO–TSO coordination is required to allocate the services according to the priorities of the power system or at least to coordinate the operation of different system operators with flexible resources available at both levels. Section 5 focuses on the second stage. The regulatory change concerning the DSO's role is expected to happen over a longer term than the first one, since the aggregation service is already allowed in some European Countries, even if it is often provided by energy suppliers.

Once all the potential participants are enabled, a final step will be the establishment of an LFM where system actors can trade their flexibility for different purposes. At this stage, three types of players are expected to trade in the LFM:

- The aggregators value the DERs' available flexibility;
- The DSO requests flexibility for grid-oriented use;
- BRPs trade flexibility for market-oriented use.

A fully-competitive LFM can be implemented, namely allowing competition on both the buyer and seller sides. The design of this market mechanism is complex, especially for the coexistence of parties requesting flexibility for different purposes (market and grid-oriented use); in other words, the DSO requests flexibility for managing the network while the BRPs request flexibility for market imbalances. At the same time, the market mechanism should be aligned with the already established markets, in particular, with the AS market managed by the TSO. The coordination is required because, e.g., aggregators are able to offer in different markets and the service offered in one market (reserve for the TSO) can worsen/improve the situation on the LFM (less/more flexibility requests from the DSO). These aspects are explored in more detail in Section 6.

4. Enabling Aggregator Trading

In the last European Directive on Energy Efficiency [22], the aggregator was essentially defined as a demand-response provider, i.e., as a means to gather short-duration customer load that is otherwise unable to participate in any organized energy market. The focus was on the demand-response as a tool to empower customers and promote energy efficiency. In [23], the aggregator is defined as "a market participant that combines multiple customer loads or generated electricity for sale, for purchase or auction in any organized energy market"; the definition has changed to include both consuming and producing units, thus considering all aspects of the DERs' flexibility.The aggregation is a commercial function of pooling decentralized generation and/or consumption to provide energy and services to actors within the system [24]. The aggregation service is the key enabler of investment and growth in the demand response and decentralized management of DERs [25]. Economies of scale can be realized since the fixed costs of market participation and communication infrastructure decrease as the number of providers increases [26]. Moreover, aggregation mitigates the risk for error forecasting in the availability of DER resources and thus limits the risk for not meeting market commitment [27]. Figure 3 indicates the possible interactions of the aggregator with existing wholesale markets. It can interact with the day-ahead (DA) market and intra-day (ID) market to providing market services and with AS markets to providing system services.

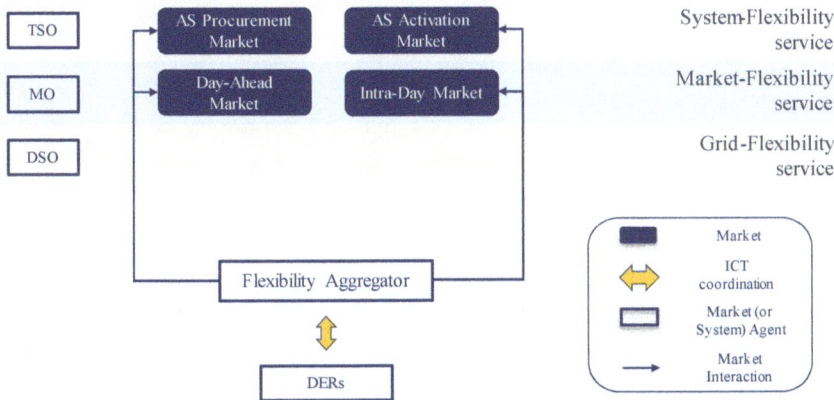

Figure 3. First stage: the aggregator should be enabled to interact with already established short-term markets to provide market services and system services.

It is worth noting the difference between an aggregator and an independent aggregator. The aggregator role can also be carried out by an energy supplier that provides the aggregation service for DERs. The energy supplier can thus exploit the flexibility of the energy resources which normally involves supplying energy and offering the end-user a lower service fee for the availability of its resources. On the other hand, an independent (or third-party) aggregator is an energy utility that can manage DERs' flexibility without having any contract with an energy supplier—the aggregator in this case has a separate contract with the DERs' owner.

4.1. Barriers to Market Access

The role of the aggregator is not even defined in most European countries, and there is no legislation regarding how this new role will be embedded in the actual framework. The aggregation service is still not completely allowed for short-term markets. In Table 1, an overview of the aggregation service for the short-term markets for 19 European countries is reported based on data from [28–30]. Each MS has different settings for the DA (e.g., closure time) and ID markets (e.g., number and duration of market windows). The same holds for reserve procurement; thus, the number or existing type of mechanism (e.g., market-based, auction-based, contract-based) and the product definition (e.g., automatic, manual, emergency reserve) depend on the type of reserve (primary, secondary or tertiary) and on the national regulation. An affirmative answer in this table means that the aggregation service is allowed in at least one of the existing market/procurement mechanisms.

Table 1. Aggregator's access to short-term markets in 19 European countries in 2017.

Country	Day-Ahead Market	Intra-Day Market	FCR [1]	FRR [2]	RR [3]
Austria	Only gen.	Only gen.	No	Yes	Yes
Belgium	No	No	Yes	No	Yes
Denmark	Yes	Yes	Yes	Yes	Yes
Estonia	No	No	No	Yes	No
Finland	Yes	Yes	Yes	Yes	Yes
France	Yes	Yes	Yes	Yes	Yes
Germany	Yes	Yes	Yes	Yes	Yes
Great Britain	No	No	Yes	Yes	Yes
Greece	No	No	No	No	No
Ireland	Yes	Yes	Yes	Yes	Yes
Italy	No	No	No	No	No
Netherlands	Yes	Yes	No	Yes	No
Norway	Yes	Yes	Yes	Yes	No
Poland	Yes	Yes	No	No	No
Portugal	No	No	No	No	No
Slovenia	No	No	No	No	Yes
Spain	No	No	No	No	No
Sweden	Yes	Yes	Yes	Yes	Yes
Switzerland	Yes	Yes	Yes	Yes	Yes

ENTSO-E terminology is used for the reserve nomenclature: [1] Frequency Containment Reserve or Primary Reserve; [2] Frequency Restoration Reserve or Secondary Reserve; [3] Replacement Reserve or Tertiary Reserve.

It can be seen that the nordic countries (Denmark, Finland, Norway, Sweden) are front runners in promoting the aggregation service. Southern countries are still lagging in allowing the aggregation of small-scale resources (Greece, Italy, Portugal, Spain). Nevertheless, even if legally open for the aggregation service, impractical requirements limit the participation of aggregators. Only a few business cases exist for aggregators managing industrial loads and generation [31]. When it comes to small-scale flexibility, the product requirements practically keep the aggregators out of the markets. Some key findings are as follows:

- Minimum bid size: the size of minimum allowable bid is a significant barrier for aggregators of small units—it should engage a significant number of customers to reach a critical size to access the market;
- Symmetric bidding requirements: in some reserve markets the bids are required to be symmetric in both upward and downward regulations; since aggregators can have unidirectional flexibility in their portfolio in specific time periods, this might be a limiting factor for the access;
- Activation time: typically designed for big generation units, the contracted reserve may be required to be online for up to 10 h, which is often not compatible with small flexible resources, even if aggregated.

4.2. Aggregator–BRP Compensation

Another strong requirement for the aggregator is the association with a BRP to access the market. The roles and responsibilities of so-called independent aggregators (e.g., not depending on any BRP) have not been clarified. Enabling independent aggregators means providing a legislation that allows aggregators to contract the end-users directly without having a pre-agreement with the energy supplier and its associated BRP [32]. Figure 4 shows the implications of a third-party aggregator having a separated contract with the prosumer.

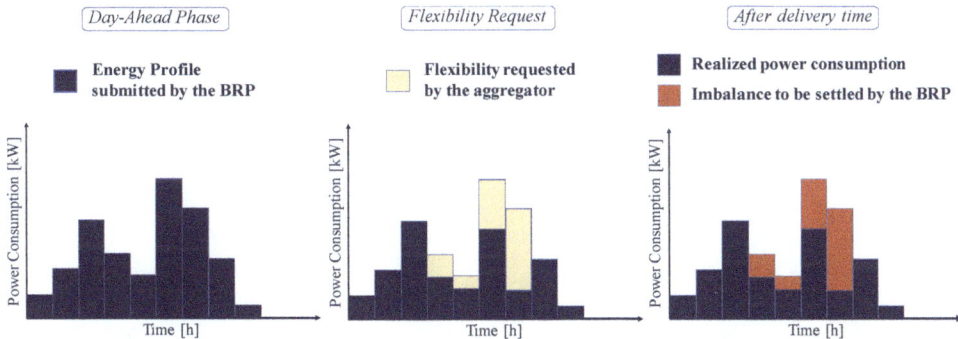

Figure 4. Impact of flexibility activation by an independent aggregator on the Balance Responsible Party's (BRP's) energy program.

The energy supplier/BRP (assuming for the sake of simplicity that they are the same company, but they can be separated entities) submits a load profile in the DA market based on its forecast. Closer to real-time operations, the independent aggregator may request a flexibility activation from the end-user. If the activation becomes effective, the resulting consumption pattern of the end-user will be different than the submitted one by the energy supplier/BRP in the DA phase. Therefore, this creates two distinct effects:

- The aggregator is getting rewarded for the offered flexibility service;
- The BRP is facing an imbalance related to the activation of the flexibility that will be settled by the TSO according to the imbalance settlement regime in place.

Clearly, a compensation payment needs to be arranged between independent aggregators and BRPs. This financial adjustment should reflect the sourcing costs for the energy/service provided and should ensure that risks and costs are associated with the party that causes them [33]. A holistic approach to the definition of roles, responsibilities, and interactions between all market parties in flexibility market has been undertaken by the Universal Smart Energy Framework (USEF). In [34], USEF contributors define different aggregator implementation models, elaborating on the interdependence of the aggregator with the BRP. This extends the work conducted in [35], where different Demand Side Response (DSR) provider models are presented based on the contractual relationship with the energy supplier.

Only France and Switzerland have defined legislation for independent aggregators. In France, three different compensation options are given [36]: (1) the contractual regime, agreed directly between the aggregator and the energy supplier; (2) the regulated regime, where the aggregator pays a fixed tariff decided by the TSO to the BRP; and (3) the corrected regime, where the aggregator is invoiced for the energy component of the energy that would have been consumed by the end-user without flexibility activation. In Switzerland, no prior agreement is needed between the indepedent aggregator and energy supplier/BRP; for the offering of balancing services, the day after the operation, the Swiss TSO corrects the position of each BRP considering all the flexibility activations that have modified its position. The BRP receives a compensation payment for the difference in consumed energy, determined by the quarter-hourly DA spot price [28]. In Ireland and in Great Britain no regulations have been established regarding independent aggregators; nevertheless, independent aggregators can directly contract with end-users without prior consensus of the BRP/supplier. In Great Britain, the independent aggregator can provide balancing services to the national TSO without compensation payment to the energy supplier; this regulation-gap is still not relevant since the number of aggregators participating in the balancing market is considerably low, but the regulator is willing to address the problem in the near future [37].

The goal of the market is to provide fair access to all potential parties without any discrimination. Thus, redesigning the market in such a way that DER aggregators are involved will act as leverage for attracting investors to create new business models [38] and accelerate the process towards the development of a smart DN through the deployment of automation, control, and hierarchical coordination of DERs. At this stage, after addressing the mentioned regulatory barriers, the aggregator would be able to access short-term markets at the wholesale level. The aggregator is expected to allocate flexibility with the objective of profit maximization; a key-challenge is to create a level-playing field with the local DSOs to contract flexibility services if local grid constraints are in place. This leads to the next stage: the change of the DSO's business model.

5. Evolution of the DSO's Role

Several policy papers claim the change in the DN due to an increasing level of distributed generation will mainly affect the figure of DSO, causing it to be no longer or only partially a regulated entity [38,39]. The need to review the regulatory framework for DSOs is widely recognized in Europe [40,41]. The biggest recognition of this trend is the article concerning the use of flexibility by the DSO included in the proposal for a directive of the European Commission regarding common rules in the internal electricity market [23]. It declares the need for a change that "allows and incentivizes distribution system operators to procure services in order to improve efficiencies in the operation and development of the distribution system, including local congestion management." Furthermore, it indicates that DSOs "shall procure these services according to transparent, nondiscriminatory, and market-based procedures". The proposed change is radical: it allows DSOs to contract services with market parties in the DN.

Research studies have already demonstrated the cost-effectiveness of using DERs' flexibility to solve local problems from the DSO perspective. The authors in [42] investigated a real-time congestion management solution by using the market-based procurement of flexibility. The flexible loads allowed the transformer overloading cost in a distribution network to be reduced by 98% with a dominant share of DERs. In [43], the use of market-based flexibility was compared to RES curtailment to assess the best option to avoid congestion at the medium and low voltage levels in a distribution grid with high RES penetration. The flexibility usage resulted in the cost reduction of congestion management while guaranteeing a high level of RES generation in the system. Finally, in [44], the market-flexibility provision was compared with three other congestion management measures, resulting both in a RES curtailment reduction (from 15 to 35%) and in a DSO congestion management cost reduction (33%).

The procurement of flexibility services for DSOs is not allowed in most European countries; however, there are a few exceptions: in Germany and Belgium, a contractual agreement can already be established. In Belgium, newly installed generation units can reduce their network tariff by giving the availability for the curtailment of active power to the local DSO. In Germany, a similar contract can be realized between the DSO and controllable thermal loads, such as space heating technologies [45]. These solutions are used for congested network zones.

With the increasing number of flexible loads and distributed generation, the DSO can use the market-based procurement of flexibility in a more systematic way to tackle network problems and to promote DER integration which is otherwise slowed down by the limited hosting capacity available with the traditional planning approach. The DSO or an independent market operator (IMO) can operate the local platform, calling for a flexibility service when needed to solve congestions in the

DA scheduling phase. This market-based mechanism has two aims: in the short-term, it is a tool for DSO to manage local problems by exploiting DER flexibility at the distribution level; in the long-term, the aim is to defer or avoid grid investments on additional transfer capacity.

One could argue whether there is a need to redesign the short-term electricity markets if the next step is the establishment of a local market mechanism where the DERs' flexibility can be fully valued. The answer is that, these two processes will develop in parallel allowing bottom-up flexibility services to be offered in any organized market at the local and wholesale levels. Figure 5 shows the potential interactions of an aggregator that is enabled to offer all three types of service.

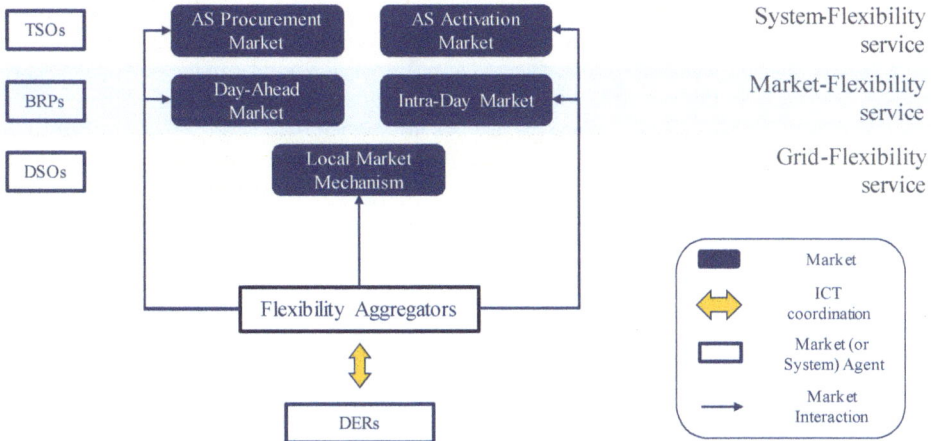

Figure 5. Second stage: The aggregator can also provide grid services since the Distribution System Operator (DSO) will be able to contract services with market parties for managing the DN.

The discussion on the appropriate time window, market time-unit, clearing mechanism, and product requirements for this local market procedure are outside the scope of this work. This is because those design requirements depend on the specific situation and on the needs of the grid operator. For example, a DSO can often encounter a congestion problem at the MV/LV substation, which can be forecasted in the DA planning. Thus, it can decide to establish the market mechanism on a DA basis. A different framework is needed to handle the voltage control problem, for which the flexibility should be procured closer to real-time dispatch and with a shorter notification time. Nevertheless, at this stage, the mechanism is oriented to procure a grid-flexibility service for the DSO, which is the only buyer, while flexibility aggregators can compete on the seller side.

5.1. Remuneration Scheme and Flexible Grid-Tariffs

Nowadays, in most European countries, the National Regulatory Authority (NRA) sets the allowed revenue (revenue cap or price cap) for a DSO in a regulatory period (whose length may vary depending on the national regulations), and the DSO sets the network tariff and connection charges accordingly. In some countries, such as France, Italy, and Spain, the NRA has even more control, also defining the level and the structure of the individual network tariff for each regulatory period [46]. Considering the penetration rate of DERs, the focus of the DSO will not be only on adjusting the network for the expected long-term peak. Additionally, there will be the need for investing in operational measures that allow a high penetration level of DERs while ensuring the quality of supply. The share between the operational expenditure (OPEX) and capital expenditure (CAPEX) is likely to change because of the DER penetration. OPEX will increase (for flexibility provision or operational grid issues) while long-term investment will have less weight with respect to the past. The cost drivers for the DSO will change and the remuneration scheme should be able to capture the process of using

OPEX (in the short-term) in order to reduce CAPEX (in the long-term), whose effects are seen in more than one regulatory period. New incentives are required to stimulate efficiency and innovation in the DSO's model, while traditionally, remuneration schemes induced DSO to invest only in grid reinforcement [27]. Through output-based incentives, the NRA can set the DSO's goals based on relevant parameters that are of interest for pursuing a particular distribution task. The minimum threshold can be fixed on performance factors such as the minimization of congested time for the network or the facilitation of low-carbon technologies. This approach is effective for tackling multiple objectives [47].

The regulatory change for the remuneration scheme will consequently affect the way that distribution costs are allocated among end-users: in other words, it can stimulate the DSO to define innovative grid tariffs to direct the consumption or production of grid-users in a system-efficient way [48]. In the past, the focus was on the reduction of energy consumption, to defer grid reinforcement for as long as possible; thus, the network tariffs were mainly energy-based or volumetric (€/kWh). These types of tariff do not consider when the power consumption is occurring or which share of the capacity line is occupying. These aspects will become more relevant in the future due to increased coincidence factors. Furthermore, DSOs are often economically exposed to consumption volumes which are continuously decreasing due to energy efficiency measures and are becoming more difficult to predict. The shift towards power-based network tariffs (€/kW) and time-dependent network tariffs are recognized as promising solutions for DERs' integration. Dynamic tariffs such as Time of Use, Real-Time Pricing or Critical Peak Pricing leading to fully-dynamic retail pricing can promote the efficient behavior of active consumers using the DN [37,38]. Previously, in dynamic tariffs, the variable part was the energy component of the retail price and the variations were not significant enough to stimulate the modification of the consumption pattern. Most of the European countries have the volumetric component as the main share in the DN tariff and this contributes one-third of the final electricity price paid by the end-users. Only the Netherlands has implemented a network tariff that is totally based on the power consumption [46]. The right economic signals can be delivered to end-users, especially at the residential level, by implementing a fully-dynamic tariff (energy plus network components) leading to the optimal use of the grid in the short-term and guiding the right investments in the long-term. Fully dynamic pricing could lead to benefits for all involved stakeholders:

- For DSO, it would mean a substantial part of the remuneration being cost reflective, as the delivered power would become an even more important cost driver with increasing DERs in the DN [49]. Thus, it would allow efficient recovery of the cost incurred in conveying efficient economic signals to end-users;
- For the end-users, on top of the remuneration for flexibility provision from the aggregator, this would lead to increased savings from shifting energy consumption from peak hours to off-peak ones;
- For aggregators and energy retailers, it would create a level playing field where they can offer a variety of flexible contracts to the end-users, depending on their willingness and availability to modify the energy consumption [50].

5.2. ICT Infrastructure

A solid ICT infrastructure is another requirement to allow the exploitation of small-scale flexibility at the distribution level. It is a key component of the future distribution system to enable data exchange and automatic control between involved parties. The European Directive 2009/72/EC, also known as the Third Energy Package, lays the foundations for large-scale deployment of the smart metering system, encouraging the European Member States (MS) to provide advanced metering systems to at least 80% of its customers by 2020 if it is economically reasonable and cost-effective in the long term [51]. The driving forces can also determine the different developments and they differ from country to country—they range from the reduction of carbon emissions and demand response promotion to consumers' requests for a clear and transparent billing system [52].

Smart metering is an enabling technology since it will allow the interoperability and connectivity of devices/premises with the overlying DN; furthermore, access to timely available data will allow improved management and control of the DN by the network operator due to an improved load and local generation forecast; consumer awareness can be increased with the possibility of accessing flexible retail contracts as well as flexibility-rewarding contracts with energy suppliers and/or independent aggregators. The aforementioned regulatory changes (review of DSO's remuneration scheme, implementation of dynamic tariffs) should be introduced for new services and new actors in order to catch all potential benefits deriving from the technology [53]. Nineteen Member States have committed to roll out up to 200 million smart meters by 2020, with a potential investment of around €40 billion [32]. Seventeen MS out of 19 will lead the change, reaching the target by 2020. The average behavior in Europe is the "Dynamic Mover" area: either the roll-out has been already initiated or there are major research projects driving the subsequent decision for deployment [32].

5.3. DSO Market Examples

Several proposals for the DSO-operated local market mechanism are currently being investigated. The FLECH market is one example [54,55]. Its aim is to integrate the grid-flexibility service within the medium-term planning during which the DSO can forecast the load scenarios with discreet accuracy and can predict how often it will need flexibility activation throughout the planning horizon. The market is structured in two phases: firstly, a reservation market, where DSO establishes contracts with aggregators that are committed to providing flexibility for a determined number of time-steps during the contracted timeframe; and secondly, an activation market where DSO can call for flexibility activation and other aggregators, which, even if they are not involved in the reservation phase, can bid at a lower price than the established price in the reservation phase (considered as a cap). The unitary cost of flexibility procurement (considering both reservation and activation) should be less than the unitary cost for increasing the power capacity of the electricity infrastructure to be cost-effective. This comparison was considered in the research conducted in [56] through an empirical framework, namely FlexMart, where the DSO tries to minimize its overall cost considering the cost of flexibility activation, the cost of line reinforcement, and the cost of energy curtailment. The consumers are rewarded for the flexibility service with a fixed benefit covering the difference between the cost of flexibility-related equipment and the benefit due to a consumption shift in off-peak hours. In this context, the relevance of dynamic network tariff is evident: it can increase the difference between off-peak and peak-hour electricity prices, consequently increasing the savings for the end-user while reducing the compensation paid by the DSO. The De-Flex Market, an interesting proposal presented by the German association of Energy Market Innovators (BNE), provides an instrument for DSO to solve local capacity constraint using DER flexibility [57]. The authors suggested the division of the network in aggregated distribution areas, subject to the different levels of restriction requirements for a certain number of 15 min operational periods. The contracting time frame with flexibility suppliers should be long enough to ensure that over the planning horizon, the DSO will get the needed flexibility. In [58], the DSO acts as a sole buyer of a reactive power market with the aim of relieving voltage problems to exploit the distributed generation available in different microgrids connected at the low voltage level. The microgrids can operate more efficiently by interacting with the utility grid, gaining profit from selling the energy generated locally and helping the utility grid when technical problems are occurring. A DA market-based congestion management is proposed in [59,60] for the trading of flexibility from microgrids, where the cost of the service is compared with the grid reinforcement. Lastly, using an agent-based approach, the flexibility market mechanism is compared with other congestion management measures in [60] showing promising results in the distribution grid with highly renewable penetration.

As the DSO will change its business model, the two main stakeholders for flexibility trading at the distribution level will be enabled. The market mechanism with DSO as the only buyer is a necessary step, since parties are emerging agents and flexibility products need to be standardized. This can be considered a 'monopolistic test' where the amount of flexibility needed and the price cap are set by the only buyer [61]. Once aggregators have reached a critical mass and the market has enough liquidity, the final step will be the shift to a fully-competitive local flexibility market (LFM) where competition is allowed on both sides.

6. Local Flexibility Market: Key Design Challenges

The DSO-operated market can be considered the first trial of a fully-competitive flexibility market where competition is ensured on both sides. Fair market access should be guaranteed to all interested market parties, including BRPs/energy suppliers that are looking at distribution level flexibility as an additional market to adjust their unbalanced positions. The complete set of interactions is represented in Figure 6.

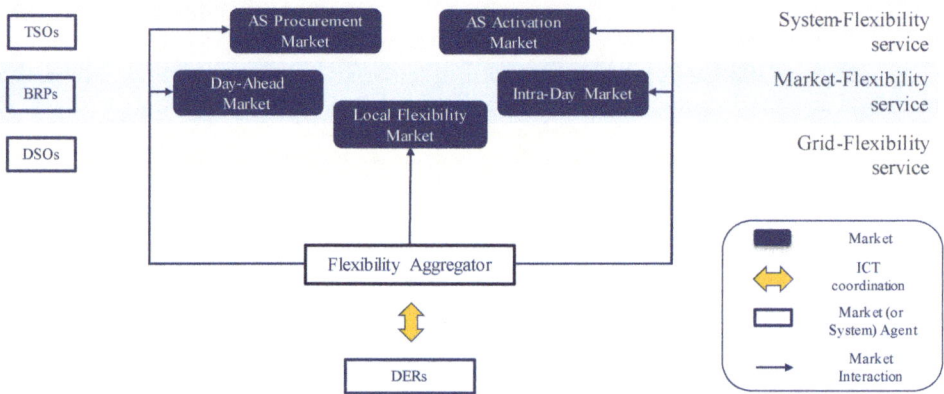

Figure 6. Third stage: Full exploitation of the DERs' flexibility is achieved with access to wholesale short-term markets and to the local flexibility market (LFM). The latter is placed in the market service layer and in the grid service layer.

The fundamental difference between the two stages is that the local flexibility market is a platform where parties can trade different services: grid-oriented services and market-oriented services. Currently, there are no practical examples of implemented LFM in Europe, but several market designs are under research or at the pilot-stage. We identified two main design challenges to be discussed in the following section: the coordination of ancillary services procurement between TSO and DSO and the coexistence of grid-oriented and market-oriented use.

6.1. TSO–DSO Coordination

If the aggregator is allowed to trade in any organized market, a situation could occur where the aggregator finds it more profitable to offer its flexibility to the TSO AS market, thus not notifying the DSO at the local level. Furthermore, it can exploit the presence of different markets (local and wholesale) with different aims by voluntarily creating a local problem and being paid afterwards to solve it. Consequently, TSO–DSO communication is essential: the local market design should consider to what extent the TSO is involved in the local market. Even if not directly involved, the TSO needs to be aware of the transactions at the local level by restoring the information flow between TSOs and DSOs which traditionally has been cut. Enhanced coordination is required to guarantee at least data exchange, sharing of balance responsibility, and grid observability of both system operators [62].

ENTSO-e states that a single marketplace may be a feasible solution to ensure that flexibility from distribution level resources is allocated to create the highest value for the system; a joint TSO–DSO AS market would reduce the number of different bidding processes for procuring ancillary services and limit the possibility of arbitrage between different market platforms [63]. In [64], the author demonstrated that the joint procurement of ancillary services can also lead to greater total system benefits compared to separate procurement. Nevertheless, a practical application of this market could result in an overwhelming burden being carried by only one central system operator. Furthermore, the needs of the TSO (balancing, frequency-related AS) and the needs of the DSO (local congestion, voltage limit violations) are extremely different: locational dependency, size, and requirements are different for addressing issues at the transmission level or at the distribution level.

A decentralized approach, consisting of several local markets operated by different DSOs, coordinated with the traditional AS markets operated by the TSO can lead to the shared computational burden and increased responsibility of the DSO in the active management of the DN. Moreover, flexibility resources that do not meet the requirements of the wholesale markets can be valued in local markets. In this case, data exchange and coordination are needed between DSO and TSO to ensure that the constraints of both transmission and DN are not violated while running different market platforms. The work in [65] provides five different coordination schemes for the TSO–DSO market procurement of ancillary services, defining valid alternatives to the joint TSO–DSO centralized market model.

6.2. Coexistence of Different Flexibility Services

Another key challenge is the coupling of grid and market-oriented use of flexibility. Apart from the system operators, market agents (BRP, energy suppliers, microgrids and even prosumers) at this last stage should be allowed to trade flexibility at the distribution level. It is important to note that the DSO, as a regulated entity, is seeking flexibility to cost-efficiently operate the grid; its request is locational dependent, meaning that it requests flexibility activation in a specific area to mitigate a problem in the same area. On the other hand, BRPs activate flexibility to solve an imbalance in their portfolio which does not have location constraints; their requests are purely profit-driven, therefore they are willing to pay a lower price for flexibility activation to avoid a higher penalty after the *ex post* imbalance settlement. On one hand, requests for grid-supporting purposes should be prioritized because the objective is to reduce the operative costs of the electric infrastructure, but on the other hand, this would mean biasing the market towards the DSO's request, violating the principle of fair competition. Different design approaches can be followed in relation to this aspect: in this work, we consider three different ways of coupling market and grid-oriented use:

- No coupling
- Conditional coupling
- Sequential coupling

This categorization comes from the analysis of the ongoing research and projects, and each solution determines the differences in the market participants, in the market operator, as well as in the level of complexity of the market design (Table 2).

Table 2. Different coordination options for grid and market flexibility services.

Coordination	Market Operator(s)	Seller(s)	Buyer(s)	References
No coupling	DSO [1] IMO [2]	Aggregators [1] BRPs, Aggregators [2]	DSO [1] BRPs, Aggregators [2]	[54–60] [1] [66,67] [2]
Conditional	IMO	Aggregators, BRPs	Aggregators, BRPs, DSO	[68,69]
Sequential	IMO	Aggregators, BRPs	Aggregators, BRPs, DSO	[65,70]

[1] DSO market, [2] Local flexibility trading platform.

Without any coupling, two separate markets will be established: one for market use and one for grid use. The DSO market will be used as a grid management measure and the examples of Section 5.3 are all valid alternatives. The market operator will be the DSO and the aggregators will be enabled to offer flexibility for grid-oriented use. In parallel, another market will run to allow the market-oriented use of flexibility managed by an IMO. In [66], the authors propose a continuous, double-sided auction for market based-control of DERs at the distribution level: prosumers and aggregators can participate, while network constraints are not considered. Each agent aims to maximize its own objective function: profit maximization for prosumers/aggregators and social welfare maximization for the IMO. A DA micro-market algorithm is proposed in [67], where prosumers and generators can trade their energy within the DN. This algorithm runs before the closure of the wholesale DA market with the aim of determining the optimal exchange with the main grid. After the clearing of the wholesale DA market, a second algorithm runs to exploit the possibility of arbitrage (the actual prices being known) by deviating from the previous program for a determined penalty cost. The technical constraints of the grid are here considered in the clearing mechanism.

The conditional coupling implies one single market where both services coexist. The fully-competitive LFM runs as long as a certain condition is satisfied; namely, the market runs without limitation if the network is not threatened by any upcoming technical problem. The market is operated by the IMO because the DSO and BRPs can request flexibility in the same market platform. This is the approach followed by the Universal Smart Energy Framework (USEF) foundation and by the German Association of Energy and Water Industries (BDEW). USEF [68] introduced four different operating regimes, depending on the grid state: the market-based mechanism (based on continuous interactions of aggregators, prosumers, and BRPs) runs without any restriction as far as the grid is in normal operation; when the grid is in a potential state of risk, the DSO starts participating in the market to procure flexibility in order to avoid any technical problem. The last two operating regimes come into play if the market-based mechanism is unable to solve the local problems, and the DSO proceeds to apply direct control over the contracted resources, managing power profiles and eventually curtailing production and shedding loads. In a similar way, BDEW introduced the Traffic Light Concept [69]: in a green state, the free local market allows trading among prosumers, aggregators, microgrids, and BRPs without any restrictions. If the DSO foresees a technical problem, it starts contracting flexibility directly in the market in the so-called 'yellow phase'; if this mechanism fails to solve the grid constraints, the grid passes into a 'red phase' in which there is a direct risk for the stability of the system, and the network operator must intervene promptly to restore the security of the supply.

Sequential coupling implies the coexistence of different types of service in the same market, but the priority is given to one use depending on the needs of the system. This market involves flexibility trade for different purposes, so again, an IMO should be in charge of managing the platform. One example [65] is a sequential market procedure where the DSO first clears the flexibility needed to solve local constraints and then passes the remaining bids to the TSO that clears the market to procure ancillary services from the distribution grid, eventually leaving the non-constrained flexibility offers for market-oriented services. Another example of such coordination can be found in [70]. In the ID clearing mechanism for flexibility trading, first, the market is cleared to match the BRPs' requests in similar and opposite directions to the DSO; after these two stages, the DSO's requests are matched with the remaining offers from flexibility aggregators. The market design, in this case, is more complicated because the criteria for the prioritization has to be defined between different uses, and bias towards one use is unavoidable. Agents might want to pay more for flexibility in an earlier stage of the clearing sequence but they are unable to due to the clearing priorities. The use of flexibility is limited for another reason: after one stage, the flexibility that is traded should respect the constraints of the previous stage or, in other words, it should not affect the resolution of constraints at previous stages.

7. Discussion and Conclusions

A comprehensive discussion on the full exploitation of flexibility deriving from Distributed Energy Resources (RES) at the distribution level is presented in this article. The establishment of a local flexibility market (LFM) is the final stage and enables the trading of the flexible part of supply and demand belonging to DERs. The available flexibility can be used for different purposes: to provide system services to the TSO, to provide grid services to the DSO, and to provide market services to the BRPs.

Starting from the actual regulatory framework in Europe, the article elaborates on the necessary changes to unleash the full potential of the DERs' flexibility, enabling the two main stakeholders of the distribution level flexibility: the aggregator and the DSO. The aggregator gathers multiple small-scale flexibility resources to allow their interaction with the short-term markets which would otherwise be unfeasible; in fact, access requirements for short-term markets are tailored around the capabilities of big generating units, which have traditionally been the only providers of flexibility. In parallel, the DSO will evolve towards the role of the active manager of the grid, and one of the key changes will be the procurement of the flexibility available in the network through a market-based mechanism to avoid or defer grid investment.

The relevant design challenges of a LFM are also discussed. TSO–DSO enhanced coordination is required to allocate flexibility to where it brings the highest benefit to the overall power system. The coordination of market and grid use of flexibility is also a crucial aspect of the LFM design. As highlighted in Section 6.2, conditional coupling of the different flexibility services is the most promising approach because it recognizes the value of the market-based flexibility procurement in the DA and in the ID planning, but it incorporates the LFM into a bigger management procedure based on the grid state, providing backup schemes when approaching the delivery time. Nevertheless, quantitative research on the value gained by using one type of coordination or another is not available, and this research gap needs to be addressed in future research. The optimal market design depends on the national regulations, specific area, the needs of the system operators, the number of active stakeholders at the distribution level, and on the amount of available flexibility. The research should define the technical, economic, and social boundaries in which the LFM can be an effective solution. Regarding the two first stages, the next research steps will focus on these challenges:

- Following the discussion in Section 4, we individuated the most critical regulatory barriers for market access. We will investigate regulatory market adjustments (lower minimum bid size, asymmetric bidding) to quantify how much they can improve market access for aggregators. A compensation payment to BRPs must be present to adequately allocate balance responsibilities among parties; on the other hand, an excessive amount will limit the profitability of new business cases for independent aggregators that are willing to enter the market. Starting from the countries (such as France and Switzerland) that have already introduced a regulation on this issue, we will compare different alternatives for compensation payment.
- In Section 5 we discussed the needed change in the DSO's remuneration scheme to introduce a more innovative tariff that can stimulate flexibility activation from end-users; we will try to individuate and quantify relevant parameters (economic or technical) that will distinguish the profitability of a local market mechanism from the procurement of flexibility from other grid management measures. The establishment of the LFM needs investments in the ICT infrastructure to allow the realization of transactions among participants. This investment decision depends on the foreseen increase in the DERs' share in the grid, on the actual state of the network, and on the needed network upgrade to accommodate the future scenarios.

Moreover, there are essential requirements for a market mechanism to be successful, namely a sufficient number of competitors and a sufficient liquidity in the commodity to trade. An insufficient number of participants or a lack of liquidity would result in an unstable market, and other mechanisms, such as a contractual-based flexibility procurement, should be preferred. Research effort needs to be directed towards the quantification of these factors to discern the preferred mechanism to establish. The final aim is the achievement of economic and effective management of the overall network infrastructure. Therefore, in light of the emerging importance of DERs as a potential flexibility source, the LFM should be considered a tool for boosting DERs' integration by creating value for bottom-up flexibility service provision.

Author Contributions: Literature Research, S.M., Conceptualization, S.M., N.H. and P.N., Methodology, S.M., Supervision, P.N., N.H. and G.P., Visualization, S.M., Writing original draft, S.M., Writing review and editing, S.M., N.H. and P.N.

Funding: This work is part of the research programme ERA-net Smart Grids Plus with project number 651.001.012, which is financed by the Netherlands Organisation for Scientific Research (NWO). (http://m2m-grid.eu/).

Conflicts of Interest: The authors declare no conflict of interest.

Abbreviations

AS	Ancillary Service
BRP	Balance Responsible Party
CAPEX	Capital Expenditure
DA	Day-Ahead
DER	Distributed Energy Resource
DN	Distribution Network
DSO	Distribution System Operator
ICT	Information and Communication Technology
ID	Intra-Day
IMO	Market Operator
LFM	Local Flexibility Market
MS	Member State
NRA	National Regulatory Authority
OPEX	Operational Expenditure
PV	Photovoltaics
RES	Renewable Energy Source
TSO	Transmission System Operator
USEF	Universal Smart Energy Framework

References

1. Hout, M.V.; Koutstaal, P.; Ozdemir, O.; Seebregts, A.; van Hout, M.; Koutstaal, P.; Ozdemir, O.; Seebregts, A. *Quantifying Flexibility Markets*; ECN-E–14-039; ECN Policy Studies: Petten, The Netherlands, 2014; p. 52.
2. Koltsaklis, N.E.; Dagoumas, A.S.; Panapakidis, I.P. Impact of the penetration of renewables on flexibility needs. *Energy Policy* **2017**, *109*, 360–369. [CrossRef]
3. Veiga, A.; Rodilla, P.; Herrero, I.; Batlle, C. Intermittent RES-E, cycling and spot prices: The role of pricing rules. *Electr. Power Syst. Res.* **2015**, *121*, 134–144. [CrossRef]
4. Batalla-Bejerano, J.; Trujillo-Baute, E. Impacts of intermittent renewable generation on electricity system costs. *Energy Policy* **2016**, *94*, 411–420. [CrossRef]
5. REN21. Renewables 2017—Global Status Report. 2017. Available online: http://www.ren21.net/wp-content/uploads/2017/06/17-8399_GSR_2017_Full_Report_0621_Opt.pdf (accessed on 1 October 2018).
6. Serițan, G.; Porumb, R.; Cepișcă, C.; Grigorescu, S. Integration of Dispersed Power Generation. In *Electricity Distribution: Intelligent Solutions for Electricity Transmission and Distribution Networks*; Karampelas, P., Ekonomou, L., Eds.; Springer: Berlin/Heidelberg, Germany, 2016; pp. 27–61.
7. Strbac, G.; Jenkins, N.; Green, T.; Pudjianto, D. Review of Innovative Network Concepts. 2006. Available online: https://ec.europa.eu/energy/intelligent/projects/sites/iee-projects/files/projects/documents/dg-grid_review_of_innovative_network_concepts.pdf (accessed on 1 October 2018).
8. Le Baut, J.; Leclercq, G.; Viganò, G.; Zenebe Degefa, M. Characterization of Flexibility Resources and Distribution Networks. 2017. Available online: http://smartnet-project.eu/wp-content/uploads/2017/05/D1.2_20170522_V1.1.pdf (accessed on 1 October 2018).
9. Parag, Y.; Sovacool, B. Perspective article: Electricity market design for the prosumer era. *Nat. Energy* **2016**, *1*, 16032. [CrossRef]
10. Teotia, F.; Bhakar, R. Local energy markets: Concept, design and operation. In Proceedings of the 2016 National Power Systems Conference (NPSC), Bhubaneswar, India, 19–21 December 2016; pp. 1–6.

11. Hvelplund, F. Renewable energy and the need for local energy markets. *Energy* **2006**, *31*, 1957–1966. [CrossRef]

12. Rosen, C.; Madlener, R. Regulatory Options for Local Reserve Energy Markets: Implications for Prosumers, Utilities, and other Stakeholders. *Energy J.* **2016**, *37*, 39–50. [CrossRef]

13. International Energy Agency. Harnessing Variable Renewables—A Guide to the Balancing Challenge. 2011. Available online: https://webstore.iea.org/harnessing-variable-renewables (accessed on 1 October 2018).

14. EDSO for Smart Grids. Flexibility: The Role of DSOs in Tomorrow's Electricity Market. 2014. Available online: http://www.edsoforsmartgrids.eu/wp-content/uploads/public/EDSO-views-on-Flexibility-FINAL-May-5th-2014.pdf (accessed on 1 October 2018).

15. Ulbig, A.; Andersson, G. On Operational Flexibility in Transmission Constrained Electric Power Systems. In Proceedings of the 2013 IEEE Power & Energy Society General Meeting, San Diego, CA, USA, 22–26 July 2012.

16. Flexibility-Enabling Contracts in Electricity Markets. 2016. Available online: https://www.oxfordenergy.org/publications/flexibility-enabling-contracts-electricity-markets/ (accessed on 1 October 2018).

17. Villar, J.; Bessa, R.; Matos, M. Flexibility products and markets: Literature review. *Electr. Power Syst. Res.* **2018**, *154*, 329–340. [CrossRef]

18. Eid, C.; Codani, P.; Perez, Y.; Reneses, J.; Hakvoort, R. Managing electric flexibility from Distributed Energy Resources: A review of incentives for market design. *Renew. Sustain. Energy Rev.* **2016**, *64*, 237–247. [CrossRef]

19. Heleno, M.; Matos, M.A.; Lopes, J.A.P. A bottom-up approach to leverage the participation of residential aggregators in reserve services markets. *Electr. Power Syst. Res.* **2015**, *136*, 425–433. [CrossRef]

20. Ohrem, S.; Telöken, D. Concepts for flexibility use—Interaction of market and grid on DSO level. In Proceedings of the 2016 CIRED Workshop, Helsinki, Finland, 14–15 June 2016; pp. 1–4.

21. GEODE. The Role of the Distribution System Operator in the Electricity Market. 2016. Available Online: http://www.geode-eu.org/uploads/GEODE%20Germany/DOCUMENTS%202016/REPORT%20DSO%20ROLE%20ELECT%20MRKT%202016%20FINAL.pdf (accessed on 1 October 2018).

22. European Parliament. Directive 2012/27/EU of the European Parliament and of the Council of 25 October 2012 on Energy Efficiency, Amending Directives 2009/125/EC and 2010/30/EU and Repealing Directives 2004/8/EC and 2006/32/EC. 2012. Available online: https://eur-lex.europa.eu/legal-content/en/TXT/?uri=celex%3A32012L0027 (accessed on 1 October 2018).

23. European Commission. Proposal for a Directive of the European Parliament and the Council on Common Rules for the Internal Market in Electricity. 2016. Available online: https://eur-lex.europa.eu/legal-content/EN/TXT/?uri=CELEX%3A52016PC0864R%2801%29 (accessed on 1 October 2018).

24. Eurelectric. Flexibility and Aggregation—Requirements for Their Interaction in the Market. 2015. Available online: https://www.usef.energy/app/uploads/2016/12/EURELECTRIC-Flexibility-and-Aggregation-jan-2014.pdf (accessed on 1 October 2018).

25. Smart Energy Demand Coalition. Enabling Independent Aggregation in the European Electricity Markets. 2015. Available online: http://smartenergydemand.eu/wp-content/uploads/2015/02/SEDC-Enabling-Independent-Aggregation.pdf (accessed on 1 October 2018).

26. Burger, S.; Chaves-Ávila, J.P.; Batlle, C.; Pérez-Arriaga, I.J. A review of the value of aggregators in electricity systems. *Renew. Sustain. Energy Rev.* **2017**, *77*, 395–405. [CrossRef]

27. Ruester, S.; Schwenen, S.; Batlle, C.; Pérez-Arriaga, I. From distribution networks to smart distribution systems: Rethinking the regulation of European electricity DSOs. *Util. Policy* **2014**, *31*, 229–237. [CrossRef]

28. Smart Energy Demand Coalition. Explicit Demand Response in Europe Mapping the Markets 2017. 2017. Available online: https://www.smarten.eu/wp-content/uploads/2017/04/SEDC-Explicit-Demand-Response-in-Europe-Mapping-the-Markets-2017.pdf (accessed on 1 October 2018).

29. Bertoldi, P.; Zancanella, P.; Boza-Kiss, B. Demand Response Status in EU Member States. 2016. Available online: http://publications.jrc.ec.europa.eu/repository/bitstream/JRC101191/ldna27998enn.pdf (accessed on 1 October 2018).

30. ENTSO-E. Survey on Ancillary Services Procurement, Balancing Market Design 2016. 2017. Available online: https://docstore.entsoe.eu/Documents/Publications/Market%20Committee%20publications/WGAS_Survey_final_10.03.2017.pdf (accessed on 1 October 2018).

31. Verhaegen, R.; Dierckxsens, C. Existing Business Models for Renewable Energy Aggregators. 2016. Available online: http://bestres.eu/wp-content/uploads/2016/08/BestRES_Existing-business-models-for-RE-aggregators.pdf (accessed on 1 October 2018).

32. European Commission. Evaluation Report on Framework of Electricity Market—Part 2. 2016. Available online: https://ec.europa.eu/energy/sites/ener/files/documents/2_en_autre_document_travail_service_part2_v2_412.pdf (accessed on 1 October 2018).

33. Smart Grid Task Force. Regulatory Recommendations for the Deployment of Flexibility—EG3 REPORT. 2015. Available online: https://ec.europa.eu/energy/sites/ener/files/documents/EG3%20Final%20-%20January%202015.pdf (accessed on 1 October 2018).

34. De Heer, H.; van der Laan, M. USEF: Work Stream on Aggregator Implementation Models. 2016. Available online: https://www.usef.energy/app/uploads/2016/12/Recommended-practices-for-DR-market-design.pdf (accessed on 1 October 2018).

35. ENTSO-E. Market Design for Demand Side Response. 2015. Available online: https://docstore.entsoe.eu/Documents/Publications/Position%20papers%20and%20reports/entsoe_pp_dsr_web.pdf (accessed on 1 October 2018).

36. RTE. Règles Pour la Valorisation des Effacements de Consommation sur les Marchés de L'énergie. 2014. Available online: https://www.cre.fr/Documents/Deliberations/Approbation/effacements-de-consommation2/consulter-l-annexe-de-rte-regles-pour-la-valorisation-des-effacements-de-consommation-sur-les-marches-de-l-energie-nebef-2.0 (accessed on 1 October 2018).

37. PA Consulting Group. Aggregators—Barriers and External Impacts. 2016. Available online: https://www.ofgem.gov.uk/system/files/docs/2016/07/aggregators_barriers_and_external_impacts_a_report_by_pa_consulting_0.pdf (accessed on 1 October 2018).

38. Boscán, L.; Poudineh, R. *Business Models for Power System Flexibility: New Actors, New Roles, New Rules*; Elsevier Inc.: Amsterdam, The Netherlands, 2016; pp. 363–382.

39. Ecorys. The Role of DSOs in a Smart Grid Environment. 2014. Available online: https://ec.europa.eu/energy/sites/ener/files/documents/20140423_dso_smartgrid.pdf (accessed on 1 October 2018).

40. Cossent, R.; Gómez, T.; Frías, P. Towards a future with large penetration of distributed generation: Is the current regulation of electricity distribution ready? Regulatory recommendations under a European perspective. *Energy Policy* **2009**, *37*, 1145–1155. [CrossRef]

41. Agrell, P.J.; Bogetoft, P.; Mikkers, M. Smart-grid investments, regulation and organization. *Energy Policy* **2013**, *52*, 656–666. [CrossRef]

42. Haque, A.N.M.M.; Nguyen, P.H.; Bliek, F.W.; Slootweg, J.G. Demand response for real-time congestion management incorporating dynamic thermal overloading cost. *Sustain. Energy Grids Netw.* **2017**, *10*, 65–74. [CrossRef]

43. Geschermann, K.; Moser, A. Evaluation of market-based flexibility provision for congestion management in distribution grids. In Proceedings of the 2017 IEEE Power Energy Society General Meeting, Chicago, IL, USA, 16–20 July 2017; pp. 1–5.

44. Kulms, T.; Meinerzhagen, A.K.; Koopmann, S.; Schnettler, A. A simulation framework for assessing the market and grid driven value of flexibility options in distribution grids. *J. Energy Storage* **2018**, *17*, 203–212. [CrossRef]

45. Copenhagen Economics and VVA Europe. Impact Assessment Support Study on: "Policies for DSOs, Distribution Tariffs and Data Handling". 2016. Available online: https://ec.europa.eu/energy/sites/ener/files/documents/ce_vva_dso_final_report_vf.pdf (accessed on 1 October 2018).

46. Eurelectric. Network Tariffs. 2016. Available online: http://www.eurelectric.org/media/268408/network_tariffs__position_paper_final_as-2016-030-0149-01-e.pdf (accessed on 1 October 2018).

47. CEER. Incentives Schemes for Regulating DSOs, Including for Innovation Consultation Paper. 2017. Available online: https://www.ceer.eu/documents/104400/-/-/f04f3e11-6a20-ff42-7536-f8afd4c06ba4 (accessed on 1 October 2018).

48. Picciariello, A.; Reneses, J.; Frias, P.; Söder, L. Distributed generation and distribution pricing: Why do we need new tariff design methodologies? *Electr. Power Syst. Res.* **2015**, *119*, 370–376. [CrossRef]

49. Mandatova, P.; Massimiano, M.; Verreth, D.; Gonzalez, C. Network tariff structure for smart energy system. In Proceedings of the 2014 CIRED Workshop, Rome, Italy, 11–12 June 2014.

50. Biggar, D.; Reeves, A. *Network Pricing for the Prosumer Future: Demand-Based Tariffs or Locational Marginal Pricing?* Elsevier Inc.: Amsterdam, The Netherlands, 2016; pp. 247–265.

51. European Parliament. Directive 2009/72/EC of the European Parliament and of the Council of 13 July 2009 Concerning Common Rules for the Internal Market in Electricity and Repealing Directive 2003/54/EC. 2009. Available online: https://eur-lex.europa.eu/legal-content/en/ALL/?uri=celex%3A32009L0072 (accessed on 1 October 2018).

52. Zhou, S.; Brown, M.A. Smart meter deployment in Europe: A comparative case study on the impacts of national policy schemes. *J. Clean. Prod.* **2017**, *144*, 22–32. [CrossRef]

53. Renner, S.; Albu, M.; van Elburg, H.; Heinemann, C. European Smart Metering Landscape Report. 2011. Available online: https://www.sintef.no/globalassets/project/smartregions/d2.1_european-smart-metering-landscape-report_final.pdf (accessed on 1 October 2018).

54. Harbo, S.; Hansen, L.H.; Heussen, K. FLECH—Market Specification Analysis. 2013. Available online: http://ipower-test2.droppages.com/Publications/FLECH%20Market%20Specification%20Analysis.pdf (accessed on 1 October 2018).

55. Zhang, C.; Ding, Y.; Nordentoft, N.C.; Pinson, P.; Østergaard, J. FLECH: A Danish market solution for DSO congestion management through DER flexibility services. *J. Mod. Power Syst. Clean Energy* **2014**, *2*, 126–133. [CrossRef]

56. Spiliotis, K.; Ramos Gutierrez, A.I.; Belmans, R. Demand flexibility versus physical network expansions in distribution grids. *Appl. Energy* **2016**, *182*, 613–624. [CrossRef]

57. Bundesverband Neue Energiewirtschaft e.V. Decentralized Flexibility Market 2.0. 2016. Available online: https://www.bne-online.de/fileadmin/bne/Dokumente/Englisch/Policy_Papers/20160704_bne_De-Flex-Market_2.0_final.pdf (accessed on 1 October 2018).

58. Madureira, A.G.; Peças Lopes, J.A. Ancillary services market framework for voltage control in distribution networks with microgrids. *Electr. Power Syst. Res.* **2012**, *86*, 1–7. [CrossRef]

59. Amicarelli, E.; Tran, T.Q. Flexibility Service Market for Active Congestion Management of Distribution Networks using Flexible Energy Resources of Microgrids. In Proceedings of the 2017 IEEE PES Innovative Smart Grid Technologies Conference Europe (ISGT-Europe), Torino, Italy, 26–29 September 2017.

60. Kulms, T.; Meinerzhagen, A.K.; Koopmann, S.; Schnettler, A. Development of An Agent-based Model for Assessing the Market and Grid Oriented Operation of Distributed Energy Resources. *Energy Procedia* **2017**, *135*, 294–303. [CrossRef]

61. Ramos, A.; De Jonghe, C.; Gomez, V.; Belmans, R. Realizing the smart grid's potential: Defining local markets for flexibility. *Util. Policy* **2016**, *40*, 26–35. [CrossRef]

62. ENTSO-E; CEDEC; GEODE; EURELECTRIC; EDSO. TSO-DSO Data Management Report. 2015. Available online: https://docstore.entsoe.eu/Documents/Publications/Position%20papers%20and%20reports/entsoe_TSO-DSO_DMR_web.pdf (accessed on 1 October 2018).

63. ENTSO-E. Distributed Flexibility and the Value of TSO/DSO Coordination. 2017. Available online: https://docstore.entsoe.eu/Documents/Publications/Position%20papers%20and%20reports/entsoe_pp_DF_1712_web.pdf (accessed on 1 October 2018).

64. Roos, A. Designing a joint market for procurement of transmission and distribution system services from demand flexibility. *Renew. Energy Focus* **2017**, *21*, 16–24. [CrossRef]

65. Gerard, H.; Rivero, E.; Six, D. Basic Schemes for TSO-DSO Coordination and Ancillary Services Provision. 2016. Available online: http://smartnet-project.eu/wp-content/uploads/2016/12/D1.3_20161202_V1.0.pdf (accessed on 1 October 2018).

66. Ampatzis, M.; Nguyen, P.H.; Kling, W.L. Local electricity market design for the coordination of distributed energy resources at district level. In Proceedings of the 2014 5th IEEE PES Innovative Smart Grid Technologies Europe (ISGT Europe), Istanbul, Turkey, 12–15 October 2014; pp. 1–6.

67. Olivella-Rosell, P.; Vinals-Canal, G.; Sumper, A.; Villafafila-Robles, R.; Bremdal, B.A.; Ilieva, I.; Ottesen, S.O. Day-ahead micro-market design for distributed energy resources. In Proceedings of the 2016 IEEE International Energy Conference (ENERGYCON), Leuven, Belgium, 4–8 April 2016; pp. 1–6.

68. USEF Foundation. USEF: The Framework Explained. 2015. Available online: http://www.globalsmartgridfederation.org/wp-content/uploads/2016/10/USEF_TheFrameworkExplained-18nov15.pdf (accessed on 1 October 2018).

69. BDEW Bundesverband der Energie- und Wasserwirtschaft e.V. Smart Grid Traffic Light Concept. 2015. Available online: https://www.bdew.de/media/documents/Stn_20150310_Smart-Grids-Traffic-Light-Concept_english.pdf (accessed on 1 October 2018).

70. Torbaghan, S.S.; Blaauwbroek, N.; Kuiken, D.; Gibescu, M.; Hajighasemi, M.; Nguyen, P.; Smit, G.J.; Roggenkamp, M.; Hurink, J. A market-based framework for demand side flexibility scheduling and dispatching. *Sustain. Energy Grids Netw.* **2018**, *14*, 47–61. [CrossRef]

energies

MDPI

Article

A Study on the Multi-Agent Based Comprehensive Benefits Simulation Analysis and Synergistic Optimization Strategy of Distributed Energy in China

Xiaohua Song [1,2], Mengdi Shu [1,*], Yimeng Wei [1] and Jinpeng Liu [1,2,*]

1   School of Economics and Management, North China Electric Power University, Beijing 102206, China;
    sxh@ncepu.edu.cn (X.S.); 1182206245@ncepu.edu.cn (Y.W.)
2   Beijing Key Laboratory of New Energy and Low-Carbon Development, North China Electric Power
    University, Beijing 102206, China
*   Correspondence: 1172206160@ncepu.edu.cn (M.S.); ljp@ncepu.edu.cn (J.L.)

Received: 17 October 2018; Accepted: 21 November 2018; Published: 23 November 2018

Abstract: With the economic and social development of China, the continuous growth of the energy demand is the trend for now and the future. As a consequence, distributed energy, especially distributed electricity power generation, has received more and more attention. Thus, the scale and utilization level of distributed energy has been continuously improved. However, due to the limitations of current technologies, resources, policies and other issues, the comprehensive benefits and synergy levels of energy sources need to be greatly enhanced. Based on the system dynamics model, this paper examines the factors affecting the comprehensive benefits of distributed energy in China, screens the key subjects, and using the literature review method, combined with the existing literature analysis, constructs a comprehensive benefit evaluation index system and evaluates the comprehensive benefits through case analysis. This paper also sorts out the distributed energy-related Chinese government policies from 2001 to 2017, and considers the scale of distributed energy development, then divides it into two development stages. The synergetic entropy is used to analyze the synergetic development degree of the two-stage distributed energy entities. The synergistic optimization strategy is proposed from the Chinese government side, power supply side, power grid side and user side, which provides theoretical methods and optimization suggestions for improving the comprehensive benefits of distributed energy and promoting sustainable development of energy.

Keywords: distributed energy; comprehensive benefits; multi-agent synergetic estimation; synergistic optimization strategy

1. Introduction

During the "13th Five-Year Plan" period and in the near future, China's economy will be in a "new normal stage" for a long term. In such circumstances, energy demand will continue to increase, and the energy supply and demand system will face prominent development problems. For instance, energy supply and demand presure continue loose, and the policy targets for energy conservation and emission reduction are continuously increasing, posing a risk to energy safety and stability. Meanwhile the system operation level has not met the requirements of marketization. The energy supply structure in China is still dominated by coal. In 2016, coal accounted for about two-thirds of the total electricity generation. In such case, innovating the development model and promoting the development of comprehensive energy sources are of great significance.

Compared with the traditional centralized energy sources, distributed energy has significant characteristics such as large development potential, flexible production, cleanliness and efficiency, as well as low pollution. The Energy Development Revolution Innovation Action Plan (2016–2030) issued by the National Development and Reform Commission of the Chinese government and the Energy Bureau proposed the high technology development plan for new energy power generation, including large-scale wind power, high-efficiency solar energy, distributed energy, energy storage technology and other related technologies. Energy technology innovation is expected to enter a golden period. The 13th Five-Year Plan for Energy Development attaches great importance to the development of distributed energy, acceleration of the construction of distributed energy projects, optimization of the solar energy development plan, and especially prioritization of the development of distributed photovoltaic power generation. With the support of the government policy, China's distributed energy development is continuously accelerating, with distributed photovoltaic power generation and biomass power generation being the most significant aspects. According to data from the China Foresight Industry Research Institute, the installed capacity of China's distributed photovoltaic power generation will double in 2017 compared with 2016. At the same time, the sustainable development goal of energy proposed by the international community provides a good opportunity for the development of distributed energy. Thus, distributed energy has great potential for development in the future.

However, despite, the fact the scale of distributed energy development has been expanding, in recent years, it is still in the initial and exploratory stage and the comprehensive benefits of distributed energy and the coordination between various entities desperately need to be enhanced. As a complex system, the entities of a distributed energy system not only include itself, but also large power grid cooprations, governments, markets and other entities have an important impact on its development. Due to the different interests and institutional mechanisms, these various entities have not reached a state of synergetic development. For instance, the power generation of distributed energy projects is self-contained, which affects the interests of power grid companies to a certain extent. The inactivity of power grid companies is inevitable, which restricts the improvement and further development of the comprehensive benefits of distributed energy system. Therefore, improving the multi-agent coordination of distributed energy is of far-reaching strategic significance for enhancing the efficiency and comprehensive benefit of energy utilization, and also promote building a safe, high-quality, economical and environmentally protective energy system, and contribute to achieving the goal of sustainable energy development proposed by the United Nations. Based on the aforementioned issues, analyzing the impact mechanism of distributed energy's comprehensive benefits and the degree of synergy between the main bodies, then discovering existing problems, exploring the potential benefit improvement potential, and optimizing multi-agent synergistic development mechanisms, are key subjects to enhance the comprehensive utilization benefits of distributed energy resources and promote large-scale investment in the distributed energy industry. At the same time, it is conducive to build a conservation-oriented society and accelerate the transformation of the energy structure.

2. Literature Review and Research Methods

2.1. Literature Review

With the growth of the total energy demand and the aim of reducing carbon dioxide emissions around the the globe, clean energy and renewable energy have attracted worldwide attention. Distributed energy systems have broadly developed prospects and great development potential due to their huge economic and environmental benefits [1–3]. Moreover, distributed energy systems have been studied in many countries as eco-friendly power systems that provide high quality power [4]. However, the development of distributed energy started late in China and still faces many difficulties. [5]. The references of this paper contain the research work on distributed energy at home and abroad that involves comprehensive benefits analysis and technical path optimization of the present situation.

In terms of the comprehensive benefits, most of the current studies are on the economic benefits and cost-benefit analysis, and there is less discussion about comprehensive benefits analysis of distributed energy. Serrano, Omu and other scholars conducted a cost-benefit analysis of distributed energy installed at a specific location [6,7]. Inamori, Zhang et al. proposed an evaluation model for the cost-benefit analysis of distributed energy [8,9].In addition, Di Somma et al. proposed the promotion of the use of low-temperature energy from the perspective of fire efficiency, analyzed the contribution of each energy device in reducing energy costs and all inputs, and considered the economic factors affecting benefits by using the Pareto boundary [10].

In terms of the multi-agent synergistic optimization research, He et al. and other scholars analyzed how to obtain reasonable profits from the perspective of users and power supply companies to promote the healthy and orderly development of the distributed energy industry [11]. Zhang, and other scholars have proposed a new method of distributed generation and load coordination control in distribution systems, aiming to promote the synergetic development of distributed energy and distribution systems [12,13]. Reinders et al. evaluated pilot projects for smart grids in residential areas in The Netherlands and Austria to better support the coordinated evolution of various stakeholders [14]. Bale et al. used the methods of subject analysis and network modeling to extract the co-evolutionary factors of the multi-participants in the transportation system and power system, and make long-term predictions on the sustainable development of transportation systems and power systems [15].

2.2. Research Methods

The research content and research route of this paper are as follows: In the first section, this paper describes and analyzes the status quo of China's distributed energy development, and puts forward the significance of comprehensive benefits analysis and multi-agent synergistic optimization.

The second section introduces the research situation of the comprehensive benefits of distributed energy and multi-agent synergy. The models and methods used in each research part of this paper are also listed.

The third section builds a causal circuit diagram based on system dynamics, analyses the factors affecting the comprehensive benefits of distributed energy, and screens the main entities for explanations. According to the current research literature, this paper refined and the index to measure the comprehensive benefit sorted out, and the three-level index system of the comprehensive benefits is constructed accordingly. Then we take the demonstration project of Dong'ao Island as an example, to analyze its comprehensive benefits.

The fourth section reviews China's key policies for distributed energy development in recent years. and divides the development stages of distributed energy. Then it constructs a synergetic entropy, analyzes the synergy relationship of multi-agents between different stages, and proposes targeted strategies for its future development from the perspective of four key entities. The fifth section presents the conclusions. The technical flow of this paper is shown in the Figure 1.

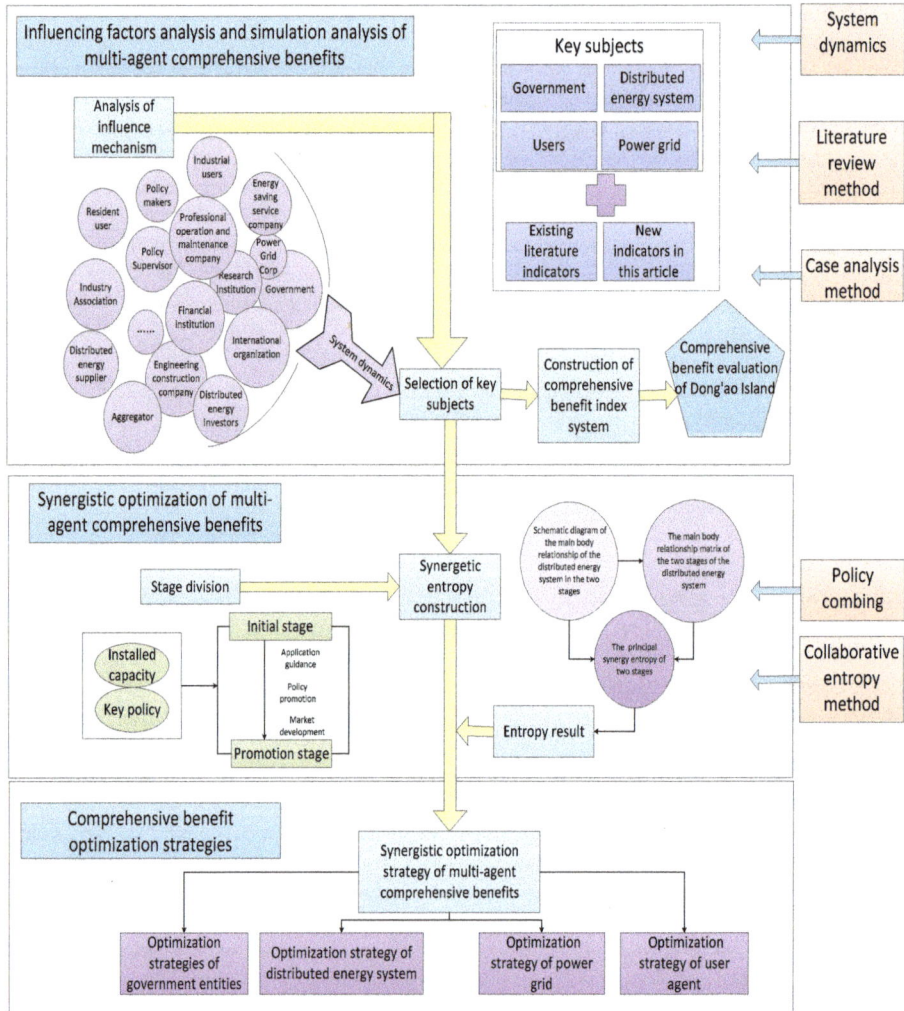

Figure 1. Schematic diagram of research content and research methods.

3. Influencing Factors Analysis and Simulation Analysis of the Distributed Energy Multi-Agent Comprehensive Benefits Based on System Dynamics Method

3.1. Analysis of Factors Affecting Comprehensive Benefits

The current development of distributed energy has already demonstrated certain comprehensive benefits. However, distributed energy systems involve different entities, including not only itself, but also large power grids, governments, users, markets, etc. Each entitiy is affected by many factors, and the interaction between each entity and the environment as well as other entities together affect the comprehensive benefits of the whole distributed energy system.

System dynamics reveals the main factors affecting system performance by revealing and analyzing the causal relationships between the internal components of the system, and then provides a basis for targeted improvement at a system operation level. The comprehensive benefits of distributed energy is a complex analysis system that needs to comprehensively consider elements regarding economic benefits, energy saving benefits, loss reduction benefits, reliability benefits, environmental benefits, and social benefits. By using system dynamics, the system can be scientifically divided into various subsystems, and the causal relationships between the system elements and the impact on the comprehensive benefits of distributed energy revealed. Based on the transmission mechanism among factors, the key factors affecting the comprehensive benefits of distributed energy can be revealed, which provides an important reference for multi-agent synergistic evaluation research. Therefore, system dynamics has applicability to the research of distributed energy comprehensive benefits transmission mechanism. Based on system dynamics analysis, it can clearly influence the main factors of distributed energy comprehensive benefits according to the direction of the conduction mechanism.

Based on the principle of system dynamics, this paper constructs a conduction mechanism diagram of the factors affecting the comprehensive benefits of distributed energy.

According to Figure 2, the interaction chain is concentrated around four main entities, including government, users, distributed energy suppliers and power grids. This paper focuses on those aforementioned entities and conduct a comprehensive benefits factor analysis.

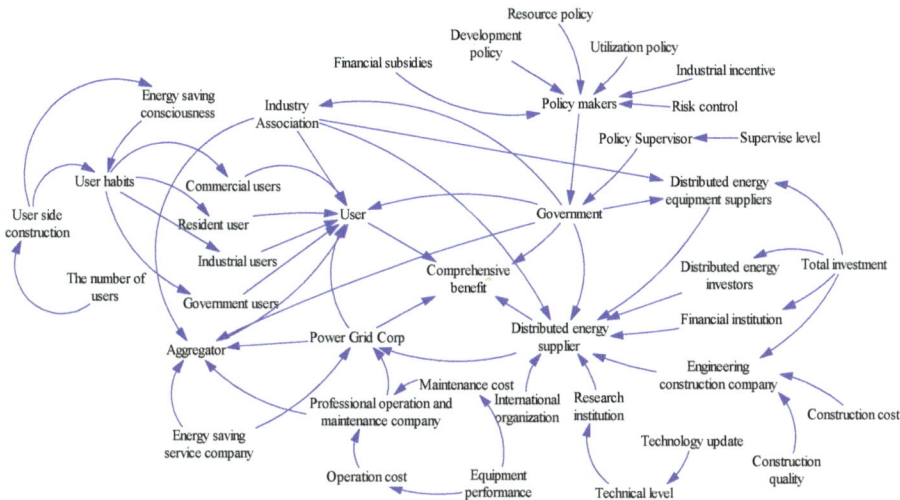

Figure 2. Conduction mechanism diagram of factors affecting the comprehensive benefits of distributed energy.

(1) Government-side analysis

It can be seen from the causal loop diagram that the influencing factors at the government level are mainly reflected in the policies issued by the government departments reponsible for distributed energy. Policy makers and policy supervisors jointly influence the promulgation and implementation of policies at different stages. Through development policies, resource policies, utilization polices and other aspects of policy support, as well as financial subsidies, industrial incentives and other positive measures formulated, policymakers affect the promotion of comprehensive benefits of the distributed energy in a certain level. At the same time, the supervision level of policy supervisors determines the implementation of distributed energy-related policies and the supervision and construction of distributed energy systems, and the comprehensive benefits will be improved when supervision is in place.

(2) Distributed energy systems analysis

The distributed energy system has the most direct impact on its comprehensive benefits, and it is deeply affected by other entities. The quality of the equipment supplied by the suppliers, the amount of funds provided by the investors and financial institutions, the construction cost of the construction companies and the quality of the projects, the support from government, the level of development of new technologies by scientific research institutions, the guidance from international organizations and trade associations affect the comprehensive benefits of distributed energy.

(3) Power grid company analysis

Distributed energy systems and power grids complement each other. The distributed energy system directly placed on the user side cooperates with the power grid to improve the reliability of the power supply. When the power of the distributed energy system is in short supply, power can be purchased from the main power grid. In the process of synergy between distributed energy systems and power grids, energy-saving service companies and professional operation and maintenance companies provide guarantees, theoretical and technical support for the integration of distributed energy systems, and affect its synergetic development, thereby affecting the improvement of its comprehensive benefits.

(4) User-side analysis

Users have a direct impact on the overall benefits of distributed energy. The user's energy consumption habits and energy-saving awareness are the key factors for selecting distributed energy generation, and the quantity reflects the situation that distributed energy replaces other energy sources. At the same time, the construction of the user terminal will directly affect the user's energy consumption convenience and energy consumption habits, thus affecting the energy efficiency and ultimately affecting the comprehensive benefits.

3.2. Distributed Energy Comprehensive Benefits Evaluation and Case Simulation Analysis

3.2.1. Distributed Energy Comprehensive Benefits Index System

This paper sorts out the indicators for measuring comprehensive benefits in the published literature, and divides the comprehensive benefits into six aspects: economy, energy saving, loss reduction, environment, reliability and society. The government side mainly involves environmental, social and economic benefits. The user side mainly involves reliability and energy saving benefits. The power grid company side mainly deals with loss reduction, environmental, reliability, economic and energy saving benefits, and the distributed energy side involves economic, energy saving, loss reduction, and reliability benefits.

In terms of the economic benefits, distributed energy shall be regarded as local peak shaving storage utility. Compared with large-scale energy storage, it not only saves economic costs but also reduces technical difficulty and has better economic benefits. Han, Dong, and other scholars measured the economic benefits from the initial investment and annual cost of the system [16,17]. Zeng used the net present value, cost-benefit ratio, and payback period to measure economic benefits [18]. He uses economic indicators and price risk indices to evaluate economic benefits [19]. Wu gave a comprehensive consideration of operating costs and pollution control costs [20]. This paper considers the comparison with conventional coal-fired power generation projects and sets the annual income increase index of the project.

In terms of the energy efficiency, renewable energy is often used in distributed energy sources, which can significantly reduce the consumption of fossil fuel and achieve energy conservation and emission reduction. Xie and other scholars use the energy-saving and emission-reduction investment costs of distributed energy units, as well as the value of fossil energy saved by conventional thermal power units, distributed energy supply equivalent energy and heat energy as a measure of energy efficiency [21]. This paper considers the ratio of renewable energy to fossil fuel energy in renewable energy units in distributed energy systems, and sets the energy replacement rate indicators.

In terms of the transmission loss reduction benefits, the energy in distributed systems can be locally consumed, reducing long-distance transportation, resulting in less electrical losses and promoting energy savings. Chen et al. used the benefits of loss reduction to measure the overall benefits [22]. Liang gave a cost-benefit analysis of micro-grids, concluding that the factors of loss reduction are the rate of loss, power generation, the number of distributed energy sources in the micro-grid [23].

In terms of the environmental benefits, distributed energy can adopt a "spontaneous use, surplus electricity online" consumption model to greatly reduce carbon dioxide emissions. Distributed energy technologies based on renewable energy can improve energy efficiency and increase the proportion of renewable energy structures. Zhang and other scholars used the emissions of major pollutants as a measure [24]. Zeng used the pollutant emission reduction and noise influence degree to calculate environmental benefits [25].

In terms of the reliability and efficiency, distributed energy can improve power supply reliability, power quality, and avoid losses caused by power outage losses and voltage drops. In the case of external power grid failures, distributed energy can be converted to independent operation mode, and continue to supply power to important loads, improving the reliability of power supply for important loads, and providing excellent power quality with other ancillary services. Mitra calculated various reliability indicators that affect the working state of distributed energy systems [26]; Gludpetch used the system average interruption frequency index, the system average interruption duration index and the unpowered energy as reliability indicators [27].

In terms of the social benefits, the advantages of distributed energy can bring benefits to society, such as economically and effectively solving the power supply issue in remote areas, achieving energy conservation and emission reduction, driving related technological progress and innovation, and stimulating employment in related industries. Tian used social saving efficiency, saving coal-fired efficiency, saving network loss benefits, reducing short-circuit current efficiency, and sustainable development benefits to optimize social resource allocation efficiency to measure social benefits [28]. This paper increases the technical update rate indicator and measures social benefits from the technical level.

According to the published literatures, after refining and processing, a three-level index system for distributed energy comprehensive benefits involving multiple entities is obtained and shown as Figure 3:

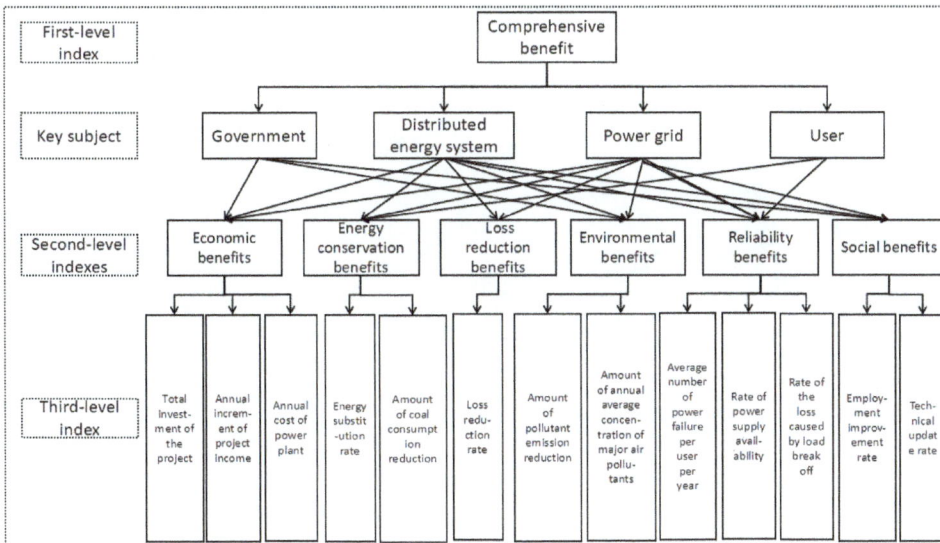

Figure 3. Multi-agent based distributed energy comprehensive benefits three-level indicator system.

3.2.2. Analysis of the Comprehensive Benefits of Typical Engineering Cases

In this paper, the comprehensive benefits analysis is carried out with the example of Dongao Island Wind-Solar-Diesel-Battery(MW grade) multi-energy complementary demonstration project.

Dongao Island is located in the south-central part of the Wanshan Island in Guangdong Province, China. Due to the unique geographical and resource conditions of the island, it is more suitable for the construction of distributed energy systems. Dongao Island's wind & solar & desiel generator and smart storage micro-grid combines wind power generation, solar power generation, diesel generation unit and battery energy storage facilitiy. This paper compares the benefits of the distributed energy accessed to Dongao Island before and after. The specific benefits are as shown in Table 1.

Table 1. Dongao Island Comprehensive Benefits Analysis Table.

Benefits	Before Accessing Distributed Energy	After Accessing Distributed Energy
Economic Benefits	The electricity cost for residents is 2.9 yuan per degree. The commercial electricity is 3.24 yuan per degree. The average annual loss of power plants is about 400,000 yuan.	The civilian use is 2.6 yuan per degree (including subsidy 0.8 yuan). The commercial price is 3.74 yuan per degree. Under the synergy of the government, the annual profit of the power station exceeds 2 million yuan.
Energy Efficiency	Diesel power generation, about 2000 tons of oil for power generation per year.	Under the synergy between the distributed energy, the power grid and the government, green energy replaces diesel power generation, and the power generation accounts for more than 80%, saving traditional energy.
Loss Reduction Benefits	The transmission line is long and the line loss is huge.	Under the cooperation of the power grid and the distributed energy, the loss of the transmission and distribution network is greatly reduced.
Environmental Benefits	Noise pollution, carbon dioxide, sulfur dioxide and other emissions seriously pollute the environment.	Under the support of the government and the power grid, the annual reduction of carbon dioxide is nearly 1500 tons, sulfur dioxide is 45 tons, dust is 40 tons and clean water is 6000 tons.
Reliability Benefits	The minimum load in the off-season is 20 kW, the maximum load in the tourist season is 600 kW, the randomness of the load is very large, the diesel engine failure rate is very high, the voltage is very unstable. There is a serious shortage of electricity in the tourist season, and users often meet with power failure.	The voltage is stable, the reliability of power consumption is greatly guaranteed. The synergy between distributed energy and users has been greatly improved.
Social Benefits	Lack of water and electricity, and residents' lives are greatly affected. Tourists are scarce and economically undeveloped.	Renewable energy has been developed, power structure has been changed, energy crisis has been prevented, and residential electricity problems have been solved. It has promoted the development of island tourism and enhance local economic level.
Comprehensive Benefits	Low energy efficiency. Poor comprehensive benefits.	Higher energy efficiency. Better comprehensive benefits.

Among the six benefits in Table 1, economic benefits, environmental benefits and energy saving benefits are the most significant and most intuitive. The improvement of the loss reduction benefits and reliability benefits also require the advancement of technology. Social benefits are more comprehensive and abstract, and affected by many external factors, so their the improvement may be slow and insignificant. Each entity has a two-way impact on the overall benefits of distributed energy. During the steady improvement of the comprehensive energy efficiency of the Dongao Island, the local government, in order to integrate it into the large power grid, cooperated with the power grid company, no longer offered any preferential treatment on the project land and policies, so the comprehensive benefits were reduced. The enhancement of the comprehensive benefits need to be further coordinated by the government, power grid, distributed energy and other main bodies.

4. Research on Synergistic Optimization Strategy of Distributed Energy Comprehensive Benefits Based on Multi-agents

In order to accurately measure the level of synergy and effectiveness of distributed energy systems, and through the synergy optimization between the subjects to further improve the overall benefits and achieve the goal of sustainable development of distributed energy resources, ensuring that user has access to affordable, reliable and sustainable modern energy, this study uses synergetic entropy to synergetic estimate distributed energy systems. Firstly, the relationship diagram of the distributed energy system is constructed as the research object. Secondly, based on the dissipation theory, the cooperative entropy index of distributed energy system is proposed, and the cooperative entropy calculation formula of distributed energy system is further established to comprehensively evaluate the evolution and the degree of synergy of distributed energy system.

The participating subjects and related relationships of distributed energy systems present different correlation characteristics and associated states at different stages of system development, and there are also hierarchical differences in the extent of driving factors. This study combines the relevant government work reports of power systems, important events of distributed energy power systems, national data and distributed energy power policies over the years to show the relationship, degree of association and type of association between the main components of distributed energy systems in the current stage.

4.1. Division of Distributed Energy Development Stage Based on Policy Analysis and Development Scale Analysis

In the 20 years of distributed energy development, the scale of China's distributed power generation has been gradually expanded under the support and guidance of relevant national policies. Natural gas distributed generation, solar photovoltaic power generation, biomass power generation, wind power generation and other related support policies have been issued. Figure 4 shows the key targeted policies for promoting distributed energy generation systems and the scale of distributed energy development represented by distributed photovoltaics and distributed biomass since 2000.

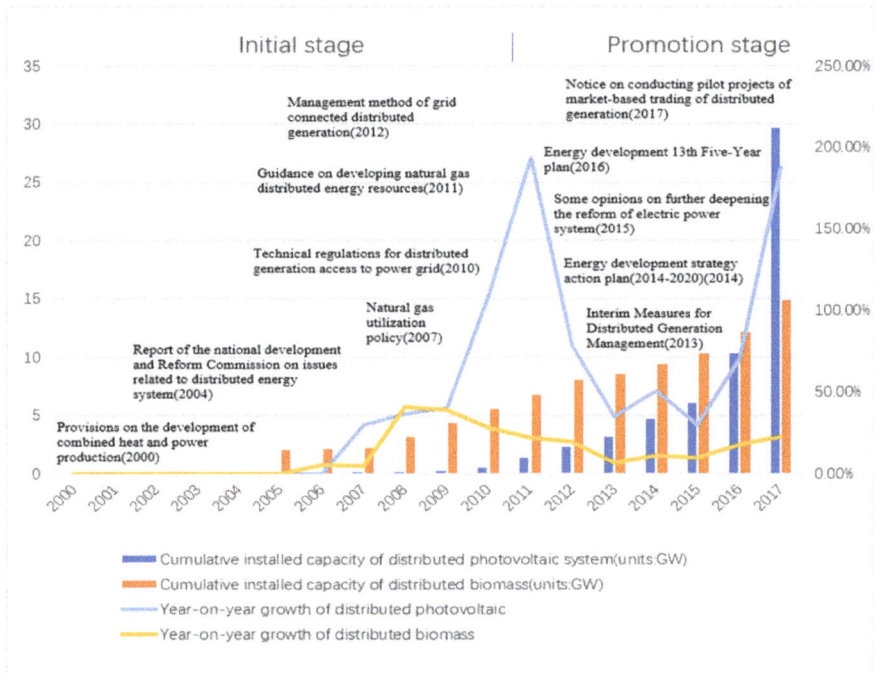

Figure 4. Distributed photovoltaic power generation, distributed biomass power generation cumulative installed capacity and year-on-year growth and key policy maps from 2000 to 2017.

On the basis of comprehensive consideration of the factors affecting the stage of China's distributed energy development, this paper divides the development time of China's distributed energy system from 2000 to 2017 according to the promulgation time of key policies and the analysis of development scale. The period from 2000 to 2011 is the initial stage, and from 2012 onwards as a promotion stage.

4.1.1. Initial Stage (2000–2011)

The initial stage is the preliminary period of distributed energy. At this stage, the term distributed energy officially appears in government documents. Most of the policies are macro-enhancement policies, and the scale of distributed energy installation is limited. During this period, a number of cogeneration projects have been initially explored, and distributed energy projects represented by natural gas-fueled distributed energy systems were gradually put into use in large cities. In 2004, the National Development and Reform Commission's Report of China on Issues Related to Distributed Energy Systems officially defined the concept of distributed energy. The pilot projects of distributed energy in economically developed areas have produced certain economic and social benefits, laying the foundation for the promotion and application of distributed energy systems into more expanded areas. However, the integration of distributed energy at this stage is still difficult, and the scale of distributed energy installation is only expanded to 10 GW, but the speed is relatively slow.

4.1.2. Promotion Stage (2012–Present)

The promotion phase is a period of substantial development of distributed energy. In this stage, policies are more targeted and the development speed of distributed energy is significantly accelerated. The policy intensity in this phase is correspondingly high. Since the "Twelfth Five-Year Plan", the development of China's distributed energy system has entered the promotion stage, and support policies have been introduced one after another, mainly involving natural gas distributed energy and distributed photovoltaics. Among them, the promulgation of the Interim Measures for Distributed Generation Management marks that China has begun to promote the development of distributed energy. In 2013, State Grid Corporation of China issued the "Opinions on Doing a Good Job of Distributed Power Grid-Connected Services" to legalize and order the grid. At this stage, there are plenty cases in which a number of individual users self-generated applications for grid connections. This resulted in the scale of installed capacity reaching a maximum of 187%, and there is a tendency to continue to expand at this rate, with a considerable future potential.

Allowing distributed energy grid-connected is the milestone in the development process of distributed energy. However, the supporting measures involved in grid-connected are not perfect, and there are obstacles in the development of distributed energy, so as the economic benefits are not significant.

4.2. Multi-Agent Synergistic Relationship Analysis of Distributed Energy Systems

Based on the characteristics of distributed energy systems and existing research results, this paper extracts the main evolutionary entities. Based on the whole process idea, the key entities are extracted from the aspects of system guarantee level, technical support level and the three process of energy planning, investment construction, operation and maintenance. The relationship between the main body and the main body of the distributed energy system in the initial stage is shown in Figure 5.

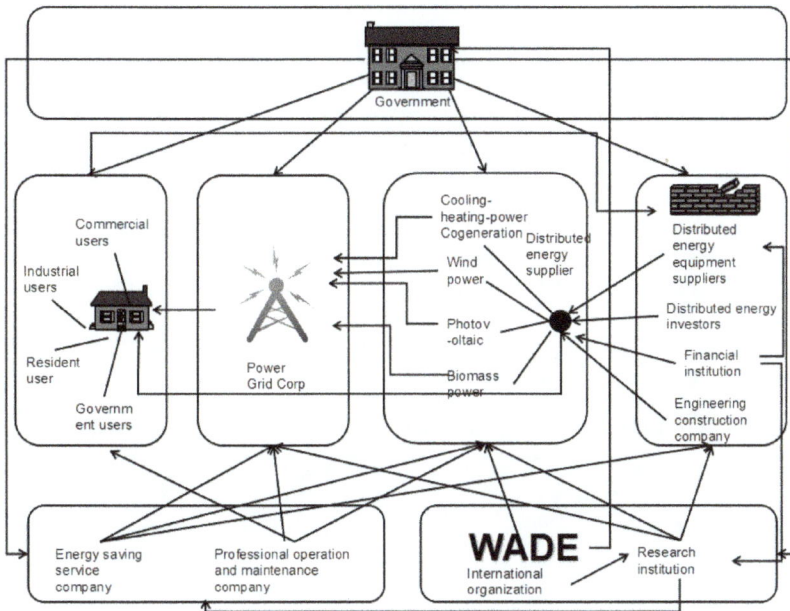

Figure 5. Schematic diagram of the main body relationship of the distributed energy system in the initial stage.

The relationship between the main body and the main body of the distributed energy system in the promotion stage is shown in Figure 6.

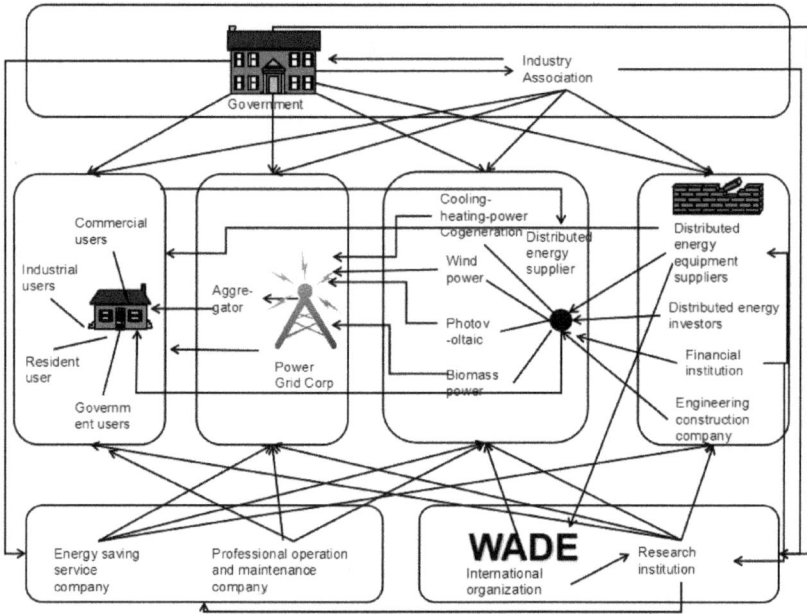

Figure 6. Schematic diagram of the main body relationship of the distributed energy system in the promotion stage.

According to the two-stage association diagram and the power-related experts' review, the study constructs the main energy relationship matrix of the distributed energy system. The degree of association between the participating entities presents different degrees of strength, and the relationship matrix between distributed energy subjects is as listed in Tables 2 and 3. Each column in the table represents the associated behavior accepted by the entity. Each row represents the associated behavior of the synergetic entities. The letters in the matrix indicate the type of association. The numbers in the matrix indicate the strength of the association. Level 1 indicates weak contact, level 2 indicates general contact, 3 indicates strong contact; P indicates policy management, C indicates capital flow, I indicates public opinion impact, E indicates basic attribute, T indicates technical support, and R indicates technology research and development.

Table 2. The multi-agent relationship matrix of the initial stage of the distributed energy system.

Main Body	A	D	F	G	H	J	K	M	N	O	Q	S
A		P2	P3	P2	P2	P3	P3	P2	P2		P2	P2
D						T3						
F						C3						
G		C1				C3			C2			
H						T3						
J							E2	E3				
K								E3				
M						C3						
N					R1	R2	R1	R2			R1	R1
O		I1				I2			I1			
Q					T2	T2	T2					
S						T2	T2	T1				

Table 3. The multi-agent relationship matrix of the promotion stage of the distributed energy system.

Main Body	A	B	D	F	G	H	J	K	L	M	N	O	Q	S
A		P2	P3	P2	P2	P2	P3	P3	P2	P2	P3		P2	P2
B	I2			I2	I2	I2	I3	I2	I2	I2	I2			
D							T3							
F							C3							
G			C1				C3				C3			
H							T3							
J								E2		E3		I2		
K									E3	E3				
L										E3				
M							C3							
N						R2	R3	R2	R2	R3			R2	R2
O	I2	I2					I2					I3		
Q						T2	T2	T2						
S							T3	T2	T1	T1				

In Tables 2 and 3, A = Government, B = Industry Association, D = Distributed energy equipment supplier, F = Distributed energy investor, G = Financial Institutions, H = Engineering construction company, J = Distributed energy supplier, K = Grid company, L = Aggregator, M = User, N = Research institutions, O = International organizations, Q = Energy service company, S = Professional operation and maintenance company.

4.3. Synergistic Entropy Construction

Entropy is one of the parameters that characterize the state of matter in thermodynamics. Its physical significance is to measure the degree of chaos of the system. In this paper, synergy entropy is used to evaluate the multi-agent synergy of distributed energy systems. The aim is to construct an effective index to measure the synergy effect of the complex multi-agent network of the whole distributed energy system. Synergistic entropy is better for dynamic evaluation than traditional methods. The calculation results can show the development trend, find the optimization direction and the sustainable development path. As a quantitative analysis method for dissipative structures, the Brusselator model also provides a theoretical basis and an operational mathematical model for studying the related problems of distributed energy system coordination. In this paper, it is applied to the synergistic analysis of distributed energy systems. Based on the existing research results, the original Brusselator model is transformed, that is, the significance represented by A, B, D, E, X and Y is transformed into the related concept of distributed energy system coevolution.

Let A and B be the components of the cooperative energy entropy of the distributed energy system, that is, A is the positive entropy generated by the synergistic participant, and B is the negative entropy formed by the synergistic participant accepting the related association behavior. D and E are A. Two possible states under the interaction with B: D is the state of non-dissipative structure, that is, the group relationship of each synergistic participant is not clear; E is the state of dissipative structure, that is, the group relationship of each synergistic participant is clear. X, Y are quantifiable indicators that affect the degree of clarity of synergistic participation subject relationships, where X represents a quantifiable positive entropy indicator and Y represents a quantifiable negative entropy indicator. Based on the above definition, this study constructs a Brusselator model of distributed energy system coordination, as shown in Equation (1):

$$A(\text{Positive entropy}) \xrightarrow{K1} X(\text{A quantifiable positive entropy indicator})$$

$$B(\text{Negative entropy}) + X \xrightarrow{K2} Y(\text{A quantifiable negative entropy indicator}) + D(\text{Non dissipative structure})$$

$$2X + Y \xrightarrow{K3} 3X \tag{1}$$

$$X \xrightarrow{K4} E(\text{Dissipative structure})$$

The study uses the synergetic entropy of the distributed energy system to represent that the distributed energy system participates in the synergistic main body and the factors affecting the synergy. In the synergistic process, the effective energy conversion efficiency decreases, and the ineffective energy consumption increases. System status coefficient changes. According to the characteristics of the entropy value, in the synergistic process of distributed energy systems, the larger the synergistic entropy value, the worse the synergistic evolution effect between entities; on the contrary, the better the synergistic evolution between entities.

Claude E. Shannon, one of the originators of information theory, expresses multiple discrete events in system S as discrete event sets. $S = \{E_1, E_2, \cdots, E_n\}$, where the probability of each event appearing randomly is $P = \{P_1, P_2, \cdots, P_n\}$, so information entropy (i.e. total amount of information) can be defined as Equation (2):

$$H(S) = -\sum P_i \log P_i, i = 1, 2, \cdots, n \tag{2}$$

Based on the above-mentioned distributed energy system synergistic Brusselator model structure (Equation 1), this study assumes that in the synergistic process of distributed energy systems, f_i is the number of the paths that the i participating synergistic entity points to the other. f_i' is the number of synergy paths for the i participants to accept other synergistic participants. Assuming that there are "n" co-participants in the distributed energy system, the total number of co-evolution paths of the distributed energy system is as follows:

$$f = \sum (f_i + f_i'), (i = 1, 2, \cdots, n) \tag{3}$$

In this article, we write P as $P = \frac{(f_i - f_i')}{f}$. According to the relationship between probability and Shannon's entropy function, the evolutionary cooperative entropy expression of distributed energy systems can be obtained:

$$CE_i = -\sum P_i \times \log |P_i|, (i = 1, 2, \cdots, n) \tag{4}$$

4.4. Synergetic Entropy Analysis of Distributed Energy Systems

According to the contents of Equations (4) and Tables 2 and 3, the out-degree value and in-degree value of each evolutionary participant and the total number of relationships associated with it in the distributed energy system are calculated. In short, taking participant A as an example, the out-degree value is the number of the relationship lines that the participant participates in the remaining participants. The in-degree value is the number of association lines that the participant accepts from other participants. The number of all associations is the sum of their out-degree and in-degree value. The evolutionary cooperative entropy value of each synergistic participant is determined by the difference between the in-degree and out-degree value and the ratio of overall relationship.

According to the above introduction, the cooperative entropy value of each participating entity in the two stages of the distributed energy system is calculated. The evolutionary cooperative entropy includes positive entropy and negative entropy. Positive entropy will generate system internal friction and increase system evolution burden. Conversely, negative entropy will neutralize the system internal friction and coordinate the evolution of the overall system; The larger the entropy value, the worse the synergy performance of the subject. The larger the negative entropy value, the better the synergy performance of the subject. Figure 7 shows in detail the changes in entropy values of distributed energy systematization participants (presenting positive entropy or negative entropy). According to the data provided in Figure 7, the change rule of the entropy value of each participant and the co-evolution effect of each stage are analyzed. Because the weight establishment process of each synergistic participant is complex, there are many driving factorsand the data collection is very difficult, this study only performs the simple addition of the entropy values of the participants in the same level.

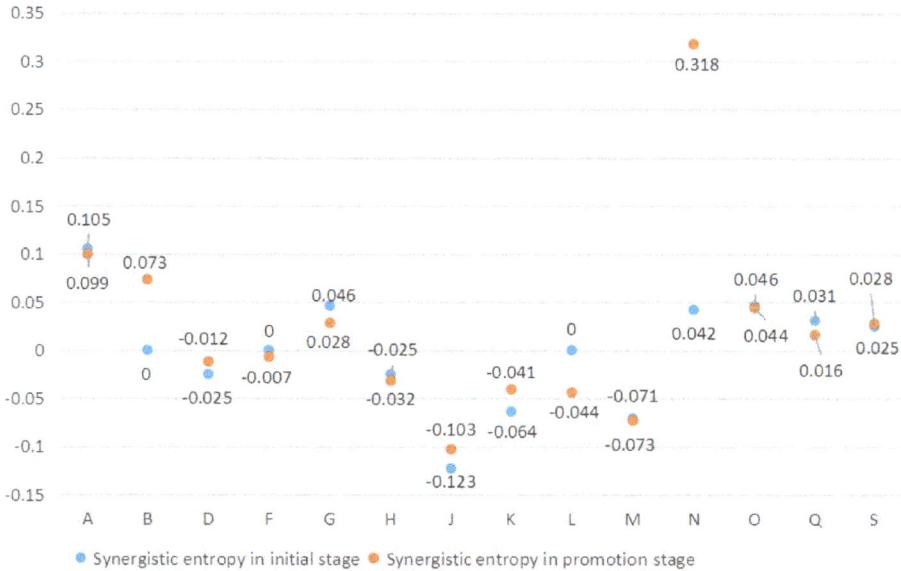

Figure 7. Distributed energy system cooperative entropy results. A = Government, B = Industry Association, D = Distributed energy equipment supplier, F = Distributed energy investor, G = Financial Institutions, H = Engineering construction company, J = Distributed energy supplier, K = Grid company, L = Aggregator, M = User, N = Research institutions, O = International organizations, Q = Energy service company, S = Professional operation and maintenance company.

Through the statistical analysis of the synergistic entropy values of the synergistic entities of the distributed energy system, this paper concludes the following: Among the 14 participants, engineering construction companies, distributed energy suppliers, distributed energy equipment suppliers, power grid companies, and users all maintain negative entropy values in both phases. Among them, the entropy value of the grid company changed from −0.064 to −0.041, the largest change, indicating that the coordination degree of the grid company is significantly improved during the promotion stage. This change is due to the serious implementation of the national energy development strategy by power grid enterprises under the background of policy promotion and the China's electricity reform policy. The grid company began to participate in distributed energy operations, and promoted the development of new energy and distributed power as an important political responsibility and social responsibility, and actively served the distributed energy system. The aggregator is a new type of participant between the user and the grid company. It appears for the first time in the promotion stage, with an entropy value of −0.044. As a participant in the power market operation, aggregators can not only buy electricity from the grid or users, but also sell their own stored power to users or power grids, with better synergy. The synergistic entropy of users changed from −0.071 to −0.073, and the synergistic ability fluctuated slightly. This is possibly due to the fact residential users began to install their own small-scale distributed energy at the promotion stage. All aspects of policy conditions are uncertain, leading to the degree of synergy appear a slight decline. The entropy values of engineering construction companies and distributed energy suppliers have not changed much, and the overall coordination situation is good.

The cooperative entropy values of government departments, industry associations, financial institutions, scientific research institutions, international organizations, and energy conservation service companies are all positive entropy values. Among them, the entropy value of international organizations has declined by a large margin, indicating that its influence is gradually increasing. This change is attributed to the increasing emphasis on the coordinated development of distributed energy systems by international organizations, which have enacted global energy policies such as the 2030 Agenda for Sustainable Development, promulgated in 2016. All countries recognize the government plays an important role in supporting power system transformation and energy system integration, and will cooperate to promote technology development and deployment in the fields of energy storage, electric vehicles and modern biomass energy, renewable energy heating, etc.,cooperate in accelerating smart grid deployment and interoperability. These policies not only provide a framework for the development of distributed energy systems in various countries, but also provide external pressure for the coordination of distributed energy systems from the perspective of international supervision. The synergy entropy of the government departments is positive and the change is not sigificant, indicating that the government's policies on distributed energy systems are not in place. At this stage, there is still a lack of technical standards and management standards at the national level.

4.5. Comprehensive Benefits Improvement Strategy of Distributed Energy System

Combined with the characteristics of distributed energy, the influencing factors of comprehensive benefits and the current synergy between various entities, guided by the sustainable development of distributed energy, this paper proposes a comprehensive benefits and multi-agent synergistic optimization strategy for distributed energy systems.

4.5.1. Synergistic Optimization Strategies of Comprehensive Benefits of Government-Side

The optimization of the government side is based on the system construction of the comprehensive energy system. Through the formulation of relevant policies and laws and regulations, the cooperation with distributed energy system and scientific research institutions should be strengthened to standardize the development of distributed energy. The development level and comprehensive benefits of distributed energy should be further improved by relying on the demonstration and promotion led by the government. Specific synergistic optimization strategies are as follows.

(1) Improve the participation at the national macro level in the multi-agent synergy of distributed energy systems. Government should co-ordinate management andlegally regulate distributed generation and energy utilization models. Meanwhile, establish a national energy comprehensive management function department that adapts to the national conditions, and formulate relevant laws and regulations on distributed power sources, and clearly define distributed energy in law; establish an integrated energy research and development institution, conduct research on major issues in the energy field, and coordinate energy long-term planning to facilitate cooperation, integration and synergetic development in the energy field.

(2) Government should continue to issue clear policy signals, provide a flexible policy environment, further clarify the development orientation of distributed energy in energy transformation, increase financial support and tax support, subsidize project construction, especially the construction of demonstration projects, and consider returning part of the value added tax. At the same time, the power price compensation mechanism shall be improved and price concessions shall be implemented to coordinate the interests of different energy supply participants. In the mean time, further refine the overall objectives and main technical indicators of distributed energy development, and provide favorable grid-connected conditions, focus on the planning of distributed energy projects, and clarify the application conditions and approval procedures. Increase the supervision of key links such as grid connection and transaction after the completion of the project.

(3) Strengthen the coordination of distributed energy market, explore the adapted trading mode of distributed energy, and promote the distributed new energy microgrid in combination with the requirements of the power system, to make the microgrid into a market entity with independent power sales rights, including the distributed energy microgrid carrier as an independent power selling entity, for direct PV supply or interaction with nearby new energy projects. Encourage grid companies to give preferential treatment to internal and external transactions of distributed energy. At the same

time, all power and energy companies are encouraged to actively participate in the construction of distributed energy projects, and arouse the enthusiasm of local capital to participate in the project to achieve win-win cooperation and benefits sharing.

(4) Enhance the degree of synergy with scientific research institutions and international organizations. The government should respond positively to the call of international organizations and integrate with the world goals. According to the future development direction, increase investment in scientific research, and propose to set up special funds for research on distributed energy technologies to cultivate professional talents. For instance, to train professional distributed energy planners. These talents must be familiar with relevant policies and regulations, understand various related technologies, and select appropriate distributed technologies according to local climate conditions and resources to achieve the best comprehensive benefits. Leaders in various energy sectors should take the lead in conducting distributed energy knowledge and technical training, enabling professionals to take the decision-making role and correctly guide the direction of distributed energy development.

4.5.2. Synergistic Optimization Strategies of Comprehensive Benefits of Distributed Energy Systems

The synergistic optimization of power side mainly consists of three aspects: energy planning, location valuation, capacity and equipment optimization. The specific optimization strategy is as follows:

(1) Energy service companies and aggregators should continuously build their core competitiveness, effectively integrate the resources of the industry chain, so that the benefits can bring together the entire industry chain and promote the synergetic development of the entire industry. In the early stage of energy planning, the planned system scope and energy load should be carefully considered, the basic types of projects should be clarified, and the development of distributed energy resources should be strengthened. Focusing on existing and future planned gas turbine power plants, the distributed energy project is developed in the controllable area of gas turbine power plants, which take advantage of the power generation and heat supply stability of the gas turbine power plant to ensure the continuous and stable operation of the system and enhance the reliability of the system operation. Actively plan integrated distributed energy supply solutions for natural gas, solar energy, wind energy, geothermal energy, biomass energy and energy storage to create an efficient and integrated energy supply model.

(2) Installing a distributed power source of appropriate capacity at a suitable location can reduce network loss and improve power quality. Methods for determining the optimal capacity and location should be explored to maximize the benefits of loss reduction. The power-side can consider actively investing in distributed energy projects in the demonstration area to strive for optimal policy support. In other areas where there is a high-quality load, a selective development project can be considered.

(3) Optimize the type and quantity of integrated energy system supply and storage equipment. Based on the overall load level of integrated energy in the planning area, the optimal combination scheme of refrigeration, heating, cooling and heat storage devices in the planning area is proposed for a variety of optimal planning objectives and the optimal operation of the whole life cycle. Secondly, the operation scheduling of the integrated energy system equipment should be optimized, and the system operation constraints and supply and demand balance constraints of energy supply and storage equipment are considered under different energy demand periods. Under the guidance of supply-demand relationship and price mechanism, various participants flexibly adjust energy supply, energy consumption and energy storage to achieve flexible interaction of integrated energy and vertical integration of supply-demand and storage, and improve the utilization efficiency of integrated energy.

4.5.3. Synergistic Optimization Strategies of Comprehensive Benefits of Power Grid Company

The optimization strategy of the power grid is based on the synergy enthusiasm of power grid enterprises. Under the condition of upgrading key technologies, coordinated planning and operation are carried out, and the pilot project is built on the premise. The specific optimization strategy is as follows:

(1) Power grid companies should consciously and actively improve the degree of synergy. Under the background of the reform of electric power, power grid enterprises have the right and obligation to actively participate in the investment, construction, operation and management of distributed energy projects, and realize the dual role mechanism transformation of distributed energy stakeholders. On the one hand, in order to avoid conflicts with the provisions of the Electricity Law, grid power companies can be used as stakeholders or project shareholders of distributed energy projects, and members of distributed energy are represented as legal representatives. On the other hand, the distributed energy enterprise thus formed can be used as a member of the power grid enterprise. Under the premise of meeting the terminal demand in the distributed energy region, according to the characteristics of the supply and demand balance of the large power grid and the function of the smart grid, the distributed generation unit's opening, stopping and load rate are adjusted and optimized to realize the switching between the two modes of operation: grid power or online power sale during peak and low valley periods.

(2) Strengthen coordination with scientific research institutions, study key technologies of distributed energy, increase independent research and development efforts, and reduce dependence on technology in developed countries. In the future, we should further study the protection and control technologies of distributed energy and the new protection principles and methods of distributed power systems to improve the security of the entire social energy supply system. Grid company should analyze the operating characteristics of various distributed power sources and microgrids, the interaction mechanism between distributed power sources, microgrids, and power distribution systems. Developing relevant theories and methods, laying the foundation for energy management and distributed generation economic dispatch. It is necessary to research on distributed power grid-connected technology for the purpose of achieving efficient and user-friendly grid-connected power generation.

(3) Research on coordinated planning methods for distributed power distribution systems. Consider establishing a distribution network design and planning theory system suitable for distributed power supply characteristics, including distribution system structure design methods that contribute to microgrid access, comprehensive performance evaluation index system including distributed energy distribution systems, and new power distribution System optimization planning theory, etc.

4.5.4. Synergistic Optimization Strategies of Comprehensive Benefits of User-Side

The user-side synergistic optimization strategy is optimized through market demand response and user-side construction. The specific strategies are as follows:

(1) Strengthen the synergy among the government, financial institution and the user, liberalize the user-side distributed power supply construction, and promote the operation mode of "spontaneous use, surplus Internet access, and power grid adjustment" to encourage enterprises, institutions, communities, and families to adjust their own conditions. Invest in the construction of various types of distributed power sources such as rooftop solar and wind energy. It can integrate many small-scale energy comprehensive utilization equipment with different forms. In addition to the traditional electric/heat/cold load, it also includes a large number of renewable energy equipment, energy storage equipment, and comprehensive energy supply equipment.

(2) Apply the energy management system on the user side to guide users to avoid the peak of power consumption and cultivate better user habits. Improve the flexibility and reliability of power supply, give priority to the use of local renewable energy or large grid through power, and encouraging new energy sources to access the power demand side management platform in the region. The energy management department should work with relevant departments to study and formulate the demand side management policy of the distributed energy source, and explore the establishment of distributed energy as a market entity to participate in service compensation mechanisms such as interruptible load peak shaving, electric energy storage peak shaving, and black starts.

(3) Increase cooperation with scientific research institutions, use the energy Internet, integrate user-side services, smart grids and distributed generation, and develop smart electricity interactive business models and intelligent power system frameworks. And consider the energy characteristics of equipment for home users and business users, and develop intelligent power technology that integrates information collection, energy efficiency assessment, equipment control, and two-way interaction to realize household energy safety monitoring, electricity consumption information and property management. Function to integrate intelligent microgrid technology with distributed energy. In terms of terminal hardware, a user-oriented intelligent interactive terminal core module shall be developed to realize energy metering and device monitoring for large-scale users.

5. Conclusions

Based on China's energy structure, this paper focuses on the development of distributed energy. Through synergistic entropy evaluation, under the guidance of system dynamics, the comprehensive benefits of distributed energy at the present stage are analyzed. By using synergy entropy concept, the synergy degree among different agents are evaluated, and the path of improving comprehensive benefits of distributed energy through synergistic optimization is found from the multi-agent level of government, power supply, power grid and users, and draw the following research conclusions:

The analysis of this paper concludes that the improvement of the comprehensive benefits of distributed energy depends on the government's policy support, the user's main needs, and the degree of synergy between the various entities. From these factors, the comprehensive benefits can be improved. Through literature analysis, a three-level indicator for measuring comprehensive benefits is proposed. Through case analysis, it is concluded that the further improvement of the comprehensive benefits of distributed energy depends on the improvement of the coordination among the main bodies. Through synergistic entropy calculation, the synergy of Power Grid Corp and distributed energy is increased fastest. The synergy of government departments, trade associations, financial institutions, scientific research institutions, international organizations and energy-saving service companies needs to be strengthened. And from the multi-agent optimization synergy level, the comprehensive benefit enhancement strategy is put forward.

In the future, distributed energy can strengthen multi-agent cooperation in government system construction, policy promulgation, demonstration and promotion, power side planning and location, equipment optimization, grid side improving technology, coordinated planning and construction pilot projects, user-side energy optimization and user-side construction, so as to improve comprehensive benefits and utilization level in an all-round way and promote the coordinated development of multi-agent and the transformation and upgrading of the energy structure.

Author Contributions: Conceptualization, X.S. and J.L.; Data curation, M.S.; Investigation, M.S.; Methodology, M.S. and Y.W.; Supervision, X.S. and J.L.; Writing–original draft, M.S.; Writing–review & editing, X.S., M.S., Y.W. and J.L.

Funding: This study is supported by the National Natural Science Foundation of China (NSFC) (71501071), Beijing Social Science Fund (16YJC064,17GLB010) and the Fundamental Research Funds for the Central Universities (2018ZD14,2017MS059).

Conflicts of Interest: The authors declare no conflict of interest.

References

1. Di Somma, M.; Graditi, G.; Heydarian-Forushani, E. Stochastic optimal scheduling of distributed energy resources with renewables considering economic and environmental aspects. *Renew. Energy* **2018**, *116*, 272–287. [CrossRef]
2. Chen, B.; Lin, S.P. Develop Distributed Energy to Promote the Energy Conservation and Emission Reduction of the State. In *Second China Energy Scientist Forum 2010*; Scientific Research Publishing Inc.: Xuzhou, China, 2010.

3.   Niu, C.H.; Li, B.J. Research on the Technical and Economic Problems of the Development of Distributed Generation. In Proceedings of the International Conference on Logistics Engineering, Management and Computer Science (LEMCS), Shenyang, China, 29–31 July 2015.

4.   Kim, H.M.; Kinoshita, T. A New Challenge of Microgrid Operation. In Proceedings of the 1st International Conference on Security-Enriched Urban Computing and Smart Grid, Daejeon, Korea, 15–17 September 2010.

5.   Zeng, M.; Ouyang, S.J.; Shi, H.; Ge, Y.J.; Qian, Q. Overall review of distributed energy development in China: Status quo, barriers and solutions. *Renew. Sustain. Energy Rev.* **2015**, *10*, 1226–1238. [CrossRef]

6.   Serrano, J.X.; Escriva, G. Simulation Model for Energy Integration of Distributed Resources in Buildings. *IEEE Lat. Am. Trans.* **2015**, *13*, 166–171.

7.   Omu, A.; Rysanek, A.; Stettler, M.; Choudhary, R. Economic, Climate Change, and Air Quality Analysis of Distributed Energy Resource Systems. *Procedia Comput. Sci.* **2015**, *51*, 2147–2156. [CrossRef]

8.   Inamori, J.; Nonogaki, M.; Hirose, K. Cost-benefit Analysis of a Microgrid System. In Proceedings of the IEEE International Telecommunications Energy Conference (INTELEC), Osaka, Japan, 18–22 Octber 2015.

9.   Zhang, Z.; Li, G.Y.; Zhou, M. Application of Microgrid in Distributed Generation Together with the Benefit Research. In Proceedings of the IEEE PES General Meeting, Minneapolis, MN, USA, 25–29 July 2010.

10. Di Somma, M.; Yan, B.; Bianco, N.; Graditi, G.; Luh, P.B.; Mongibello, L.; Naso, V. Operation optimization of a distributed energy system considering energy costs and exergy efficiency. *Energy Convers. Manag.* **2015**, *103*, 739–751. [CrossRef]

11. He, Y.X.; Xu, Y.; Xia, T.; Zhang, J.X. Business Impact and Policy on the Major Players in the Market of the Development of Distributed Energy in China. *Math. Prob. Eng.* **2016**, *10*. [CrossRef]

12. Zhang, L.; Chen, X.Y.; Chen, K.; Ding, X.H.; Chen, X.Y.; Liao, Y.C.; Yu, K. Coordinated Control Strategy of Distributed Photovoltaic Generation and Load. *Appl. Mech. Mater.* **2014**, *457–458*, 1266–1271. [CrossRef]

13. Li, M.J.; Yan, J.Y.; Lu, T.Q.; Liu, S.N.; Liu, T.; Li, D.X. Research on Distribution Network Comprehensive Integration Technology in Micro-Grid. In Proceedings of the International Conference on Sensing, Diagnostics, Prognostics, and Control (SDPC), Shanghai, China, 16–18 August 2017.

14. Reinders, A.; de Respinis, M.; van Loon, J.; Stekelenburg, A.; Bliek, F.; Schram, W.; van Sark, W.; Esterl, T.; Uebermasser, S.; Lehfuss, F. Co-evolution of smart energy products and services: A novel approach towards smart grids. In Proceedings of the Asian Conference on Energy, Power and Transportation Electrification (ACEPT), Singapore, 25–27 October 2016.

15. Bale, C.S.; Varga, L.; Foxon, T.J. Energy and complexity: New ways forward. *Appl. Energy* **2015**, *138*, 150–159. [CrossRef]

16. Han, Z.H.; Qi, C.; Xiang, P.; Liu, M.H.; Wang, S. Benefit analysis and comprehensive evaluation of distributed energy system. *Therm. Power Gener.* **2018**, *2*, 31–36. [CrossRef]

17. Dong, F.G.; Zhang, Y.; Shang, M.M. Research on multi index comprehensive evaluation of distributed energy system. *Proc. CSEE.* **2016**, *12*, 3214–3223. [CrossRef]

18. Zeng, M.; Xie, B.; Yan, B.J.; Lin, X.; Zhang, Y.N.; Xue, S. Comprehensive benefits evaluation of micro network based on multi factor analysis. *Water Res. Power* **2013**, *31*, 247–249, 256. (In Chinese)

19. He, J.; Zhou, K.P.; Xin, J.H.; Chen, K.; Deng, C.H. New Metrics for Assessing Fuel Price and Technological Uncertainty in Microgrid Power Planning. In Proceedings of the 10th IEEE Conference on Industrial Electronics and Applications, Auckland, New Zealand, 15–17 June 2015.

20. Wu, H.B.; Liu, X.Y.; Ding, M. Dynamic economic dispatch of a microgrid: Mathematical models and solution algorithm. *Int. J. Electr. Power Energy Syst.* **2014**, *63*, 336–346. [CrossRef]

21. Xie, X.; Cao, Y.; Yuan, Y.; Guo, S.Q. Benefit analysis of energy saving and emission reduction for micro grid diesel engine based on ladder peak valley tariff. *Autom. Electr. Power Syst.* **2014**, *8*, 1–6.

22. Chen, B.S.; Liao, Q.F.; Liu, D.C.; Wang, W.Y.; Wang, Z.Y.; Chen, S.Y. Comprehensive evaluation index and method of regional integrated energy system. *Autom. Electr. Power Syst.* **2018**, *4*, 174–182.

23. Liang, H.S.; Cheng, L.; Su, J. Cost–benefit analysis of microgrid. *Proc. CSEE* **2011**, *S1*, 38–44. [CrossRef]

24. Zhang, T.; Zhu, T.; Gao, N.P.; Wu, Z. Research on optimal design and multi index comprehensive evaluation method of distributed thermoelectric energy system. *Proc. CSEE* **2015**, *14*, 3706–3713. [CrossRef]

25. Zeng, C.H. Analysis on environmental benefits and environmental problems of distributed energy planning in a certain area of Beijing. *J. Shenyang Inst. Eng. Nat. Sci.* **2014**, *4*, 305–308. [CrossRef]

26. Xu, X.F.; Mitra, J.; Wang, T.T.; Mu, L.H. Reliability Evaluation of a Microgrid Considering its Operating Condition. *J. Electr. Eng. Technol.* **2016**, *11*, 47–54. [CrossRef]
27. Gludpetch, S.; Tayjasanant, T. Optimal Placement of Protective Devices For Improving Reliability Indices in Microgrid System. In Proceedings of the IEEE PES Asia-Pacific Power and Energy Engineering Conference (APPEEC), Kowloon, China, 8–11 December 2013.
28. Tian, S.X.; Cheng, H.Z.; Chang, H.; Qi, Q.R.; Liu, L.; Hong, S.Y. Social benefit analysis and evaluation method of UHV power grid. *Electr. Power Autom. Equip.* **2015**, *35*, 145–153. [CrossRef]

energies

MDPI

Article

Fault-Tolerant Temperature Control Algorithm for IoT Networks in Smart Buildings

Roberto Casado-Vara [1,*], Zita Vale [2], Javier Prieto [1] and Juan M. Corchado [1]

[1] BISITE Digital Innovation Hub, University of Salamanca. Edificio Multiusos I+D+i, 37007 Salamanca, Spain; javierp@usal.es (J.P.); corchado@usal.es (J.M.C.)
[2] GECAD—Research Group on Intelligent Engineering and Computing for Advanced Innovation and DevelopmentInstitute of Engineering—Polytechnic of Porto (ISEP/IPP), 4249-015 Porto, Portugal; ZAV@isep.ipp.pt
* Correspondence: rober@usal.es

Received: 15 November 2018; Accepted: 5 December 2018; Published: 7 December 2018

Abstract: The monitoring of the Internet of things networks depends to a great extent on the availability and correct functioning of all the network nodes that collect data. This network nodes all of which must correctly satisfy their purpose to ensure the efficiency and high quality of monitoring and control of the internet of things networks. This paper focuses on the problem of fault-tolerant maintenance of a networked environment in the domain of the internet of things. Based on continuous-time Markov chains, together with a cooperative control algorithm, a novel feedback model-based predictive hybrid control algorithm is proposed to improve the maintenance and reliability of the internet of things network. Virtual sensors are substituted for the sensors that the algorithm predicts will not function properly in future time intervals; this allows for maintaining reliable monitoring and control of the internet of things network. In this way, the internet of things network improves its robustness since our fault tolerant control algorithm finds the malfunction nodes that are collecting incorrect data and self-correct this issue replacing malfunctioning sensors with new ones. In addition, the proposed model is capable of optimising sensor positioning. As a result, data collection from the environment can be kept stable. The developed continuous-time control model is applied to guarantee reliable monitoring and control of temperature in a smart supermarket. Finally, the efficiency of the presented approach is verified with the results obtained in the conducted case study.

Keywords: control system; fault-tolerant control; algorithm design and analysis; IoT (Internet of Things); nonlinear control

---

1. Introduction

The advances in communications techniques, network topologies and control methods, have contributed to the development of Networked Control Systems (NCSs), expanding their possibilities. As a result, in the last several decades, NCSs have received considerable attention form the scientific community, mainly due to their wide-ranging application possibilities [1]. Once an Internet of Things (IoT) network is formed by multiple IoT nodes, controller or actuator nodes, it is feasible for them to capture data from a large range of existing structures. However, when the accuracy of IoT nodes is reduced, the data they capture is faulty and causes inappropriate decisions. Therefore, it is critical to increase the ability of the IoT network to detect IoT nodes which are not operating properly [2]. This work introduces a new predictive temperature control algorithm for fault tolerant detection of a large number of IoT nodes, providing an efficient temperature control. The implementation of a system to control and monitor the precision states of the IoT nodes will ensure reliability of the data captured by the IoT network. The discrete time control focuses on system efficiency at a discrete time range rather than a continuous time range. The discrete-time control issues, such as linear systems have been investigated. Amato et al. deal with the finite-time stabilization of continuous-time linear systems is considered. The main result provided is a sufficient condition for the design of a dynamic output feedback controller which makes the closed loop system finite-time stable [3,4]. Therefore, Polyakov et al. consider the control design problem for finite-time and fixed-time stabilizations of linear multi-input system with nonlinear uncertainties and disturbances, so the robustness properties of the network are improved [5]. The works presented above show that the quality of any linear control algorithm is estimated by different performance indices such as robustness with respect to disturbances. Although these authors make their study in discrete time, the algorithm we have developed is an important starting point. Meanwhile, the studies on the discrete-time control of nonlinear system have also been carried out for triangular systems [6] or nonlinear dynamical networks [7]. These two papers have a different approach to the problem of discrete-time control. Korobov et al. solve the issue of global stabilization in finite-time for a general class of triangular multi-input multi-output ($MIMO$) systems with singular input–output links combining the controllability function method with a modification of the global construction. Hui et al. focus on the analysis of semistability and stability in finite time and on the synthesis of systems with a continuous equilibrium. These two approaches address the problem of control in nonlinear systems in very specific cases of triangular and semi-stable systems. Although these are two rather limited case studies, they give a very good focus on how to deal with nonlinear control problems. Discrete-time control techniques have been applied for many practical applications, for instance, multi-agent systems [8] and secure communications [9]. Both works present a new adaptive fuzzy output feedback control approach composed for a type of nonlinear single input and single output feedback control systems with unmeasured status and input saturation. In these two works, we can see that fuzzy control is a good approach to the problem of nonlinear control, but the authors think that, for this case, it is an invalid technique, since all the control functions of the system are known. Feedback nonlinear systems representing a class of nonlinear control systems have been widely considered [10,11]. The problem we address is the topic of predictive maintenance of IoT networks in continuous-time, with the aim of increasing the monitoring and control reliability of IoT networks, as it is done in continuous-time. By using continuous-time Markov chains to predict the future accuracy states of sensors, IoT networks will collect quality data because their nodes will always work in an optimal state.

Motivated by the above observation, this paper proposes a new feedback control algorithm to improve predictive maintenance of the IoT networks. The algorithm finds the IoT nodes that do not function correctly and collect false data. To optimize the monitoring and control processes of the IoT network, a novel application of the continuous-time Markov chains is used. We predict the future accuracy states of the IoT nodes and, in case it is predicted that a sensor will become faulty after the time control period has expired, the controller sends a signal that this IoT node has to be replaced. Moreover, if an IoT node has to be replaced, the control algorithm creates a virtual sensor in that position. This virtual sensor estimates the temperature of that sensor based on the temperature of its neighboring nodes. In this way, the IoT network collects data in continuous-time range without any loss of reliability in the data due to malfunctions in the IoT devices.

The problem of data quality and the detecting of incorrect data has been extensively studied [12]; these works search the quality of data applying different techniques as game theory [13] or other types of metrics [14]. These articles provide a solid design of how to increase the quality of data; in our opinion, these works are focused on homogeneous data and discrete time; even so, they are an excellent support for our research. The above-mentioned studies on data quality and detection of incorrect data concern discrete time, and the outputs for continuous time systems are quite limited. Actually, continuous time control systems have been applied in a large range of fields, such as feedback control of nonlinear systems [15,16]. These papers deal with the stability of discrete-time networked systems with multiple sensor nodes under dynamic scheduling protocols. In fact, this is a great advancement for the stability of nonlinear systems because it addresses dynamic systems with multiple nodes.

In our research, we are using similar techniques for improve fault tolerant control with multiple IoT nodes. Although the work of these authors is in discrete time, the techniques they use are very sophisticated and useful in the field of control theory. Decision-support is an important topic in control theory. Automated trading plays a crucial role in supporting decision-making in bilateral energy transactions [17,18]. In fact, a proper analysis of the past actions of opposing traders can increase the decision-making process of market players, allowing them to choose the most appropriate parties with whom to trade in order to increase their performance. Demand–response aggregators were developed and deployed around the world, and more in Europe and the United States. Aggregator involvement in energy markets increases the access of a small resource to them, enabling case studies to be presented for flexibility of demand [19,20]. Real-time simulations [21,22] have applications to control theory. In fact, this work analyzes the way in which the players' features are modeled, particularly in their small-scale performance, thus simplifying the simulations while preserving the quality of the results. Authors also carried out a comparative analysis of the real values of the electricity market with the market results obtained from the scenarios generated. In [23,24], Zhang et al. proposes a new time-delay communications algorithm based on adapted control. Although in our research we have used a control algorithm based on feedback, we think that a possible improvement of our proposal is that the control algorithm is adaptive. This article is a good example of how to use adaptive control to stabilize a system. In addition, control theory has several applications in the field of demand response. In [21,25], the authors propose an algorithm to predict demand response based on a simplex optimization method. Although this is a nice approach to solve this kind of problem, we think that this approach can be optimized for its application to control theory. However, some problems related with the above topics can be solved using neural networks [26]. In other areas such as supply chain [27,28], fraud detection [29] and edge/fog computing architectures [30], control techniques are beginning to be applied to optimize processes. Control algorithms face the following challenges in the field of temperature data quality and predictive maintenance of IoT networks.

1. For the fault tolerant control in continuous time, solving differential equations with complex conditions and boundaries that change in every loop is needed.
2. Algorithms that improve data quality and detect incorrect data can lead to false positives. It is essential to differentiate between a hot (cold) temperature point and a faulty IoT node.

In this paper, we address research gaps in the supervision and control of continuous time networked systems with multiple IoT devices. Our goal is to present an optimized control algorithm to achieve maximum efficiency in fault tolerant control. A unified model of a continuous time hybrid control system is presented along with a data quality and incorrect data recognition algorithm and a feedback control algorithm to provide prediction of the accuracy status of the IoT nodes. The output of the data quality algorithm is the input of the predictive feedback control algorithm. The main contribution of this paper can be summarized as follows:

1. To the best of our knowledge, the suggested method provides efficient feedback control for the continuous time system model regarding detection of incorrect data or malfunction of IoT devices.
2. A new way of predicting IoT node accuracy states from error measurements and, through the Markov continuous time chains, algorithm predict future IoT node accuracy states in continuous time.
3. A novel control algorithm capable of integrating the above contributions to provide an innovative IoT network temperature control mechanism.

The efficiency of the presented approach is illustrated by a numerical case study. Preliminary results on the improvement of data quality and detection of wrong date in WSNs have been presented in the work of Casado et al. [13].

The rest of the paper is organized as follows. Section 2 shows the procedure of the control algorithm design in this paper. A case study is shown in this section and simulation studies are performed in Section 3. Finally, Section 4 concludes this paper.

2. System Model

This section presents the control algorithm that we have developed. The control algorithm is a hybrid of two other algorithms: (1) Cooperative control algorithm (Section 2.1). This algorithm receives the data collected by the IoT network and increases the quality of the data by searching and self-correcting false data. The output variables of this algorithm are the input variables of the following algorithm; (2) accuracy state prediction algorithm (Section 2.2). This algorithm implements a predictive maintenance system to make the IoT network more robust. Figure 1 shows the model described in this paper, where ϵ is the measurement error that temperature IoT node are allowed to have. $u^{(t+k)}$ is the controller function, this function detects if an IoT node is faulty or operates correctly at time $t + k$ (i.e., t is the current algorithm step time, while k is a time interval that we want to control. In this way, $t + k$ is the time interval that elapses from the current time t). $z^{(t+k)}$ is the prediction accuracy states function; this function predicts the accuracy state of IoT nodes in the time window $t + k$ (i.e., we know the accuracy state of the IoT nodes at time t, so this function gives us the most probably precision state in time $t + k$). $f^{(t)}$ is the feedback function at time t.

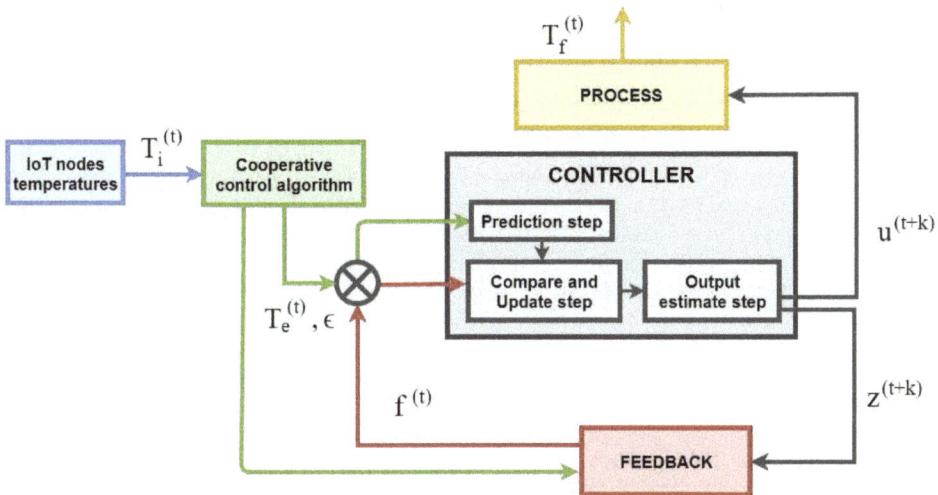

Figure 1. This algorithm improves the fault tolerance of the IoT network via the designed control algorithm in the time interval $(t, t + k)$, where k is the interval of time that we want to control.

The algorithm proposed in this paper controls the temperature of a smart building. For this purpose, data collected in time t from the IoT network is the input of the algorithm (i.e., $T_i^{(t)}$ in blue block). The cooperative control algorithm forms coalitions of neighboring IoT nodes to detect incorrect data and thus auto-correct temperature. This first part of the proposed algorithm calculates the difference between the collected temperature collected by the IoT network and the optimal output temperature of the cooperative control algorithm. Then, this calculated error (i.e., $T_e^{(t)}$) in time t is sent to the controller as input of the prediction step. The prediction step resolves the Markov strings in continuous-time resulting in the probability that the IoT nodes have the same error that in time t or this error will change. Forecasts of the accuracy state of the IoT nodes are sent to the actuator (i.e., thermostats) to set the process (i.e., smart building) temperature. From the controller, there are two send signals: (1) Since t is the current time in the current loop, assume that k is the time interval to be determined; $z^{(t+k)}$ predicts the accuracy of the IoT nodes at the end of the time interval $t + k$. (2) The second signal that comes out of the controller $u^{(t+k)}$ determines which IoT nodes need to be repaired and which are operating correctly. The process sends the final temperature coming out of the algorithm to the feedback function that compares the prediction of the accuracy states with the new temperature inputs of the algorithm and corrects the error in the predictions for the next step of the algorithm.

2.1. Cooperative Control Algorithm

The cooperative control algorithm is located in the reference input. The cooperative control algorithm requires the data to be in a matrix. The input of this algorithm is the temperature collected by the IoT network of the smart building. This data has a transformation process until it is in the correct form so that the algorithm can process it. The IoT nodes collect the data as follows, IoT node places in: $s_{(i,j)}$ have the following temperature: $ts_{(i,j)}$. The other IoT nodes behave in a similar way. Therefore, the first transformation that data has is to place them in an ordered mesh from point $(1,1)$ to point (n,n) so that each of these points matches the position of the smart nodes. It is easy to create a matrix from the mesh and apply the cooperative control algorithm to it. If we have a mesh with n sensors ordered from $(1,1)$ to (n,n), a matrix shown in Equation (1) is created without loss of generality:

$$T_{n,n} = \begin{pmatrix} ts_{1,1} & \cdots & ts_{1,n} \\ \vdots & \ddots & \vdots \\ ts_{n,1} & \cdots & ts_{n,n} \end{pmatrix}. \tag{1}$$

2.1.1. Mathematical Description of the Algorithm

Let $n \geq 2$ be the amount of players in the game, ordered from 1 to n, and let $N = \{1, 2, ..., n\}$ be the group of players. A coalition, S, is formed to be a subgroup of N, $S \subseteq N$, and the group of the whole coalitions is called by $\mathbb{S}$. A cooperative game in N is a function u (characteristic function of the game) that applies to every coalition $S_i \subseteq \mathbb{S}$ a real number $u(S_i)$. Moreover, one of the conditions is that $u(\emptyset) = 0$. In this case, the game will be non-negative (the outputs of the characteristic function are always positive), monotonous (if there are more players in the coalition, the expected characteristic function value does not change), simple and 0-normalized (the players are required to cooperate with one another, as each player will obtain no profit on his own).

In this case, the group of players is the group of organized IoT nodes S and the characteristic function u is denoted as:

$$u : 2^n \longrightarrow \{0,1\} \tag{2}$$

so that, for every coalition of nodes, $u = 1$ or 0 according to a particular coalition can vote or not, respectively (see Equantions (2) and (3)):

$$\mathbb{S} \ni S_i \longrightarrow u(S_i) = \{0,1\} \in \mathbb{R}, \tag{3}$$

where $\mathbb{R}$ are the Real numbers.

2.1.2. Cooperative IoT Nodes Coalitions

The potential for IoT nodes to form coalitions will be restricted by their location, i.e., coalitions can only be composed of neighbouring IoT nodes. Let us consider the matrix of the IoT nodes and a pair of IoT nodes $s_{i,j}$ and $s_{k,m}$ will be in the same neighbourhood if and only if:

$$\| (i-k)^2 - (j-m)^2 \| \leq 1; \tag{4}$$

in other words, if every IoT node to which the game is applicable is the centre of a Von Neumann neighborhood, its neighbors are those who are at a Manhattan range (in the matrix) equal to one. In addition, authorized coalitions have to meet the following conditions:

1. Coalition of IoT nodes have to be in the same neighborhood as presented in Equation (4).
2. Coalitions cannot be formed by a single IoT node.

2.1.3. A Characteristic Function to Find Cooperative Temperatures.

In the suggested game, we want to decide in a democratic way the temperature of the current IoT nodes. To do this, the IoT nodes will create coalitions that will determine the final temperature of the IoT nodes, which will be decided by whether or not they can vote in the election process. From the characteristic function defined in Equation (2), if the value is 1(0), the coalition can vote (not vote) respectively. Assume that s_i is the master IoT node with its related temperature t_{s_i}, the characteristic function is built in the following way:

1. First, the average temperature of all the IoT node is calculated:

$$T_{s_i}^k = \frac{1}{V} \sum_i^V t_{s_i}.$$ (5)

Here, $T_{s_i}^1$ represents the average temperature of the IoT node' neighbourhood s_i (including it) in the first iteration of the game and V is the amount of neighbours in the coalition.

2. The next iteration is to compute an absolute value for the temperature difference between the temperatures of each IoT node and the average temperature:

$$\overline{T}_{s_i}^k = \left(\frac{1}{V} \sum_i^V | t_{s_i} - T_{s_i}^k |^2 \right)^{\frac{1}{2}}.$$ (6)

3. Using the differences in temperature values with regards to the average temperature $\overline{T}_{s_i}^k$ (see Equation (6)), a confidence interval is created and defined as follows:

$$I_{s_i}^k = \left(T_{s_i}^k \pm t_{(V-1, \frac{\alpha}{2})} \frac{\overline{T}_{s_i}^k}{\sqrt{V}} \right).$$ (7)

In Equation (7), we use the Student's t-distribution with a significance level of $\alpha = 1\%$.

4. In this step, we use a hypothesis test. If the temperature of the sensor lies in the interval $I_{s_i}^k$, it belongs to the voting coalition; otherwise, it is not in the voting coalition. Once the confidence interval is calculated, the algorithm runs the characteristic function of the game (u^k) to find which elements will be in the voting coalition:

$$u^k(s_1, \dots, s_n) = \begin{cases} 1, & \text{if } t_{s_i} \in I_{s_i}^k, \\ 0, & \text{if } t_{s_i} \notin I_{s_i}^k. \end{cases}$$ (8)

5. The characteristic function will repeat this process iteratively (k is the number of the iteration) until all the IoT nodes in that iteration belong to the voting coalition. In the cooperative game theory, the Payoff Vector (PV) is the outcome of cooperative actions carried out by coalitions (i.e., the output of applied the characteristic function to the coalitions). At each iteration k, the following PV of the coalition is available S_j (with $1 \le j \le n$ where n is the number of sensors in the coalition) in step k ($PV(S_j^k)$):

$$PV(S_j^k) = (u^k(s_1), \dots, u^k(s_n)) \text{ where } \sum_i^n u^k(s_i) \le n.$$ (9)

The stop condition of the game steps is $PV(S_j^k) = PV(S_j^{k+1})$, at which the algorithm ends. That is, let $PV(S_j^k) = (u^k(s_1), \dots, u^k(s_n))$ and let $PV(S_j^{k+1}) = (u^{k+1}(s_1), \dots, u^{k+1}(s_n))$. The step process ends when both payoff vectors contain the same elements. This process is shown in the following equation:

$$
\begin{cases}
u^k(s_1) = u^{k+1}(s_1) \\
\vdots \\
u^k(s_n) = u^{k+1}(s_n).
\end{cases}
\tag{10}
$$

Then, the game can find the solution that is shown in the following subsection.

2.1.4. Solution of the Cooperative Game

Once the characteristic function has been applied to all IoT nodes involved in this iteration of the game, a payoff vector is available in iteration k (see Equation (9)). Since the proposed game is cooperative, the solution is a coalition of players that we have called game equilibrium (GE). The GE of the proposed game is defined as the minimal coalition with more than half of the votes cast. Let n be the amount of players in this iteration of the game. The winning coalition has to comply with the following conditions:

1. The sum of the elements of the coalition PV must be higher than half plus 1 of the votes cast:

$$
\sum_i^n u^k(s_i) \geq \frac{n}{2} + 1.
\tag{11}
$$

2. The coalition is maximal (i.e., coalition with the greatest number of elements, different from 0, in its payoff vector $PV(S_j^k)$).

Therefore, the solution to the proposed game is the coalition, from among all possible coalitions that are formed at each step k of the game, that satisfies both conditions.

2.1.5. Temperatures of the Winning Coalition

Once the characteristic function finds which is the winning coalition, it is possible to compute the temperature of the main IoT node. Let $\{s_1, \ldots, s_j\}$ be the winning coalition's IoT node and $\{t_{s_1}, \ldots, t_{s_j}\}$ be their related temperature.

The temperature that the game has voted to be the main IoT node's temperature (MST) is computed as follows:

$$
MST = \max_{j \in |S_{winner}|} \{j \cdot t_{s_i}\}_{s_i \in S_{winner}},
\tag{12}
$$

where $|S|$ is the amount of elements in the winning coalition. Therefore, the MST will be the maximum temperature that has the highest involved frequency. In the case of a draw, it is resolved by the Lagrange criterion.

2.1.6. Diffuse Convergence

There is a temperature matrix at each game iteration (see Equation (1)). Hence, we define a sequence of arrays $\{M_n\}_{n \in \mathbb{N}}$ where the M_i element corresponds to the temperature matrix in step i of the game. Therefore, it can be said that the sequence of matrices is convergent if:

$$
\forall \epsilon > 0, \text{ there is } i_0 \in \mathbb{N} \text{ such that } |M_{i-1} - M_i| \leq \epsilon \ \forall i \in \mathbb{N}.
\tag{13}
$$

That is, if the element $m_{n,m}^{i-1} \in M_{i-1}$ and the element $m_{n,m}^i \in M_i$ are set and the convergence criterion is applied, we have:

$$
\forall \epsilon_{n,m} > 0 \text{ there is } N \in \mathbb{N} \text{ such that } |m_{n,m}^{i-1} - m_{n,m}^i| \leq \epsilon_{n,m}
$$
$$
\forall i \in \mathbb{N}, \forall i \geq i_0 \text{ and } m_{n,m}^{i-1} \in M_{i-1}, m_{n,m}^i \in M_i.
\tag{14}
$$

Therefore, by applying the criterion of convergence in Equation (14) to all the elements, a new matrix is obtained; it calculates the difference in the temperatures obtained in the game's previous step and those obtained in the current step:

$$
\begin{pmatrix}
|m_{1,1}^{i-1} - m_{1,1}^{i}| & \cdots & |m_{1,m}^{i-1} - m_{1,m}^{i}| \\
\vdots & \ddots & \vdots \\
|m_{n,1}^{i-1} - m_{n,1}^{i}| & \cdots & |m_{n,m}^{i-1} - m_{n,m}^{i}|
\end{pmatrix}. \tag{15}
$$

For the succession of matrices to be convergent, each of the sequences of elements that are formed with the $|m_{n,m}^{i-1} - m_{n,m}^{i}|$ must be less than the fixed $\epsilon > 0$. In this work, it is established that $\epsilon = 0.01$. With the definitions provided above, we are now ready to define the diffuse convergence of the game. The game is diffuse convergent if at least 80 % of the elements of the matrix are convergent; then, the game reaches the equilibrium.

2.2. Accuracy State Prediction Algorithm

In this subsection, we propose a new feedback control algorithm for predictive fault tolerant control to improve the monitoring and control of the IoT networks. Section 2.2.1 presents the accuracy state categories of IoT nodes. The predictive algorithm is based in the continuous-time Markov chains, and, in our model, we compute the solution of this equation in Section 2.2.2. We provide the theoretical solution of the Markov chains (i.e., the transition matrix). Finally, in Section 2.2.3, the elements of the algorithm are shown (i.e, controller, feedback and process).

2.2.1. Initial Accuracy State

Initially, it is necessary to define a scale of accuracy degradation expressed in percentages. This is done according to the data obtained by the algorithm that we had developed in previous research [13]. This scale will be the discussion universe of the random variable X_n that defines the current state of precision of the system related to the error of the sensors. Therefore, the sensors' possible states are $X_n = \{A = high\ accuracy,\ B = accurate,\ C = low\ accuracy,\ F = failure\}$. Below, Table 1 has the selection made for each variable.

Table 1. Accuracy state of sensors.

X_n	IoT Nodes Accuracy State	Error (%)
A	High accuracy	$e \leq 10$
B	Accurate	$10 < e \leq 20$
C	Low accuracy	$20 < e \leq 35$
F	Failure	$e \geq 35$

Let $T_i^{(t)}$ be the matrix of initial temperatures collected by the WSN, and let $T_f^{(t)}$ be the final temperatures, obtained after applying the data quality algorithm. Then, the accuracy error matrix of the sensors, according to the data quality algorithm, is given by the following equation:

$$
T_e^{(t)} = |T_f^{(t)} - T_i^{(t)}|, \tag{16}
$$

where the coefficients e_{ij} of the matrix $T_e^{(t)}$ are the differences between the initial and final temperature in absolute value for each sensor.

Given the $T_e^{(t)}$ matrix, we now apply the error correction given by the allowed error margin ϵ, and adjust the error matrix:

$$
T_e^{(t)} = |T_e^{(t)} - Id \cdot \epsilon|. \tag{17}
$$

Now, let's centralize these measures to calculate the states of the sensors. To this end, we calculate the average of the elements of the array m_ϵ and the maximum of the array $T_\epsilon^{(t)}$ that we call max_ϵ. Therefore, the centralizing measure is defined as:

$$\delta = m_\epsilon + max_\epsilon. \tag{18}$$

This measure is applied to the matrix $T_\epsilon^{(t)}$ to calculate the percentages associated with each error and therefore calculate the states of each sensor:

$$T_\delta^{(t)} = \begin{pmatrix} t_{1,1}^\delta = \frac{(t_{1,1} \cdot 100)}{\delta} & \cdots & t_{1,n}^\delta = \frac{(t_{1,n} \cdot 100)}{\delta} \\ \vdots & \ddots & \vdots \\ t_{n,1}^\delta = \frac{(t_{n,1} \cdot 100)}{\delta} & \cdots & t_{n,n}^\delta = \frac{(t_{n,n} \cdot 100)}{\delta} \end{pmatrix}. \tag{19}$$

Then, one can define the following function in order to estimate the accuracy state of the sensors in time t. For this purpose, we use the Solution of Kolmogorov's differential equations to design this function:

$$g^{(t)} : M_{n,n}(\mathbb{R}) \longrightarrow M_{n,n}(\{X_n\}) = T^{g(t)} \tag{20}$$

defined as follows:

$$g^{(t)}(t_{i,j}^\delta) = \begin{cases} A, & if & t_{i,j}^\delta \le 10\%, \\ B, & if & 10\% < t_{i,j}^\delta \le 20\%, \\ C, & if & 20\% < t_{i,j}^\delta \le 35\%, \\ F, & if & t_{i,j}^\delta \ge 35\%, \end{cases} \tag{21}$$

where $t_{i,j} \in T_\delta^{(t)}$, and let $T^{g(t)}$ be the matrix with the accuracy states of the sensors at time t.

2.2.2. Transition Matrix

Let λ_A be the time the sensor remains in state A (exponential distribution). λ_B and λ_C are defined in a similar way. In addition, let ξ_A be the time the sensor that remains in state A. Let μ_A (μ_B, μ_C) be the probability that a sensor in state A (B, C) at time t shifts to state F in the time interval $(t, \Delta t + t)$. Thus, if the sensor was in state A at time t_i, the probability of the sensor remaining in state A at time t_{i+1} is given by the following equation:

$$p_{AA} = P(\xi_A > t + \Delta t | \xi_A > t) = \frac{e^{-\lambda_A(t+\Delta t)}}{e^{-\lambda_A t}} = e^{-\lambda_A \Delta t} = 1 - \lambda_A \Delta t + o(\Delta t). \tag{22}$$

Similarly, the probability that a sensor in state A at the beginning will shift to state B is given by the following equation:

$$\begin{aligned} p_{AB} = P(\xi_B > t + \Delta t | \xi_A > t) &= 1 - ((1 - \lambda_A \Delta t + o(\Delta t)) - (\mu_A \Delta t + o(\Delta t))) \\ &= (\lambda_A - \mu_A)\Delta t + o(\Delta t). \end{aligned} \tag{23}$$

In this way, we can build the transition matrix between t and $t + \Delta t$, where the coefficients of the transition matrix are the probabilities of the sensors' switching states (e.g., p_{AF} is the probability that a sensor in state A at the beginning will eventually shift to state F in the interval $(t, \Delta t + t)$).
In this way, the transition matrix $P(t)$ is built:

$$P(t) = \begin{pmatrix} P(\xi_A > t + \Delta t | \xi_A > t) = p_{AA} & \cdots & p_{AF} \\ \vdots & \ddots & \vdots \\ P(\xi_A > t + \Delta t | \xi_F > t) = p_{FA} & \cdots & p_{FF} \end{pmatrix}. \tag{24}$$

2.2.3. Predictive Control Algorithm

Here, we describe how the control algorithm works. This algorithm is used by the sensor control system to monitor and control the accuracy of the sensors. In Figure 1, the set point (green arrow) with the reference inputs contain the following variables: (1) The accuracy error matrix, T_e (see Equation (16)). This matrix has the precision errors of the mesh of sensors. For each step of the algorithm at every time t, this matrix is introduced to update the data of the algorithm. (2) The allowed error ϵ. This parameter enters the flow in each of the steps of the algorithm.

Controller

The first action performed by the controller is the prediction step. In this stage of the algorithm, the transition matrix of the developed model is used (see Equation (24)). Let $z^{(t)} : T^{g(t)} \longrightarrow z^{(t)}(T^{g(t)}) = T^{z(t+k)}$ be the prediction function of accuracy states (i.e., Prediction step) for each time t and let $t + k$ where $k \in \{1, 2, \cdots\}$ be the predicted time. Given $t_{i,j}^{\delta} \in T^{\delta}$, the controller function u is defined as follows:

$$z_{ij}^{(t+k)}(t_{i,j}^{g}) = max\{\mathbb{P}_{t_{i,j}^{g(t+k)}A}, \mathbb{P}_{t_{i,j}^{g(t+k)}B}, \mathbb{P}_{t_{i,j}^{g(t+k)}C}, \mathbb{P}_{t_{i,j}^{g(t+k)}F}\}. \tag{25}$$

Let $z^{(t)}(T^g) = T^{z(t+k)}$ be the matrix of the states of accuracy given by the prediction function. The output of this function is the accuracy state of the sensors at time t.

The next step of the algorithm is to compare the measurements with the feedback function in order to update them. Let $x^{(t)} : T^{z(t)} x T^{f(t-k)} \longrightarrow x^{(t)}(T^{z(t)}) = T^{x(t)}$ be the comparison function defined by the following numerical values $\{A = 1, B = 2, C = 3, F = 4\}$ as follows:

$$x^{(t)}(t_{i,j}^{z(t)}, t_{i,j}^{f(t-k)}) = w_{x_1(t)} t_{i,j}^{z(t)} + w_{x_2(t)} t_{i,j}^{f(t-k)}, \tag{26}$$

where $w_{x_n(t)}$ with $n \in \{1, 2\}$ are the weights given for each of the coordinates of the function x.

Let $y^{(t)} : T^{x(t)} \longrightarrow y^{(t)}(T^{x(t)}) = T^{y(t)}$ be the update function defined as follows:

$$y^{(t)}(T^{x(t)}) = \begin{cases} 1 & if \quad 0 \leq t_{i,j}^{x(t)} \leq 1.5, \\ 2 & if \quad 1.5 < t_{i,j}^{x(t)} \leq 2.5, \\ 3 & if \quad 2.5 < t_{i,j}^{x(t)} \leq 3.5, \\ 4 & if \quad t_{i,j}^{x(t)} \geq 3.5. \end{cases} \tag{27}$$

The update function refreshes the accuracy states of the prediction function with the results obtained from the comparison function.

Let $u : T^{y(t)} \longrightarrow u^{(t)}(T^{y(t)}) = T^{u(t)}$ be the controller function (i.e., output estimate step) and let $T^{u(t)}$ be the system controller matrix at time t. Then, this function finds sensors that are in faulty state (F). In this way, the system creates a virtual sensor to maintain system monitoring. In addition, it will send a request to the service staff to replace the malfunctioning sensor. Given $t_{i,j}^{y(t)} \in T^{y(t)}$, u is defined as follows:

$$u(t_{i,j}^{y(t)}) = \begin{cases} 1 & if \quad t_{i,j}^{y(t)} = F, \\ -1 & if \quad t_{i,j}^{y(t)} \neq F. \end{cases} \tag{28}$$

Thus, if $u(y^{(t)}) = 1$, the system creates a virtual sensor in the position (i, j) and requests maintenance.

Feedback

Let $h^{(t)} : T^{g(t)} x T^{g(t+k)} x T^{z(t+k)} \longrightarrow h^{(t)}(T^{z(t+k)}) = T^{h(t)}$ be the auxiliary feedback function. Given $k \in \{1, 2, \cdots\}$ and the accuracy states in numerical values are $\{A = 1, B = 2, C = 3, F = 4\}$, h is defined as follows:

$$h^{(t)}(t_{i,j}^{g(t)}, t_{i,j}^{g(t+k)}, t_{i,j}^{z(t+k)}) = w_{h_1(t)} t_{i,j}^{g(t)} + w_{h_2(t)} t_{i,j}^{g(t+k)} + w_{h_3(t)} t_{i,j}^{z(t+k)}, \tag{29}$$

where $w_{h_n(t)}$ with $n \in \{1, 2, 3\}$ are the given weights for each of the coordinates of the function h.

Let $f^{(t)} : T^{h(t)} \longrightarrow f^{(t)}(T^{h(t)}) = T^{f(t)}$ be the feedback function defined as follows:

$$f^{(t)}(T^{h(t)}) = \begin{cases} A & if & 0 \le t_{i,j}^{h(t)} \le 1.5, \\ B & if & 1.5 < t_{i,j}^{h(t)} \le 2.5, \\ C & if & 2.5 < t_{i,j}^{h(t)} \le 3.5, \\ F & if & t_{i,j}^{h(t)} \ge 3.5 \end{cases} \tag{30}$$

The feedback function returns the accuracy state of the sensor (i, j) back to the flow. In this way, it is verified that the controller is working correctly and that virtual sensors are not created for the repair of sensors that are working properly.

Process

The process matrix $T^{p(t)}$ shows when sensors need maintenance. The process matrix is defined as follows:

$$T^{p(t)} = T^{u(t-1)} + T^{u(t)}. \tag{31}$$

Thus, when the coefficient of the matrix corresponds to a particular sensor, it means that it has to be replaced $t_{(i,j)}^{p(t)} \ge 0.5\% t_{max}$ time periods with $t_{(i,j)}^{p(t)} \in T^{p(t)}$ (i.e., assuming that $t_{max} = 5$ years, then a sensor has to be replaced if $t_{(i,j)}^{p(t)} \ge 9$ days).

Then, the controller function sends a signal to the process which sends back the matrix of final virtual temperatures at time t (i.e., $T_{vf}^{(t)}$). When the controller sends the signal that a sensor is in the state of failure, the process creates a virtual sensor in that position and simulates the temperature so that the monitoring and control of the building does not lose efficiency. Let $\{T_f^{(t)}\}_{t \ge 0}$ be the matrix succession with the final temperatures at time t given by the algorithm described in Casado et al. [13]. Moreover, let $VS_{i,j}^{(t)}$ be the virtual sensor in the position (i, j) at time t. Then, the temperature of the $t_{i,j}^v$ is provided by the temperature $t_{i,j} \in T_f^{(t)}$.

3. Results

In this section, we present the case study and the results obtained during the experiment. The control algorithm gets data collected by the IoT nodes and auto-corrects the faulty data. Furthermore, in case the controller predicts that an IoT node will be in fault state, it will create a virtual temperature sensor in order to keep the reliability of the IoT network. In this way, the monitoring and control efficiency of the IoT network is improved. This section is organized as follows: In Section 3.1, we provide the solution of the continuous-time Markov chain and its transition matrix $(P(t))$ for every t. Section 3.2 shows the experimental details of the case study (i.e., hardware, temperature collected, etc.). Finally, Section 3.3 presents the results of the application of the control algorithm in the case study and the error decrease in the IoT nodes.

3.1. Case Study Experimental Setup

This case study supposes that the IoT nodes (i.e., temperature sensor) can undergo four accuracy states throughout their useful life (A = high accuracy, B = accurate, C = low accuracy, F = failure). The probability that a sensor in state A at instant t shift to state F in the time interval $(t, t + \Delta t)$ is $0.1\Delta t + o(\Delta t)$, if it is in state B it is $0.2\Delta t + o(\Delta t)$ and if it is in state C it is $0.5\Delta t + o(\Delta t)$. In this simulation, we assume that the time during which the sensors remain in state A is an exponential time of 2.1 in state A and 1.2 in state B.

From A in a time interval $(t, t + \Delta t)$, the sensor can pass to F with probability $0.1\Delta t + o(\Delta t)$. If ξ is the time the sensor stays at A, you have:

$$P(\xi > t + \Delta t | \xi > t) = \frac{e^{-2.1(t+\Delta t)}}{e^{-2.1t}} = e^{-2.1\Delta t} = 1 - 2.1\Delta t + o(\Delta t). \tag{32}$$

Therefore, Equation (32) is the probability of remaining in state A at instant t_{i+1} if it was in A at instant t_i. Then, the probability of shifting to B between t and $t + \Delta t$ is

$$1 - ((1 - 2.1\Delta t + o(\Delta t)) - (0.1\Delta t + o(\Delta t))) = 2\Delta t + o(\Delta t). \tag{33}$$

In the successive stages, we finally reach a calculation in which the transition matrix is between t and $t + \Delta t$, as shown in Table 2.

Table 2. In this simulation, we have assumed that state F is absorbent. That is, for the sensor to move from F to any other state, it needs to be repaired by a maintenance worker.

	A	B	C	F
A	$1 - 2.1\Delta t + o(\Delta t)$	$2\Delta t + o(\Delta t)$	$o(\Delta t)$	$0.1\Delta t + o(\Delta t)$
B	0	$1 - 1.2\Delta t + o(\Delta t)$	$\Delta t + o(\Delta t)$	$0.2\Delta t + o(\Delta t)$
C	0	0	$1 - 0.5\Delta t + o(\Delta t)$	$0.5\Delta t + o(\Delta t)$
F	0	0	0	1

Thus, the derivative of the matrix in the zero is:

$$P'(0) = \begin{pmatrix} -2.1 & 2 & 0 & 0.1 \\ 0 & -1.2 & 1 & 0.2 \\ 0 & 0 & -0.5 & 0.5 \\ 0 & 0 & 0 & 0 \end{pmatrix}, \tag{34}$$

which may be expressed using the Jordan matrix form for the whole period of time t as follows:

$$P(t) = \begin{pmatrix} 1 & 1 & 2 & 1 \\ 1 & 0.8 & 0.9 & 0 \\ 1 & 0.56 & 0 & 0 \\ 1 & 0 & 0 & 0 \end{pmatrix} \begin{pmatrix} 1 & & & \\ & e^{-0.5t} & & \\ & & e^{-1.2t} & \\ & & & e^{-2.1t} \end{pmatrix} \begin{pmatrix} 0 & 0 & 0 & 1 \\ 0 & 0 & \frac{0.9}{0.504} & \frac{-0.9}{0.504} \\ 0 & \frac{0.56}{0.504} & \frac{-0.8}{0.504} & \frac{0.24}{0.504} \\ 1 & \frac{-1.12}{0.504} & \frac{0.7}{0.504} & \frac{-0.084}{0.504} \end{pmatrix}. \tag{35}$$

For example, the term $p_{AF}(t)$ represents the probability that a sensor that begins its useful life at stage A functions incorrectly at time t, so:

$$P(\text{Life span} \leq t) = p_{AF} = 1 - \frac{0.9}{0.504}e^{-0.5t} + \frac{0.48}{0.504}e^{-1.2t} - \frac{0.084}{0.504}e^{-2.1t}. \tag{36}$$

In Figure 2, the graphical representation of the Markov chain is presented. Probabilities of changes in the accuracy states of the sensors are shown in Table 2. The instance simulation presented in this section demonstrates that sensors in any of the precision states (i.e., A,B,C) can move to the fault state

(*F*)—while from state *A* it goes to state *B*, and from state *B* to state *C*. This is so, since, in this example, we assume that the sensor from any of its precision states can fail, while we assume that a high accuracy sensor (*A*) has to go through the precise state (*B*) before moving to the low accuracy state (*C*).

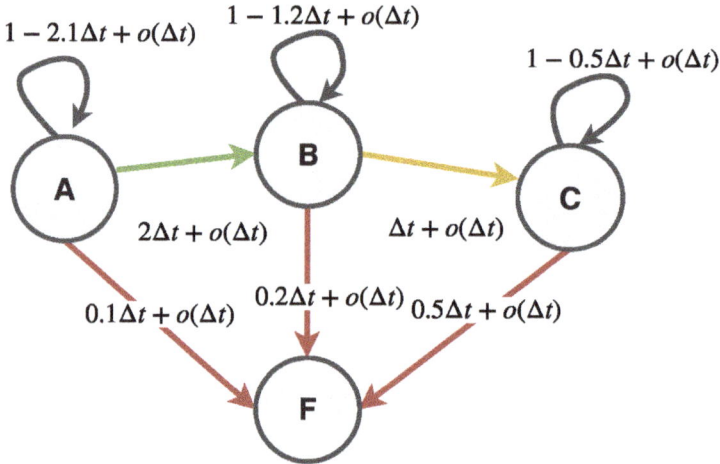

Figure 2. Graphical representation of the Markov chain of the solution of the Kolmogorov differential equations of the proposed simulation.

Given the Markov chain used for this simulation with transition matrix given by Equation (35), the stationary paths given by the probabilities of change of precision state of the sensors are shown in Figure 2. This figure illustrates the probability that a sensor's initial accuracy, state *A*, will shift to a different state in time *t*. Let's assume that $t_{max} = 5$ years (i.e., lifespan of the sensor is five years), then, at $t = 0$, the probability that the sensor remains in state *A* is 1, while, at $t \geq 0$, the probability that the sensor remains in state *A* decreases. Thus, the greater the value of *t* , the greater the probability that a sensor changes to state *B*, *C* and *F*, respectively. For $t \longrightarrow \infty$, the accuracy state *F* of the sensor has a probability of 1 (i.e., the sensor is in failure state) [31].

3.2. General Description of the Experiment

To validate the proposed algorithm, we have selected a smart building. In the moment that the IoT nodes measured the temperature, the actuator (i.e., thermostat) in the selected building showed 23 °C. A grid was applied to locate the IoT nodes on the ground. With the assistance of laser measurements, the IoT nodes were vertically positioned in each section of the building. A total of 25 IoT nodes were deployed.

A combination of the ESP8266 microcontroller in its commercial version "ESP-01" was the type of sensor deployed in the building and a DHT11 temperature and humidity IoT nodes (Figure 2). The sum of the two allows us more versatility in data gathering and adaptation to the case study, as the DHT11 sensor is specifically designed for indoor environments (it has an operating range of 0 °C to 50 °C) according to its datasheet [32]. The microcontroller obtains the data of this IoT nodes through the onewire process and transmits it to the surroundings through Wi-Fi by using HTTP protocols and GET/POST petitions. The ESP-IDF scheduling system supplied by the microcontroller maker was used to schedule the device.

The temperature sensors had been collecting data at 15 minute intervals, for an entire day. For the analysis, we selected the data collected by the sensors in the following time interval 2018-11-02T08:30:00Z and ended on 2018-11-02T21:30:00Z. A particular point in time has been chosen because the game is static and not dynamic (in other words, the game does not handle data in a time period). Below, a mathematical overview of the measured values with the IoT nodes is provided in Table 3.

Table 3. Statistical table of measurements of the IoT nodes.

Timestamp Start	Total Timestamp	Min Temp	Max Temp	Mean	Standard Deviation
2018-11-02T09:00Z	13:00:00Z	20.1 °C	24.6 °C	22.91 °C	0.71 °C

We assume that $t = 5$ years, so if we want to find an interval of one day, we have to do some transformation in t. In this experiment, we have considered the next time interval $(t, t + \Delta t)$, a year has 365 days, and 5 years has $(365 \cdot 5)$ days, so an interval of one day in five years is written as follows: $\Delta t = \frac{1}{365 \cdot 5}$ (i.e., a day). To validate the model, we applied the accuracy state prediction model to the data collected by the sensors placed in the building.

3.3. Case Study Results

In this case study, we have tested the proposed model to increase the efficiency of monitoring and control of an IoT network. This is achieved by improving the quality of data collected by the IoT nodes and the predicted maintenance of these nodes. In this way, the reliability of the data is increased and the energy efficiency of the smart building is increased. The temperature collected by the IoT nodes is the input of the control algorithm. In Figure 3, the evolution of the temperature can be found from its initial state (i.e., data collected by the IoT nodes) until the control algorithm sends the data to the process to set the regulators that control the temperatures of the building sections. The building temperature is slightly warmer in areas where there are large temperature differences. The control algorithm finds these zones and self-corrects if necessary these temperatures to reach the equilibrium in which the temperatures are consistent in the whole building.

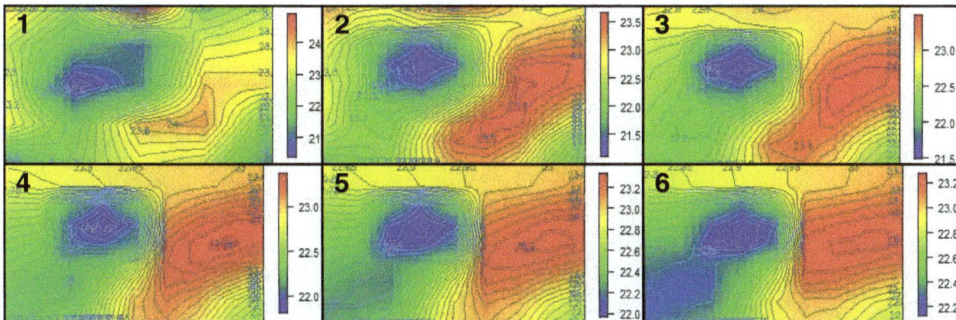

Figure 3. Graphic representation of the matrix of initial temperatures, the evolution of the temperature and the final temperatures in this case study. In Figure 3 (1) can be found the temperatures collected by the IoT nodes. In addition, the measurements that the control algorithm will find as false data can be found in the same figure. Also, the evolution of the controlled temperature is shown in Figure 3 (2)–(5). Final temperatures after the control algorithm is executed are shown in Figure 3 (6).

The suggested algorithm performs an efficient transformation in the ETL system. We can implement our approach as a process step included in the ETL system for the creation of new temperature data, which are self-corrected and ready-to-use. A major part of the thermal noise caused by the data arriving from the IoT node is removed (noise is generated when the IoT node is faulty or non-accurate). Figure 4 provides the amount of IoT nodes (in percents) containing thermal noise for every step of the game. It can be remarked that, when changing the accuracy of the IoT nodes from 0.05 °C to 0.1 °C, the results achieved are quite distinct.

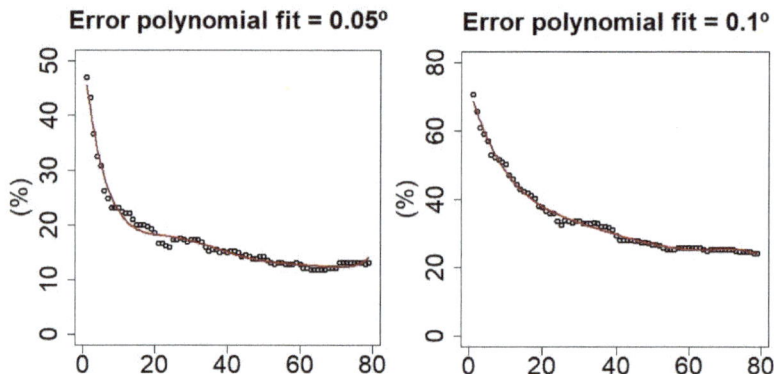

Figure 4. Board with thermal noise reduction in the progression of the algorithm with several confidence intervals from 0.05 °C to 0.1 °C. In the display board, the noise of the % in the temperature matrix is shown opposite the amount of steps. For every one of them, the permitted error range for the temperature collected by the IoT nodes is variable. As the allowable error range is increased, the thermal noise in the temperature matrix also is increased.

However, if it changes to 0.05 °C, 45% of the IoT nodes had thermal noise, and, in a few (<10) steps, the noise was decreased to less than 15%. When the relative permitted error was incremented, the percentages of IoT nodes that had a bit of thermal noise also incremented. For instance, with 0.1 °C of relative error, 70% of the IoT nodes had thermal noise and as the step increment was decreased below 25 percent. However, at a certain point, the noise began to freeze. These IoT nodes will keep having some noise for the selected error (Table 4).

Table 4. Table showing the possible errors and % of noise both during and after applying the game.

Allowed Error (° Celsius)	IoT Nodes with Noise at the Beginning (%)	IoT Nodes with Noise at the End (%)
0.05	47.06	13.25
0.1	70.59	24.22

There are also two useful implementations of our current approach: (1) Identifying the IoT nodes that supply incorrect data and setting up the new IoT node by inserting them in the IoT network; (2) Smart detecting of incorrect data in an IoT network is a major issue, as it allows fault tolerant control of the IoT network and a high quality of data. Furthermore, predictive maintenance allows the good operation of the IoT network. As faulty IoT nodes are detected, the maintenance cost is significantly decreased, as the service technician can focus only on faulty nodes.

4. Conclusions

This paper has addressed the problem of fault tolerant control of IoT nodes in continuous-time NCSs. The feasibility of the proposed approach was verified with a case study in which the closed-loop system was modeled as a continuous-time feedback system with the continuos-time Markov chains to improve the quality of the data collected by the IoT nodes. Through a newly constructed feedback control-based algorithm, an improved control system has been created. It allows for deriving a smart building's maximum allowable energy efficiency such that the resulting closed-loop system improves the control of an IoT network. A numerical case study illustrates the efficiency of our model in Section 3. Figure 3 shows a graphic representation of the evolution of the temperatures and how the fault tolerant control algorithm works. In this figure, one can find how the incorrect data are self-corrected by the control algorithm, improving the monitors and controls of the IoT network. In addition, in Figure 4, we present the percentage of IoT nodes that are collecting incorrect data and how the control algorithm decreases the amount of malfunction IoT nodes. This claim is also supported by Table 4. In it, you can find that, after applying the control algorithm, the amount of malfunctioning IoT nodes is greatly reduced. However, in many real scenarios, the ability to detect an imprecise or malfunctioning IoT node from a hot (cold) spot is limited. In a future work, we will try to solve this problem with artificial intelligence.

Author Contributions: Literature Research, R.C.-V., Conceptualization, R.C.-V., Z.V., J.P. and J.M.C., Methodology, R.C.-V., Supervision, Z.V., J.P. and J.M.C. and G.P., Visualization, R.C.-V. and Z.V., Writing original draft, R.C.-V., Writing review and editing, R.C.-V., Z.V., J.P. and J.M.C.

Funding: This paper has been partially supported by the Salamanca Ciudad de Cultura y Saberes Foundation under the Talent Attraction Programme (CHROMOSOME project).

Acknowledgments: This paper has been partially supported by the Salamanca Ciudad de Cultura y Saberes Foundation under the Talent Attraction Programme (CHROMOSOME project).

Conflicts of Interest: The authors declare no conflict of interest.

References

1. Hespanha, J.P.; Naghshtabrizi, P.; Xu, Y. A survey of recent results in networked control systems. *Proc. IEEE* **2007**, *95*, 138–162. [CrossRef]
2. Mo, Y.; Garone, E.; Casavola, A.; Sinopoli, B. False data injection attacks against state estimation in wireless sensor networks. In Proceedings of the 2010 49th IEEE Conference on Decision and Control (CDC), Atlanta, GA, USA, 15–17 December 2010; pp. 5967–5972.
3. Amato, F.; Ariola, M.; Cosentino, C. Finite-time stabilization via dynamic output feedback. *Automatica* **2006**, *42*, 337–342. [CrossRef]
4. Amato, F.; Ariola, M.; Cosentino, C. Finite-time control of discrete-time linear systems: Analysis and design conditions. *Automatica* **2010**, *46*, 919–924. [CrossRef]
5. Polyakov, A.; Efimov, D.; Perruquetti, W. Robust stabilization of MIMO systems in finite/fixed time. *Int. J. Robust Nonlinear Control* **2016**, *26*, 69–90. [CrossRef]
6. Korobov, V.I.; Pavlichkov, S.S.; Schmidt, W.H. Global positional synthesis and stabilization in finite time of MIMO generalized triangular systems by means of the controllability function method. *J. Math. Sci.* **2013**, *189*, 795–804. [CrossRef]
7. Hui, Q.; Haddad, W.M.; Bhat, S.P. Finite-time semistability and consensus for nonlinear dynamical networks. *IEEE Trans. Autom. Control* **2008**, *53*, 1887–1900. [CrossRef]
8. Khoo, S.; Xie, L.; Zhao, S.; Man, Z. Multi-surface sliding control for fast finite-time leader–follower consensus with high order SISO uncertain nonlinear agents. *Int. J. Robust Nonlinear Control* **2014**, *24*, 2388–2404. [CrossRef]
9. Soares, J.; Ghazvini, M.A.F.; Borges, N.; Vale, Z. A stochastic model for energy resources management considering demand response in smart grids. *Electr. Power Syst. Res.* **2017**, *143*, 599–610. [CrossRef]
10. Li, Y.; Tong, S.; Li, T. Composite adaptive fuzzy output feedback control design for uncertain nonlinear strict-feedback systems with input saturation. *IEEE Trans. Cybern.* **2015**, *45*, 2299–2308. [CrossRef]
11. Li, Y.X.; Yang, G.H. Event-triggered adaptive backstepping control for parametric strict-feedback nonlinear systems. *Int. J. Robust Nonlinear Control* **2018**, *28*, 976–1000. [CrossRef]
12. Pipino, L.L.; Lee, Y.W.; Wang, R.Y. Data quality assessment. *Commun. ACM* **2002**, *45*, 211–218. [CrossRef]
13. Casado-Vara, R.; Prieto-Castrillo, F.; Corchado, J.M. A game theory approach for cooperative control to improve data quality and false data detection in WSN. *Int. J. Robust Nonlinear Control* **2018**, *28*, 5087–5102. [CrossRef]
14. Wang, R.Y. A product perspective on total data quality management. *Commun. ACM* **1998**, *41*, 58–65. [CrossRef]
15. Liu, K.; Seuret, A.; Fridman, E.; Xia, Y. Improved stability conditions for discrete-time systems under dynamic network protocols. *Int. J. Robust Nonlinear Control* **2018**, *28*, 4479–4499. [CrossRef]
16. Zhang, X.; Lin, Y. Adaptive output feedback tracking for a class of nonlinear systems. *Automatica* **2012**, *48*, 2372–2376. [CrossRef]

17. Lezama, F.; Soares, J.; Hernandez-Leal, P.; Kaisers, M.; Pinto, T.; do Vale, Z.M.A. Local energy markets: Paving the path towards fully Transactive energy systems. *IEEE Trans. Power Syst.* **2018**. [CrossRef]
18. Rodriguez-Fernandez, J.; Pinto, T.; Silva, F.; Praça, I.; Vale, Z.; Corchado, J.M. Context aware q-learning-based model for decision support in the negotiation of energy contracts. *Int. J. Electr. Power Energy Syst.* **2019**, *104*, 489–501. [CrossRef]
19. Faria, P.; Spínola, J.; Vale, Z. Reschedule of Distributed Energy Resources by an Aggregator for Market Participation. *Energies* **2018**, *11*, 713. [CrossRef]
20. Fotouhi Ghazvini, M.A.; Soares, J.; Morais, H.; Castro, R.; Vale, Z. Dynamic Pricing for Demand Response Considering Market Price Uncertainty. *Energies* **2017**, *10*, 1245. [CrossRef]
21. Silva, F.; Teixeira, B.; Pinto, T.; Santos, G.; Vale, Z.; Praça, I. Generation of realistic scenarios for multi-agent simulation of electricity markets. *Energy* **2016**, *116*, 128–139. [CrossRef]
22. Santos, G.; Pinto, T.; Praça, I.; Vale, Z. An interoperable approach for energy systems simulation: Electricity market participation ontologies. *Energies* **2016**, *9*, 878. [CrossRef]
23. Zhang, X.; Lin, Y. Adaptive output feedback control for a class of large-scale nonlinear time-delay systems. *Automatica* **2015**, *52*, 87–94. [CrossRef]
24. Zhang, J.X.; Yang, G.H. Fault-tolerant leader-follower formation control of marine surface vessels with unknown dynamics and actuator faults. *Int. J. Robust Nonlinear Control* **2018**, *28*, 4188–4208. [CrossRef]
25. Ghazvini, M.A.F.; Soares, J.; Abrishambaf, O.; Castro, R.; Vale, Z. Demand response implementation in smart households. *Energy Build.* **2017**, *143*, 129–148. [CrossRef]
26. Wang, H.; Liu, K.; Liu, X.; Chen, B.; Lin, C. Neural-based adaptive output-feedback control for a class of nonstrict-feedback stochastic nonlinear systems. *IEEE Trans. Cybern.* **2015**, *45*, 1977–1987. [CrossRef] [PubMed]
27. Casado-Vara, R.; González-Briones, A.; Prieto, J.; Corchado, J.M. Smart Contract for Monitoring and Control of Logistics Activities: Pharmaceutical Utilities Case Study. In Proceedings of the 13th International Conference on Soft Computing Models in Industrial and Environmental Applications, San Sebastian, Spain, 6–8 June 2018; Springer: Cham, Germany, 2018; pp. 509–517.
28. Casado-Vara, R.; Prieto, J.; De la Prieta, F.; Corchado, J.M. How blockchain improves the supply chain: Case study alimentary supply chain. *Procedia Comput. Sci.* **2018**, *134*, 393–398. [CrossRef]
29. Casado-Vara, R.; Prieto, J.; Corchado, J.M. How Blockchain Could Improve Fraud Detection in Power Distribution Grid. In Proceedings of the 13th International Conference on Soft Computing Models in Industrial and Environmental Applications, San Sebastian, Spain, 6–8 June 2018; Springer: Cham, Germany, 2018; pp. 67–76.
30. Casado-Vara, R.; de la Prieta, F.; Prieto, J.; Corchado, J.M. Blockchain framework for IoT data quality via edge computing. In Proceedings of the 1st Workshop on Blockchain-enabled Networked Sensor Systems, Shenzhen, China, 4 November 2018; pp. 19–24.
31. Mailund, T. Continuous-Time Markov Chains. In *Domain-Specific Languages in R*; Apress: Berkeley, CA, USA, 2018; pp. 167–182.
32. DHT11 datasheet. Available online: http://www.micropik.com/PDF/dht11.pdf (accessed on 6 December 2018).

energies

MDPI

Article

Object-Oriented Usability Indices for Multi-Objective Demand Side Management Using Teaching-Learning Based Optimization

Mayank Singh * and Rakesh Chandra Jha

Department of Electrical and Electronics Engineering, Birla Institute of Technology Mesra, Ranchi 835215, India; rcjha@bitmesra.ac.in
* Correspondence: mayank2626@gmail.com; Tel.: +91-829-426-7125

Received: 23 November 2018; Accepted: 21 January 2019; Published: 24 January 2019

Abstract: This paper proposes Object-Oriented Usability Indices (OOUI) for multi-objective Demand Side Management (DSM). These indices quantify the achievements of multi-objective DSM in a power network. DSM can be considered as a method adopted by utilities to shed some load during peak load hours. Usually, there are service contracts, and the curtailments or dimming of load are automatically done by service providers based on contract provisions. This paper formulates three indices, namely peak power shaving, renewable energy integration, and an overall usability index. The first two indices indicate the amount of peak load shaving and integration of renewable energy, while the third one combines the impact of both indices and quantifies the overall benefit achieved through DSM. The application of the proposed indices is presented through simulation performed in a grid-tied microgrid environment for a multi-objective DSM formulation. The adopted microgrid structure consists of three units of diesel generators and two renewable energy sources. Simulation has been done using MATLAB software. Teaching-Learning-Based Optimization (TLBO) is adopted as the optimization tool due to its simplicity and independency of algorithm-specific control parameters. Five different cases of renewable energy availability with results validate the efficiency of the proposed approach. The results indicate the usefulness in determining the suitable condition regarding DSM application.

Keywords: optimization; DSM; microgrid; solar; wind; teaching-learning

---

1. Introduction

1.1. Motivation

DSM can be considered as the coordinated reduction in load during a specific time so as to maximize the energy usage from renewable energy and minimize the generation cost. Most of the research on DSM have tried to explore the technical benefit of DSM in design and operation area, recently some studies have tried to quantify the economic benefits of DSM. Thus motivation of the work is to quantify both the technical and economic benefits of DSM using OOUI.

1.2. Literature Review

Currently, the power industry is facing numerous challenges because of the fast-changing structure of the power network. The integration of small-size renewable energy sources with the conventional grid has become a significant challenge. In addition to this, the uncertainty of power availability from these renewable energy sources is also a serious concern. The search for a suitable scheme which provides a solution for technical and commercial challenges associated with the above-discussed problems is the main area of research for system operators. DSM has appeared as a potential solution for many of these problems.

Any scheme that involves alteration in operation of a complex system should be reliable to a certain level. A reliability study of a power system with DSM is presented in [1]. Since a plan for load curtailment can only be prepared when load forecasting is reliable, stochastic optimization and Gray Wolf Optimization (GWO)-based load forecasting and subsequent DSM programs are presented in [2,3]. Load shifting is also an alternative method of performing DSM. In this aspect, a corrective load shifting program is found to be a better option to conventional preventive load shifting [4,5]. In another work, loads are categorized in rigid and flexible types, and DSM is performed keeping in mind the adaptability of the flexible loads [6]. A novel application of DSM to smoothing the peaks of load curve in a small demand area in the Sultanate of Oman is described in [7]. The key contribution was load profile prediction based upon which the DSM program was prepared. The DSM program developed here handles the issue of variable power generation from solar energy-based generation. In a work, the uncertainty due to wind power generation and its impact on the DSM scheme has been discussed [8]. The methodology discusses the fulfillment of two key objectives of DSM, i.e., minimization of emission and cost. Battery storage and diesel-based energy sources are also employed with solar or wind generators to improve the reliability of the system. A DSM program for the hybrid photovoltaic system has been also proposed in the literature [9]. The target of DSM was set to minimize the system component size and extension of battery life. This indicates that the system operation and its reliability can also be improved through DSM.

A unique presentation of DSM using a spatial and temporal approach has been proposed in [10]. Spatial and temporal DSM is a mathematical improvement carried over the conventional DSM mathematical system. In this improved scheme, a power diagram for each load bus is prepared to accomplish the optimality for a complete system. With the proposed DSM method, the overall performance of the system increases with reduced operating cost and increased voltage quality. A Genetic algorithm (GA)-based DSM program has been presented in [11]. With accurate prediction, the DSM program presented successfully improves the economic dispatch. The impact of solar and wind energy variability has been depicted through sensitivity, and its impact on the DSM program has been analyzed [12]. An optimal sizing problem has been investigated with and without DSM, and a solution with DSM employed was found to be more suitable while sizing the sources for optimal operation [13].

In some of the recent works, DSM has been explored for energy management of the microgrid structure as well. The energy planning and management can be done in a certain advance time duration. This duration may be one day (day ahead scheduling), one hour (hour ahead scheduling), or it can be real-time management (real-time scheduling). In one recent work, demand response-based hierarchical energy management of a microgrid was proposed using a scenario-based optimization scheme [14]. In an hourly scheduling study, it was found that the payment for community aggregated electricity consumption can also be reduced through optimal hourly scheduling of electrical loads [15]. Following the same line, work has also been carried out for consumption management. The electrical appliances can be scheduled for optimal consumption. An informatics solution has been proposed for optimal scheduling which interacts with different appliances and utilities and coordinates the optimal scheduling of loads [16]. In a microgrid, multiple sources operate simultaneously. An operation algorithm has recently been proposed for proper scheduling of individual sources. The algorithm is based on the interaction between different sources and operates a demand response for energy management [17]. The load reduction or load profile flattening has also been done based on a shifting optimization algorithm. This method successfully lowers the electricity consumption and bill payment [18].

A prominent challenge for a demand response program is the probabilistic nature of renewable sources. The conventional numerical methods to optimize the system often fail to reach the best possible solution. Adaptive schemes like genetic algorithm- and artificial intelligence-based optimization have been tried upon to obtain the best possible solution in such adverse mathematical environment. A DSM program based on the artificial immune network has been proposed for the peak load problem [19]. Results show that the adaptive technique enabled the DSM program to keep the peak load within one percent of the desired limit. The Artificial Neural Network (ANN)-based decision-making system has also been proposed in the literature to manage the energy storage system and solar PV-based generator to reduce the consumer electricity cost [20]. The adaptive techniques have not only improved the stress issue and design aspect, but also helped in obtaining better economic performance through DSM. Using the load problem of the individual customer, the appliance usage is so adjusted that a minimum bill for each consumer is obtained [21].

To implement the DSM program, an optimization technique is required as a tool for problem-solving. There are conventional techniques used to solve the DSM problem in the literature, such as linear programming [22,23] and dynamic programming [24], which give a globally optimal solution. However, with an increase in the size and complexity of the problem the solution search space increases. These approaches required more computational effort and time with the addition of more variables and variety of constraints. To resolve this problem, modern approaches, such as simulated annealing and GA, have been proposed as alternatives [25,26].

The evolutionary techniques are advanced optimization schemes where the optimization problem is solved with the help of specific control parameters. For example, GA uses the mutation rate and crossover rate, and particle swarm optimization uses the inertia weight, and social and cognitive parameters. In all these evolutionary methods, the tuning of parameters has a significant effect over the end result. The TLBO algorithm has an advantage over other methods in this aspect. It need only the fewer parameters, thus reducing the burden of tuning the parameters. Thus, the TLBO algorithm is simple and effective [27–29]. Due to these advantages, TLBO has been adopted in this work.

All works discussed here so far have tried to explore the technical benefit of DSM in design and operation areas. However, there are very few works available in the literature where an attempt has been made to quantify the economic benefit of DSM. An index-based assessment of DSM program is discussed in the literature to quantify the technical and economic benefits that can be obtained through the application of DSM. A DSM quality index quantifies the technical gain, whereas a DSM appreciation index quantizes the economic benefits of DSM program [30]. The work quantifies the benefits of DSM. However, the index includes only a single outcome, i.e., load current reduction, and is based only on per unit energy cost reduction. The other advantages of DSM, such as greater integration of renewable energy sources and peak load shaving, are not included in the index calculation. Therefore, need arises for the quantification of DSM usability which includes the desired objects of DSM. This paper proposes an OOUI which can be used for deciding the usability of the DSM program for a certain application.

1.3. Contribution and Paper Organization

Major contributions of this work are as follows: (1) This work proposes three indices, namely, peak power shaving index, renewable energy integration index and an overall usability index; (2) the proposed indices help the operator determine conditions during which DSM is relatively more beneficial in term of peak power shaving and renewable energy integration; (3) the economic benefit of demand response can also be quantified using the proposed index mechanism; and (4) OOUI can provide aid to the system operator in policy-related decisions.

The rest of this paper is organized as: Section 2 presents the mathematical modeling of the system under study. Section 3 presents a microgrid-DSM formulation. Section 4 introduces the proposed indices and their analysis. Section 5 presents the TLBO-based optimization scheme and its adaption for the demand response program. Section 6 presents the case studies based on MATLAB simulation, results, and discussion. Section 7 provides the conclusion of the proposed research work, followed by the references.

2. Microgrid Modeling

The schematic diagram of the microgrid is given in Figure 1. This microgrid consists of three units of diesel generators (DG1, DG2, DG3) and two renewable energy sources: one wind and one solar generator. The switching logic block in Figure 1 takes care of scheduling all six energy resources. Three different types of customer are assumed to be connected in this system using advanced metering infrastructure (AMI). AMI provides all the required information of the customer load to the utility. A scheduling interval of 24 h is used in this paper. Renewable energy sources are coupled with DGs and the main grid through a solid state interface converter which allows power flow from the grid and distributed energy resources (diesel, wind, solar energy resources) to loads connected at the point of common coupling. The maximum interchangeable power allowed between the main grid and microgrid is 4 kW.

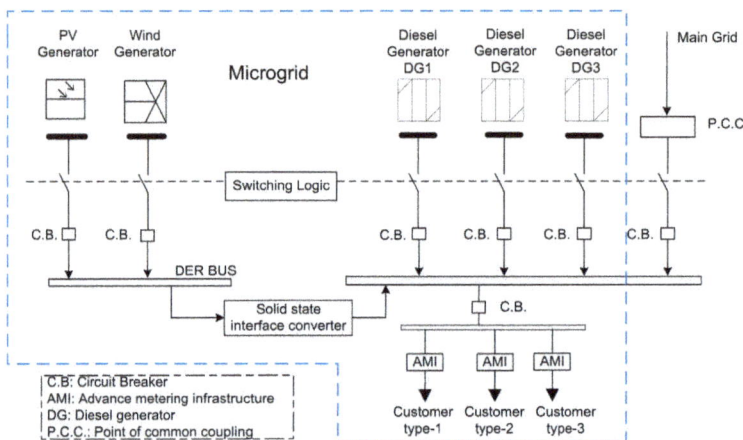

Figure 1. Schematic: grid-integrated microgrid.

Figure 2a,b presents the availability of electrical energy from wind- and solar-based power plants in our microgrid system from Figure 1. Figure 3 shows the initial hourly demand of three customers. Figure 4 shows the hourly power interruptibility cost($\lambda_{j,t}$) in dollars. Descriptions of the dieselgeneration fuel cost function with its parameters are as follow:

$$C_l(P_{l,t}) = a_l P_{l,t}^2 + b_l P_{l,t} \tag{1}$$

where $P_{l,t}$ is the power output and a_l, b_l, are the fuel cost coefficients related to the lth DG at any time instant t; for example, if the cost function of DG2 is $C_2(P_{2,t}) = a_2 P_{2,5}^2 + b_2 P_{2,5}$ then here a_2 and b_2 are parameters related to DG2 and $P_{2,5}$ is the output power of DG2 during the 5th hour. Information about the conventional DG parameters are tabulated in Table 1, providing all the cost parameter data of the diesel generators.

Table 1. Data for diesel generator cost parameters.

Parameter	Abbreviations	DG1	DG2	DG3
First fuel cost coefficient ($/kW)	a_l	0.06	0.03	0.04
Second fuel cost coefficient ($/kW)	b_l	0.5	0.25	0.3
Minimum output power limit (kW)	$P_{l,min}$	0	0	0
Maximum output power limit (kW)	$P_{l,max}$	4	6	9
Ramp up rate(kW/hour)	DR_l	3	5	8
Ramp down rate(kW/hour)	UR_l	3	5	8

(a)

(b)

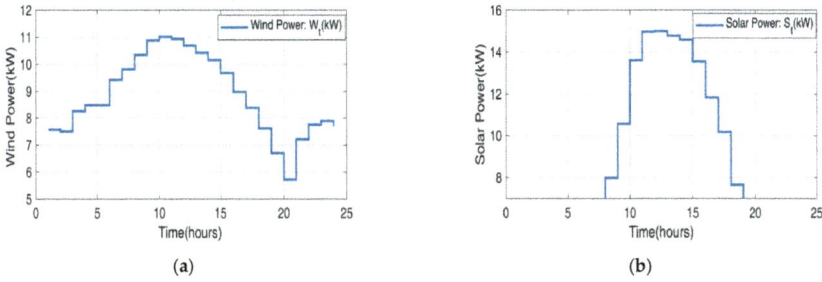

Figure 2. (**a**) Wind power availability curve; and (**b**) solar power availability curve.

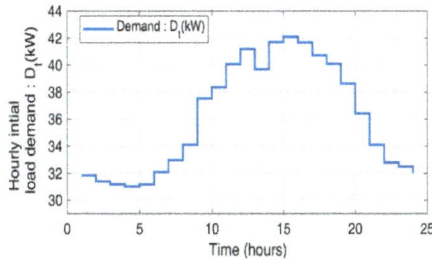

Figure 3. Load curve.

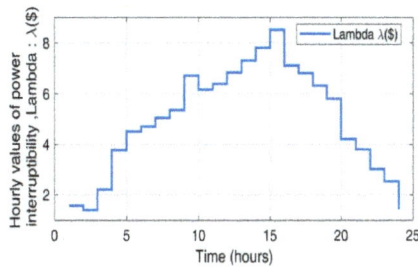

Figure 4. Power interruptibility curve.

The solar and wind modeling presented here is based on a complex and detailed model. For the detailed model please referred to [12,31].

Wind energy conversion units develop electrical energy from wind speed using the mathematical relation of Equation (1a), referred to in [12,31]:

$$E_t = 0.5 \, \eta \, \rho_{air} C_p A \, V^3 \tag{1a}$$

where η is the efficiency of the wind generator, A is the swept area of the wind turbine rotor; V is the velocity of the wind at the hub height, C_p is the power coefficient of the wind turbine, and ρ is the air density.

The PV generator output per hour is given below [12,31]:

$$E_{pv} = \eta A_c I_{pvt} \tag{1b}$$

where η is the efficiency of the solar pv generator, A_c is the area of the solar pv array, and I_{pvt} is the per hour incident of solar irradiation (kWh/m^2) on the solar array.

Load modeling is an important part of the demand response program as the main target of the program is to reschedule the consumer loads. The loads in this paper are taken as aggregate loads as individual customers. Unlike some other load modeling where loads are presented in terms of individual load types (i.e., air conditioners, refrigerators, etc.), or presented as characteristic impedance equations (i.e., constant impedance loads, non-linear loads, etc.), this paper considers loads as cumulative consumer units. The flexibility of the load is presented through a term θ_i which indicates the willingness of individual customers to participate in the rescheduling process. The willingness value varies from 0 to 1, which indicates zero percent willingness to one-hundred percent willingness. The details of customer-related parameters are discussed in Section 3.2 and tabulated in Table 2.

Table 2. Customer cost function related data.

Description	Abbr.	Customer Type1	Customer Type2	Customer Type3
Customer type	i	1	2	3
1st coefficient related to customer cost function	$K_{1,i}$	1.079	1.378	1.847
2nd coefficient related to customer cost function	$K_{2,i}$	1.32	1.63	1.64
User Type	θ_i	0	0.45	0.9
Curtail limit (kW)	CM_i	80	85	90

The microgrid conceptually utilizes the potential of solar and wind energy generation units. The small-scale generation units based on renewable energy sources are normally interfaced into the conventional grid at the low-voltage distribution level. Such microgrid structures lower the cost, negating traditional economies [32]. The studied system adopted in this paper is a microgrid structure in grid-tied mode. The adopted microgrid structure consists of a solar-based generator, wind-based generator, and three conventional diesel generators at the source end.

The mathematical model describing the structure and behavior of individual components and the complete system is prepared and presented in the next sections.

3. Microgrid DSM Problem Formulation

This section presents the microgrid-DSM formulation. Optimization results of this formulation are used in the upcoming section to show the application and usability of the proposed indices.

Adopted microgrid DSM problem formulation has two objectives, namely, minimization of fuel cost function $G(X)$ for diesel generators by maximum utilization of renewable energy sources, and maximization of utility benefit function using DSM. The formulation of the multi-objective function involves these two objectives, as shown in Figure 5.

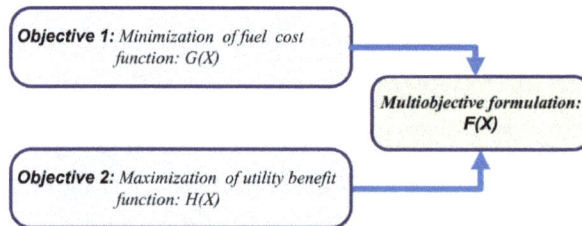

Objective 1: *Minimization of fuel cost function: G(X)*

Multiobjective formulation: *F(X)*

Objective 2: *Maximization of utility benefit function: H(X)*

Figure 5. Formulation of multi-objective function $F(X)$.

3.1. Objective1: Minimization of the Fuel Cost Function, G(X)

In this mode the reduction of the fuel cost of microgrid generators is the main objective. In this work, a power exchange scheme is assumed to exist between the main grid and microgrid. This scheme is needed to cater to the intermittent nature of renewables. When the microgrid system is not able to complete the load demand then power has to be taken from the main grid, and if the microgrid supply is a surplus with respect to its need then the excess power is given back to the main grid. Thus, this power exchange cost needs to be considered during modeling of the function to minimize the fuel cost of conventional generators of the microgrid. Equation (3) represent the required mathematical model of this cost function where $C_r(Pr_t)$ is the cost of transferable power. "γ_t", (dollars per kWh) is the rate of power exchange at a specific bus [12], whereas $(Pr_t) > 0$ and $(Pr_t) < 0$ reflect the power transfer between the microgrid and main grid, while the condition $(Pr_t) = 0$ reflects there is no power exchange between the grid and microgrid. Positive and negative values reflect monetary gain and loss. The function related to minimization of fuel cost in this mode is as follows:

$$\min G(X) = \sum_{t=1}^{T}(Cost_t) = \min\left\{\sum_{t}^{T}\sum_{l}^{L}C_l(P_{l,t}) + \sum_{1}^{T}C_r(Pr_t)\right\} \tag{2}$$

$$\text{where} \quad C_r(Pr_t) = \begin{cases} (Pr_t)(\gamma_t) & if \ (Pr_t) > 0 \\ zero \ if \ (Pr_t) = 0 \\ (Pr_t)(-\gamma_t) & if \ (Pr_t) < 0 \end{cases} \tag{3}$$

where l, t, T, L hold integer values only:

l: Variable to represent conventional generating units. Its value ranges from $1 \leq l \leq L$;

t: Dispatch interval, $1 \leq t \leq T$; in present work, is expanded over $T = 24$ time horizon.

$G(X)$: Operating cost function.

$Cost_t$: Total cost at time t to deliver power;

$P_{l,t}$: Conventional generator, number of units;

$C_l(P_{l,t})$: Fuel cost function for conventional generators.

$C_r(Pr_t)$: Transferable power cost;

Pr_t: Transferable power;

γ: Location marginal prices [33];

L: Total number of the conventional generating unit.

Equation (2) subjected to following constraint:

$$\left\{\sum_{l=1}^{L}(P_{l,t})\right\} + \{Ps_t + Pw_t + Pr_t\} + \left\{\sum_{i=1}^{n}(x_{j,t})\right\} - D_t = 0 \tag{4}$$

$$P_{l,\min} \leq P_{l,t} \leq P_{l,\max} \tag{5}$$

$$0 \leq Ps_t \leq S_t \tag{6}$$

$$0 \leq Pw_t \leq W_t \tag{7}$$

$$-Pr_{\max} \leq Pr_t \leq Pr_{\max} \tag{8}$$

$$-DR_l \leq P_{l,t+1} - P_{l,t} \leq UR_l \tag{9}$$

where:

i: Customer number;

n: Total number of customers;

$x_{i,t}$: Curtailed power by customer number i, at tth time interval

W_t: Forecast maximum wind power obtainable by available solar energy generators;

S_t: Forecast maximum solar power obtainable by available wind energy generators;

Pw_t: Wind generator power availability (during time t)

Ps_t: Solar generator power availability (during time t)

D_t: Load demand at time t;

$Pr_{\max}$: Denotes maximum allowable exchange between grid and microgrid.

Equation (4) signifies the power balance. Collective generation of a conventional microgrid must fulfill all demand inside the microgrid. A small three-feeder radial low voltage scheme is taken in the present study. Losses for the model under consideration are assumed as having very low numeric value. The rest of the constraints are for power generation capacity of conventional and renewable power generators. The constraint, defined by Equation (5), is for generation limits of the conventional generators. This ensures that limits are not violated. Constraints defined using Equations (6) and (7) stand for renewable energy generators to ensure that the optimal values for renewable energy generators are within the forecast range. The fourth constraint, Equation (8), is to restrict the exchange of power. The sixth constraint, Equation (9), takes care of the maximum allowed ramp up and down rates.

3.2. Objective2: Maximization of Utility Benefit Function, H(X), Using Demand Response

In general, demand response is viewed as the involvement of the end-user in the power system. This involvement is in the form of cost variation with time and obtained incentive. Demand response can be divided into two classes: the first one based on incentive, and the second based on a time-based rate program [34,35]. In this work we have adopted the first one.

The main interest of the utility is to maximize its benefit. The mathematical model of the function to represent this interest may be modeled as Equation (10). In the present work, we categorized electricity consumers in three different categories, and the following equations demonstrate the modeling for their behavior.

In this work, the quadratic outage cost function for more than one customer is used [36]. Equation (12) defines the customer cost function, where the "Z" term sorts users according to θ, whereas θ is giving information about the user type and this information is used to categorize the customers [36–39]. Details of parameters related to the customer cost function are tabulated in Table 2. Customer benefit is defined as the difference between the incentive received by him and the cost incurred by him. Utility benefit can be stated as the difference between the cost of the load curtailed by user and the incentive obtained by the user. The mathematical model of the benefit function is as follows:

Benefit function of the utility:

$$V_2 = \lambda x - y \tag{10}$$

Equation (10) denotes the benefit function of the utility where term "λx" is the total hourly benefit to the utility when it does not deliver "x" kW power to a specific customer at the rate of "λ" dollars per kWh [12]. The term y is compensation in terms of incentive which is paid to customer for reducing its load by "x" kW.

Benefit function of the customer:

$$V_1 = y - c \tag{11}$$

where:

V_1: Customer benefit in \$/kWh. It must be $V_1 \geq 0$ for user participation.
V_2: Utility benefit in \$/kWh.
x: Curtailed power by customer in kW;
y: Monetary compensation the customer receives in \$/kWh;
λ: Cost of power.

Equation (11) denotes the benefit function (V_1) of the customer, where c is the monetary loss to the user who reduces his power requirement by "x" kW when DSM is applied and "y" is monetary incentive value that the customer receives. The customer would only participate if $V_1 \geq 0$, meaning he gets some benefit.

The customer cost function for the ith customers can be modeled as:

$$c_i = A + B - Z \tag{12}$$

$$A = K_{1,i}x_i^2, B = K_{2,i}\,x_i, Z = K_{2,i}x_i\theta_i \tag{13}$$

and $K_{1,i}$, $K_{2,i}$ are coefficients related to the costincurred by customer i.

The main interest of the utility is to maximize its benefit (V_2). If the number of users is "i" then it can be mathematically expressed as $\max(\lambda_i x_i - y_i)$. A block diagram representation of this expression is shown by Figure 6.

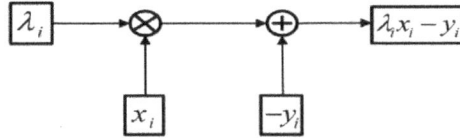

Figure 6. Block-diagram model of utility objective function.

The above demand response model maximizes the expected benefit for the utility when implemented over the total optimization period of twenty-four hours. With considering maximum powertarget and the total budget as practical constraintsit can be modeled as:

$$\max H(X) = \max \sum_{t=1}^{T=24} \sum_{i=1}^{n} [\lambda_{i,t}x_{i,t} - y_{i,t}] \tag{14}$$

subject to:

$$\sum_{t=1}^{T} [y_{i,t} - \{(K_{1,i}x_{i,t}^2) + (K_{2,i}x_{i,t}) - (K_{2,i}x_{i,t}\theta_i)\}] \geq 0 \tag{15}$$

$$\sum_{t=1}^{T} [y_{i,t} - \{(K_{1,i}x_{i,t}^2) + (K_{2,i}x_{i,t}) - (K_{2,t}x_{i,t}\theta_i)\}]$$
$$\leq \sum_{t=1}^{T} [y_{(i-1),t} - \{(K_{1,(i-1)}x_{i,t}^2) + (K_{2,(i-1)}x_{i,t}) - (K_{2,(i-1)}x_{(i-1),t}\theta_{(i-1)})\}] \tag{16}$$

$$\sum_{t=1}^{T} \sum_{i=1}^{n} y_{i,t} \leq TB \tag{17}$$

$$\sum_{t=1}^{24} x_{i,t} \leq CP_i \tag{18}$$

where i, n, t, T hold integer values only, and i ranges between $1 \leq i \leq n$, in Equation (15) and $2 \leq i \leq n$, in Equation (16), and "TB" denotesthe total budget of the utility. The daily limit of power interruptibility for customer i is denoted by CP_i. Equation (15) denotes the constraint which restricts the incentive of an end-user on a daily basis. The constraint of Equation (16) is to restrict the customer benefit, which must be greater for greater curtailed customer power. The constraint denoted by Equation (17) limits the total utility payment to end-users as an incentive within its budget. The constraint represented by Equation (18) controls the individual curtailed customer power (on total time horizon).

3.3. Multi-Objective Formulation

The DSM problem in this paper involves two objectives, fuel cost minimization and utility benefit maximization. Putting Equations (2) and (14) together represents the DSM problem as a multi-objective optimization problem. However, the optimization tool adopted in this paper is TLBO, which is a single-objective optimization algorithm.

Thus, to implement TLBO both of the functions need to be converted into the same type. We converted the objective represented by Equation (14) from a maximization problem to a minimization problem by multiplying it with (-1). This transformation is represented in Equation (19).

The two objectives represented by Equations (2) and (19) are scalarized by pre-multiplying each of them with a weighing factor "w" and "$(1 - w)$" such as the total weight: $w + (1 - w) =$ The overall single-objective optimization problem is represented by Equation (20):

$$\min \sum_{t=1}^{T(=24)} \sum_{i=1}^{n} \left(y_{i,t} - \lambda_{i,t} x_{j,t} \right) \tag{19}$$

$$\min F(X) = \min \left[(w) \left\{ \sum_{t=1}^{T} \sum_{i=1}^{n} C_l(P_{l,t}) + \sum_{1}^{T} C_r(Pr_t) \right\} + (1 - w) \left\{ \sum_{t=1}^{T(=24)} \sum_{j=1}^{n} \left(y_{i,t} - \lambda_{i,t} x_{i,t} \right) \right\} \right] \tag{20}$$

where $x_{i,t}, y_{i,t}, P_{w,t}, P_{s,t}, P_{r,t}, P_{l,t}$ are decision variables.

In this work we transformed the constrained objective function into the unconstrained objective by adding penalty terms for each constraint violation (Equations (4)–(9) and (15)–(18)). In the present work we a have given equal weighting to both objectives.

There are many methods to handle the constraint of an objective function in an optimization problem. In the present work, we designed the constraints of our objective function using an exterior penalty approach in which infeasible design variables are penalized, but feasible design variables are not penalized [40].

4. Teaching-Learing-Based Optimization (TLBO) Algorithm

Under this section TLBO is first introduced, and then its adaptation for multi-objective formulation has been presented.

In [29] six different meta-heuristic algorithms including TLBO are compared. These algorithms are, namely, Whale Optimization, Fire Fly, PSO, Differential Evaluation, and GA. This compression is done mainly in terms of selecting the best suitable optimization method for cost minimization of the micro-grid. In terms of convergence characteristics analysis among these methods TLBO reported the fastest convergence. Search agents of TLBO are found more effective to explore the feasible search space. It is reported in the [29] that TLBO search agents require the least effort. Among the above-mentioned six meta-heuristic techniques TLBO is reported as the best in most of the conditions.

The microgrid DSM is a multi-objective problem, and TLBO is a promising method for optimization. TLBO is proposed by [27,28]. The algorithm is divided in two phases, namely, teacher and learner. This phase of TLBO is explained below.

A. Teacher Phase:

In this phase a teacher tries to improve the average result of the class in his subject. In this identification of the best solution from the population is done on the based on the objective function value.

If $X_{k,i}$ is the fittest solution at any iteration i then it can be represented as $X_{j,kbest,i}$. Calculation of the mean result of the learner, in a subject j will be done in the next step. The effort of the teacher to improve the mean result of the class in his subject depends on the capability of the teacher. This improvement in the existing mean result is modeled as:

$$Difference\_Mean_{j,k,i} = r_{j,i}(X_{j,kbest,i} - T_f M_{j,i}) \tag{21}$$

where:

$X_{j,kbest,i}$: Result of the best learner in subject j;

$r_{j,i}$: Random number in the range [0, 1]; and

T_f: Teaching factor. It decides the value of mean to be changed the value of T_f is selected in random way with equal probability as:

$$T_f = round[1 + rand(0, 1)\{2 - 1\}] \tag{22}$$

updated solution in the teacher phase:

$$X'_{j,k,i} = X_{j,k,i} + Difference\_Mean_{j,k,i} \tag{23}$$

where:

$X_{j,k,i}$: A value in the solution,

$X'_{j,k,i}$: Updated value of $X_{j,k,i}$. Accept $X'_{j,k,i}$ if it gives better value of the function

j: jth design variable. Denotes subject chosen by the learners. $j = 1, 2, \ldots, m$;

k: kth member of population. Denotes learner. $k = 1, 2, \ldots, n$;

i: ith iteration, $i = 1, 2, \ldots, G_{max}$,

G_{max}: Denotes maximum iterations.

All the values of the function accepted in this phase are reserved and treated as the learner phase input.

B. Learner phase:

In this phase learners improve their knowledge by interaction. A population size of n is considered. At any iteration, random compression between learners is done. Random selection of two learners is conducted, namely, P and Q. For these two randomly-selected learners $X'_{P,i} \neq X'_{Q,i}$. Updated values of the first phase:

$$X''_{j,P,i} = \begin{cases} X'_{j,P,i} + r_{j,i}(X'_{j,P,i} - X'_{j,Q,i}), & f(X'_{P,i}) < f(X'_{Q,i}) \\ X'_{j,P,i} + r_{j,i}(X'_{j,Q,i} - X'_{j,P,i}), & f(X'_{P,i}) > f(X'_{Q,i}) \end{cases} \tag{24}$$

$X''_{j,P,i}$ is accepted if it provides a better function value. Accepted function values at the end of the learner phase are reserved and treated as the input of the next iteration.

The steps involved in programming for TLBO implementation in presented paper and its flow chart are given in below section.

TLBO Steps Adopted to Optimize Multi-Objective DSM Function F(X)

In this section step needed to adopt TLBO to optimize $F(X)$ is presented first and followed by its flow chart illustration as given in Figure 7.

Step1: Initialize population, design variables (X) of DSM program, and termination criterion:

$$X = [P_{l,t} P_{r,t} P_{w,t} P_{s,t} x_{i,t} y_{i,t}] \tag{25}$$

Step2: Calculate the mean of each design variable:

$$DiffrenceMean = r\left(X_i(s) - T_f \overline{X}(s)\right) \tag{26}$$

In the above equation $\overline{X}(s)$ mean the result while "r": random number, $\in [0, 1]$. The value of T_f determined randomly as:

$$T_f = round[1 + rand(0, 1)\{2 - 1\}] \tag{27}$$

Step3: Determine the fittest solution.

Step4: Improve the solution according to the fittest solution:

$$X'_i(s) \leftarrow X_i(s) + r\left(X_i(s) - T_f \overline{X}(s)\right) \tag{28}$$

Step5: Determine the better solution between X'_i and X_i.

Step5(a): If the above condition is not satisfied then reject the new result and keep the previous one and go to the next step.

Step5(b): If the condition is satisfied then accept the new one by replacing the old and go to step6.

Step6: Choose two random solution X'_i and X'_k.

Step7: Determine fittest solution between X'_i and X'_k.

Step7(a): If X'_i is better than X'_k then perform the update below and go to Step8:

$$X'_i(s) + r\left(X'_i(s) - X'_k(s)\right) \rightarrow X''_i(s) \tag{29}$$

Step7(b): If X'_i not better than X'_k than perform the below update and go to the next step.

$$X'_i(s) + r\left(X'_k(s) - X'_i(s)\right) \rightarrow X''_i(s) \tag{30}$$

Step8: Is solution X''_i better than X'_i?

Step8(a): If the condition is not satisfied then reject X''_i, keep X'_i and go to step9.

Step8(b): If the condition is satisfied then accept X''_i, replace X'_i and go to the next step.

Step9: Is the termination criteria satisfied? If the answer is yes then go to the next step; otherwise start the whole process again from step2.

Step10: Report the optimum solution.

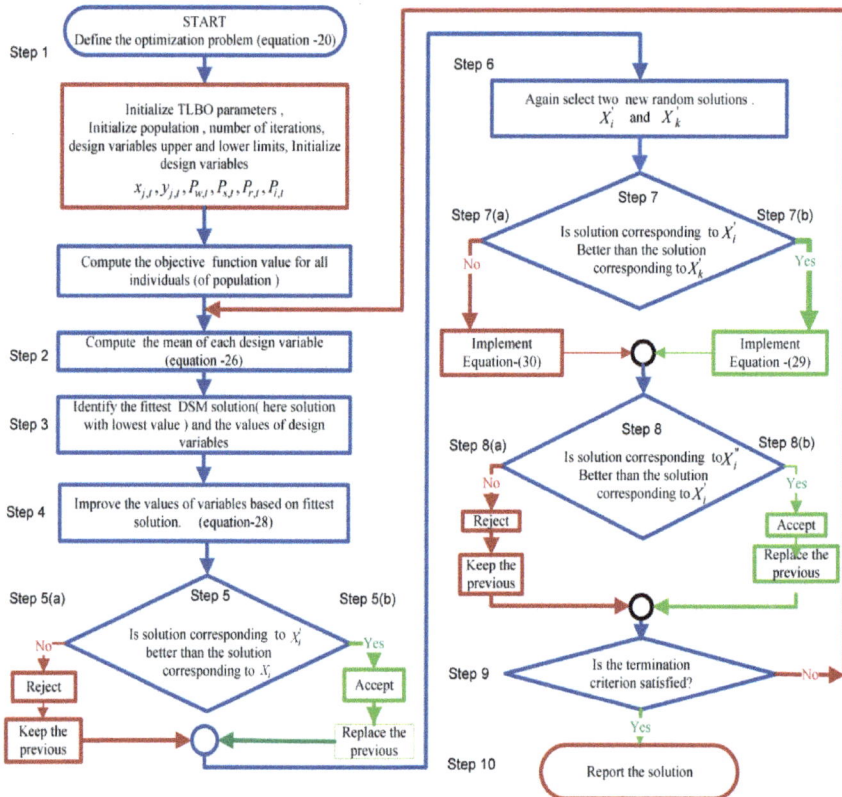

Figure 7. Flow chart of TLBO implementation to optimize $F(X)$.

5. Object-Oriented Usability Indices (OOUI)

Under this section the proposed indices are defined and discussed.

Peak load shaving and increased integration of renewable sources were identified as the main objectives of a demand-side management program. A usability index is expected to present the fulfillment of these two objectives in any specific case study via quantified numerical factors. The outcome of any demand response program does not contain any such indices directly based on which the relative fulfillment of these objectives could be identified. In the absence of such indices, the policy decision of going with demand-side management may not produce the optimized result all the time. Therefore, OOUI are proposed in this section to indicate the relative fulfillment of the two main objectives.

In a power network, the electricity demand varies with time which is presented through the load profile. The power operator faces a challenge of balancing load and generation when peak load conditions appear. Demand-side management, when appropriately performed, reduces some load in peak hours and relieves the operator's stress. Load factor, which is the ratio of the average load on the generator over a period of time to the peak load in the same time interval, indicates the variability of the load. A low load factor indicates that the load is highly variable. A high load factor is desirable for the economic feasibility of plant.

If a demand response program performs as per the requirements, it should reduce the load in peak hours and, therefore, the load factor should improve after allying DSM. A peak power shaving factor is defined as:

$$m_{pps} = \frac{LF_{DSM}}{LF_{WDSM}} \tag{31}$$

where LF_{DSM} is the load factor of the power network when DSM is applied and LF_{WDSM} is the load factor of the power network when it operates without demand response. If DSM successfully shaves the load in peak hours, then the load factor will improve and the peak power shaving index will be greater. For better peak power shaving operation, m_{pps} should be as high as possible.

Renewable energy generators are often small in size and unpredictable in terms of availability. Therefore, the generation support from these sources may or may not always be helpful for the power operator. A good demand response program, in terms of renewable energy integration, is one which ensures a good amount of load satisfied with renewable energy sources. A renewable energy integration index indicates the percentage utilization of electrical energy from renewable-based sources. This index is given as:

$$m_{rei} = \frac{P_{gr}}{P_{gt}} \tag{32}$$

where P_{gr} active power is supplied by renewable energy sources and P_{gt} is the total active power supplied.

An overall usability index which indicates the fulfillment of both the objectives of DSM presented by OOUI. Depending upon the specific system architecture and policy-related decisions, the relative weighting of these two indices may vary. Therefore, the overall usability index also includes the individual weighting and is defined as:

$$m_{dsmuf} = w_{pps}m_{pps} + w_{rei}m_{rei}; \text{ where } w_{pps} + w_{rei} = 1; \tag{33}$$

Above, w_{pps} and w_{rei} are the relative weighting of the peak power shaving index and renewable energy integration index, respectively.

Theoretical Analysis of the Proposed Indices

This section presents the theoretical analysis of proposed indices. First of all, the analysis of the peak power shaving index is presented, and then the renewable energy integration index is analyzed. Finally, the overall usability index is analyzed.

Let load demand to utility over a time interval T is given as:

$$d_t = [d_1, d_2, d_3, \ldots \ldots, d_T], \ t I e T \tag{34}$$

where $d_1, d_2, d_3, \ldots \ldots, d_T$, are loads during 1st, 2nd, 3rd ... and Tth, dispatch intervals (in the presented work hourly load variation with dispatch intervals of one hour are considered (Figure 3a). Average load demand to utility over time period T is calculated using the expression:

$$\frac{1}{T} \sum_{t=1}^{T=24} d_t \tag{34a}$$

Now the load factor with DSM (LF_{DSM}) and without DSM (LF_{WDSM}) is calculated using the following expression:

$$LF_{WDSM} = \left(\frac{1}{T} \sum_{t=1}^{T} d_t / d_{max} \right) \text{and } LF_{DSM} = \left(\frac{1}{T} \sum_{t=1}^{T} d'_t / d'_{max} \right) \tag{35}$$

where:

t: dispatch interval.

d_t: load demand to utility when no DSM is applied.

d_{max}: maximum load over the time period T when no DSM is applied

d'_t: load demand to utility when DSM is applied.

d'_{max}: maximum load over the time period T when DSM is applied

m_{pps} is defined as the ratio of LF_{DSM}/LF_{WDSM}. This can be represented as:

$$m_{pps} = \left(\frac{1}{T} \sum_{t=1}^{T} d'_t / d'_{max} \right) / \left(\frac{1}{T} \sum_{t=1}^{T} d_t / d_{max} \right) \tag{36}$$

Any index proposed as an indicator of DSM usability should quantify the obtained benefit. One of the main purposes of DSM is to reduce the stress on the operator. The sudden high demand from the load in some hours of operation puts stress over the system. Therefore, the index which shows the benefit in terms of reduction of load should be an indicator.

Peak power index analysis can be done under the following points:

1. Peak power shaving index greater than one that is $m_{pps} > 1$.
2. Peak power shaving index less than one that is $m_{pps} < 1$.
3. Peak power shaving index equal one that is $m_{pps} = 1$.
4. Comparative result of peak power shaving index obtained for DSM solutions.

The first condition, $m_{pps} > 1$, will arise if the load factor with DSM (LF_{DSM}) is higher than the load factor without DSM (LF_{WDSM}). In this condition the value of the peak power shaving index will be higher. Here, the value of m_{pps} indicates how much improvement is achieved in terms of load factor after implementing DSM.

The second condition, $m_{pps} < 1$, will arise if the load factor with DSM (LF_{DSM}) is lower than the load factor without DSM (LF_{WDSM}). In this condition the value of the peak power shaving index will be lower than 1. The value of m_{pps} in this condition indicates a decrease in the load factor value with respect to the without-DSM case.

The third condition, $m_{pps} = 1$, will arise if the load factor with DSM (LF_{DSM}) is equal to the load factor without DSM (LF_{WDSM}). This also indicates that the load factor without DSM is equal.

If the peak power shaving index is calculated and presented for different solutions of DSM, then its relative higher value indicates a better load factor compared to the DSM outcome in other cases.

Energies **2019**, *12*, 370

Coincidently, if all DSM solutions are equally good considering load reduction, then the higher value of the peak power shaving index in a specific DSM program can give a ranking of achieved relative peak shaving.

The following example gives an illustrative view about the application of the peak power shaving index. Let us assume that there is a DSM scheme, $F(X)$, for a grid-tied microgrid. This DSM is adopted for five different availability conditions of renewable energy. These conditions are named as case 1, case 2, case 3, case 4, and case 5. The proposed peak power shaving index is calculated for each case and their values are presented as 'm_{pps}'.

If the operator wants to know that, among all five cases, in which case maximum peak shaving has been achieved, the answer can be provided by the values of m_{pps}. To demonstrate the working of m_{pps}, a bar chart is plotted. Indicator values as given in Figure 8 represent that in case 3 the peak power shaving index hold maximum value which indicate that the percentage contribution of peak power shaving is the maximum in case 3 relative to other DSM cases. Additionally, the higher value of case 3 indicates that the load factor in this case with respect to other cases is much better.

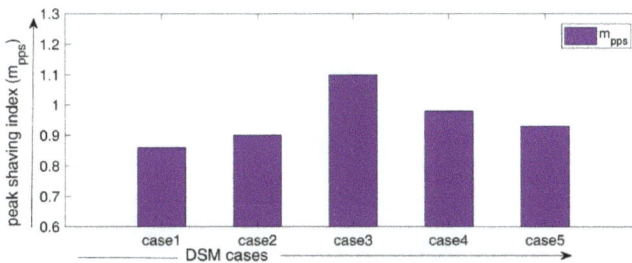

Figure 8. Illustrative graph for peak power shaving index.

The ratio of active power from renewables (P_{gr}) to total active power (P_{gt}) quantifies the involvement of renewable energy. Suppose if the value of this index is 0.26, then it means renewable energy contribution in this case to fulfill total demand is 26%. The lowest possible value for this index can be zero, which will indicate the condition of no-contribution of renewable energy in fulfilling the demand. If the value of m_{rei} is equal to one than it mean 100% renewable energy is used to fulfill all the demand.

If the operator wants to know that, among all cases, which one achieves relatively major integration of renewable energy sources while implementing DSM, m_{rei} can help here by providing a numerical value in the form of an index. The DSM case with a higher value of m_{rei} involves a greater renewable energy percentage to fulfill load demand.

An overall usability index indicates DSM usability combined with peak load shaving and renewable energy integration. Depending upon the specific system architecture and policy-related decisions, the individual weighting of m_{pps} and m_{rei} may vary.

In extreme cases, if the weight associated with m_{rei} is set at 0 and the weight associated m_{pps} is set at 1, then the overall usability index (represented by m_{dsmuf}) becomes equal to the peak power shaving index. Thus, it indicates that the desired aim of DSM is only to take care of peak power shaving. Similarly, if this extreme setting is interchanged for DSM, it can be concluded that the DSM scheme is only focusing on renewable energy integration and has zero interest in achieving load factor improvement or peak power shaving.

6. Case Studies, Simulation Results, and Discussion

Under this section we first present the different renewable energy availability cases. Simulation results of the multi-objective formulation of the DSM using the TLBO method in the MATLAB environment is presented next. Finally, determination of OOUI, followed by a discussion, is conducted.

Figure 9 represents the approach adopted in the presented work with respect to the application and usability of the proposed indices. First of all, a microgrid multi-objective DSM formulation ($F(X)$) is conducted. In the next step the TLBO algorithm is adopted to solve $F(X)$, and the optimized results are obtained. From these solutions the proposed OOUI are calculated and made available to the operator. A MATLAB-based simulation has been performed to obtain all of the simulation results.

It is considered that the participation of all loads is equally shared, and each type of load has equal right to make a decision to get involved in the demand response program. The microgrid operator has the information about the interruptible energy limit of the user on per day bases and this information is used to index the end users in order of increasing interest to diminish their energy need. Additionally, the outage cost function coefficients ($K_{1,i}$ and $K_{2,i}$) of involved users are considered to be known to the operator. The daily utility budget is $500.

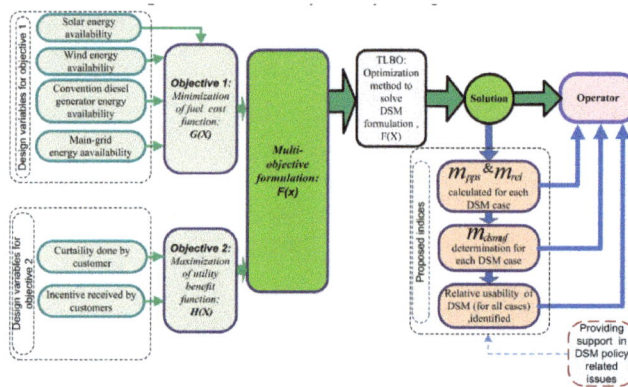

Figure 9. Schematic of the proposed indices application in a grid-tied microgrid.

6.1. Case Studies, Simulation Results

Solution of the multi-objective formulation of DSM is obtained using TLBO for five different scenarios of renewable energy availability in the microgrid. These scenarios are presented as cases 1–5. Case 1 deals with the condition of average wind and average solar availability; in this case total availability of renewable power is 346.24 kW. Case 2 considers the maximum wind and average solar availability and total renewable power availability in this case is 400.5 kW. Case 3 is when wind availability is at a minimum, and solar availability is average and overall availability of renewable sources is 275.03 kW. Average wind and maximum solar availability are termed as case 4; in this case renewable energy availability is 362.76 kW. Finally, case 5 deals with the generation scenario when the wind is average and solar has minimum availability, and the total availability of renewable energy is 296.01 kW. The solar power availability and wind power availability under different cases are represented in Figures 10 and 11; these figures present the hourly variation of solar and wind energy availability in different cases.

Under all of these conditions the DSM program is run, and the iteration vs. fitness value plots are given in Figure 12. The results indicate that the optimization target is achieved around 2000 iterations in all cases. As a result of the optimization new load profiles are obtained for each case. These new load profiles are in accordance with the two objectives of $F(X)$. These load curves for different cases are plotted and presented in Figure 13, with and without the DSM load curve. Values of parameters obtained from these load curves are listed in Table 3. Total, average, and peak demand when DSM is not implemented is 864.9 kW, 36.0375 kW, and 42.09 kW, and the load factor is 0.84. These parameters are needed to calculate OOUI values.

Wind energy curves

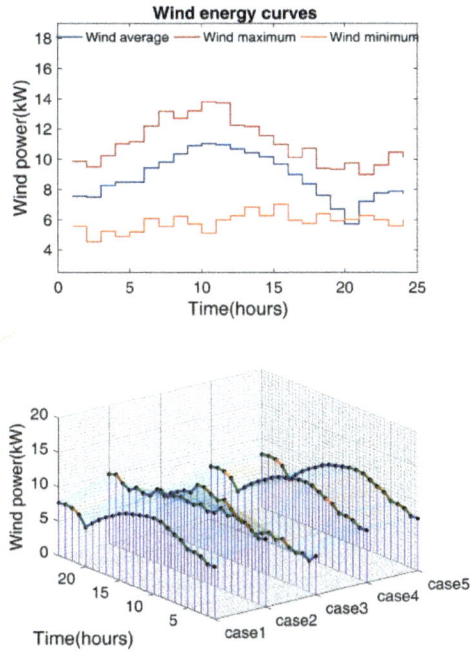

Figure 10. Energy availability from wind.

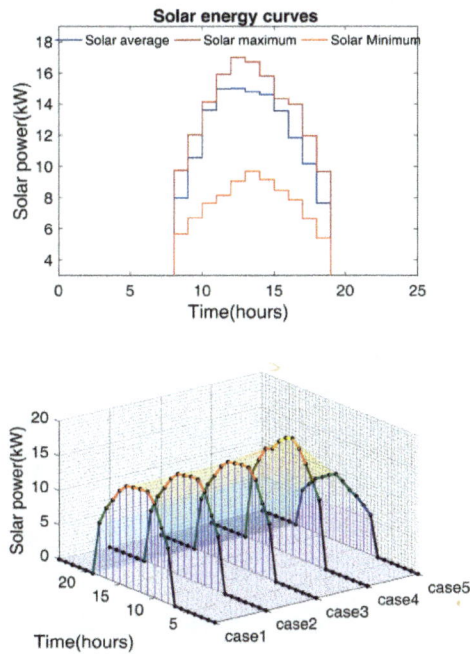

Solar energy curves

Figure 11. Energy availability from solar.

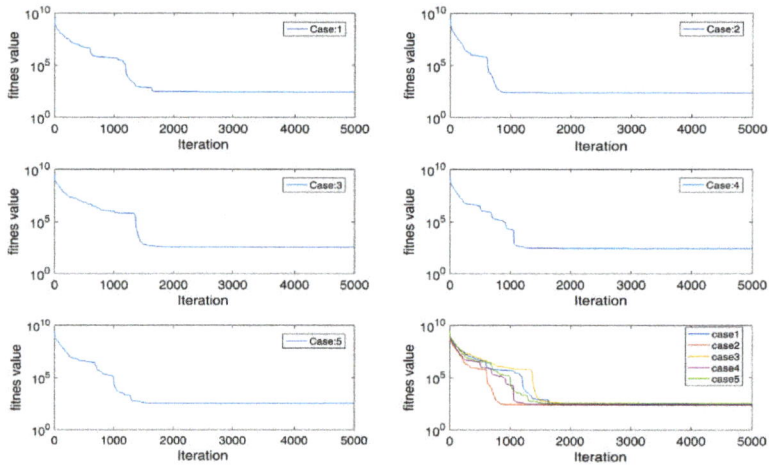

Figure 12. Convergence curves of TLBO in the case of the $F(X)$ objective.

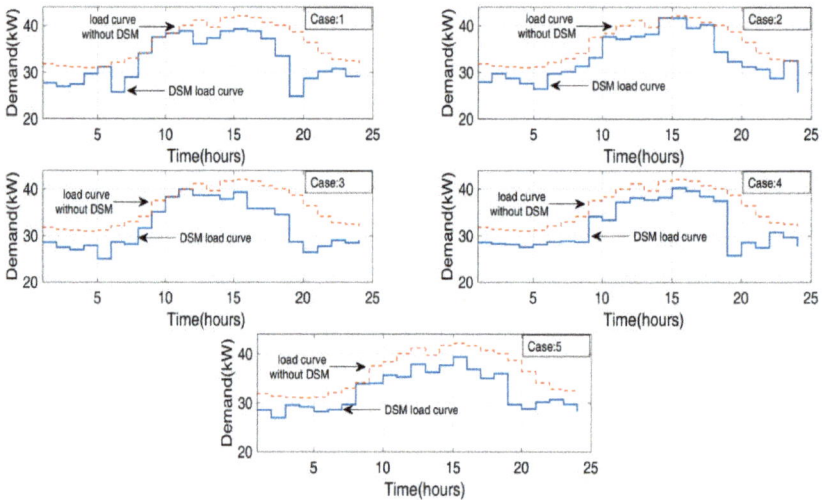

Figure 13. Load curves obtained for cases 1–5 with DSM.

Table 3. Load curve data for DSM cases 1–5.

Cases	Total Demand (kW)	Avg. Demand (kW)	Peak (kW)	Load Factor	Renewable Energy Availability	
					(kW)	(%)
case 1	780.236	32.50983	39.25754	0.828117	346.24	34.62
case 2	793.9236	33.08015	41.66873	0.793884	400.45	40.04
case 3	767.8551	31.99396	39.95539	0.800742	275.03	27.50
case 4	771.3937	32.1414	40.25102	0.798524	362.76	36.27
case 5	775.5615	32.31506	39.35548	0.821107	296.01	29.60

6.2. Calaulation of Proposed Indices

The calculations of the proposed indices are based on the mathematical relation as given in Section 5. Table 4 presents the calculated values of these indices under all five cases studied.

The renewable energy integration index presents how much the participation of renewable sources has been achieved. The peak power shaving index indicates how much peak load shaving has been achieved. The overall usability index is calculated with the help of the weighted sum of these two indices.

This work considers three different weighting scenarios named as scenario 1, scenario 2, and scenario 3. In scenario 1, equal weighting $(w_{pps} = w_{rei} = 0.5)$ is given to both objectives. In the second scenario, the renewable energy integration has been given higher weighting ($w_{rei} = 0.9$). In the third scenario, the peak power shaving remains the preferred objective and has higher weighting ($w_{pps} = 0.9$). Table 4 presents the calculated indices and overall index under all five cases of energy availability and weighting scenarios. The peak power shaving targets are achieved with the maximum amount in case 1 when wind and solar energy availabilities are average.

Table 4. OOUI for DSM.

Cases	W(kW)	S(kW)	m_{pps}	m_{rei}	m^{s1}_{dsmuf}	m^{s2}_{dsmuf}	m^{s3}_{dsmuf}
					($w_{pps} = 0.5, w_{rei} = 0.5$)	($w_{pps} = 0.1, w_{rei} = 0.9$)	($w_{pps} = 0.9, w_{rei} = 0.1$)
Case 1	Average	Average	0.986	0.444	0.715	0.498	0.932
Case 2	Maximum	Average	0.945	0.504	0.725	0.548	0.901
Case 3	Minimum	Average	0.953	0.358	0.655	0.418	0.894
Case 4	Average	Maximum	0.951	0.470	0.711	0.518	0.903
Case 5	Average	Minimum	0.978	0.382	0.680	0.441	0.918

W and S: Wind power availability forecast, while S denote Solar power availability forecast. m_{rei}: Renewable energy integration index; m_{pps}: Peak power shaving index. w_{pps}: Weighting of m_{pps}: and w_{rei} for m_{rei}. m_{dsmuf}: Overall usability index. m^{s1}_{dsmuf}: It is for scenario 1. In this scenario weighting of two factors are $w_{pps} = 0.5$ and $w_{rei} = 0.5$. m^{s2}_{dsmuf}: It is for scenario 2. In this scenario weighting of two factors are $w_{pps} = 0.1$ and $w_{rei} = 0.9$. m^{s3}_{dsmuf}: It is for scenario 3. In this scenario weighting of two factors are $w_{pps} = 0.9$ and $w_{rei} = 0.1$.

The next best case is case 4 when wind availability is average and solar energy is at maximum availability. This indicates that a significant amount of peak power shaving can be achieved when at least an average solar availability is ensured. The reason can be attributed to the solar availability in peak load duration. With average or maximum availability, solar provides additional support to conventional generators and the grid in easing the peak load conditions. Renewable energy utilization is another objective of DSM. The renewable energy integration index indicates the fractional quantity of renewable generation concerning a conventional generator. The calculated values indicate that maximum integration of renewable energy is achieved with DSM in case number 4, where wind availability is average and solar availability is the maximum. In general, the condition where both have maximum availability should be the best scenario for renewable energy utilization.

For such condition to occur, a combination of suitable weather conditions is required. Therefore, the next best situation is when one source generates the maximum and the other generates average energy. These two conditions occur in case 2 and case 4. However, as the values of the renewable energy integration index indicate, solar energy has more impact on fulfilling this target. When both the indices are combined to obtain a final index, which presents the object-oriented DSM usability, the condition for the maximum benefit of DSM is obtained. As can be seen from Table 3, the overall maximum benefit from DSM can be achieved in case 3, when wind provides minimum energy and solar provides average energy. This conclusion remains valid for any weighting given to either objective of DSM. Whether either objectives have equal weighting, or one of them has higher weighting, the best result can be obtained only in case number 3. This conclusion can help in policy-making while deciding to adopt the DSM for any specific system. Figures 14 and 15 present the different indices in plots where the conclusion can be easily identified and established.

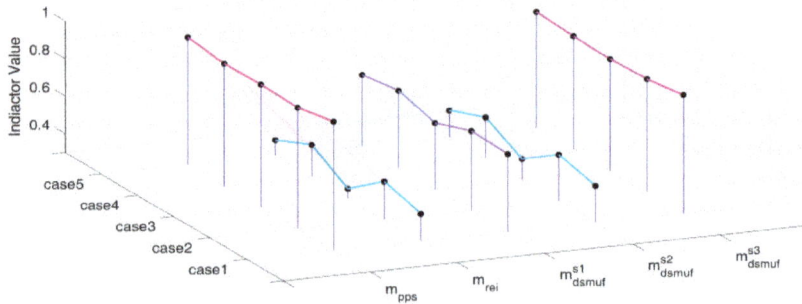

Figure 14. Plot of OOUI for DSM.

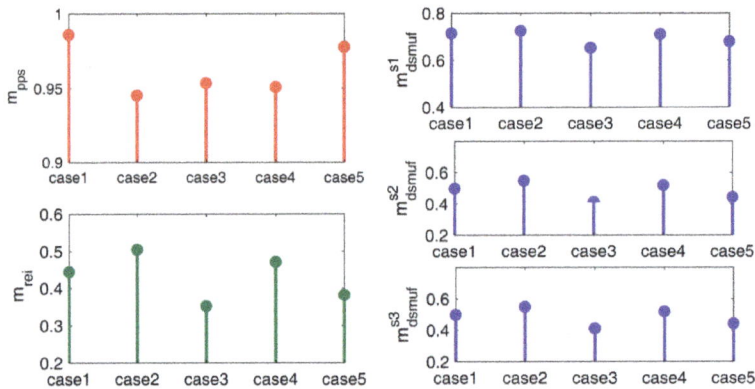

Figure 15. Plot of OOUI for DSM in different cases and scenarios as described in Table 4.

6.3. Discussion of the Results with Comparative Analysis

This section is divided into the following two parts: (A) The findings of the present work are discussed keeping the focus on the obtained values of the proposed indices for DSM; and (B) comparative discussion with similar studies published by other authors to show the uniqueness of the present work.

(A) Observation of Tables 3 and 4, focusing on the numeric values of peak power shaving index, reveals two things: (1) regarding peak shaving and (2) the load factor. A close look of Tables 3 and 4 shows that the peak shaving actor holds adirect relation with the load factor and a reciprocal relation with peak shaving. Observation of Table 5 reveals that case 1 holds rank 1 when ranking is done on the basis of the peak power shaving index. This indicates that when DSM is applied to case 1 then it achieves maximum peak shaving in obtaining its objectives. Case 2 is ranked 5th on the basis of the peak shaving index value, which means that peak reduction is the minimum in this case. In case 2 the maximum peaks are 41.67 kW (Table 3) which is very close to without DSM load curve peak (42.09 kW). Figure 13 (case 2) shows that in this case DSM also reduces the overall demand and achieves its objectives, but is poor in terms of peak shaving. In terms of higher peak reduction, cases 1–5 can be arranged as follows: case 1 > case 5 > case 3 > case 4 > case 2.

Observation of Tables 3 and 4, while focusing on the numeric values of renewable energy index, reveals that its value is the maximum in case 2 and the minimum in case 3. This indicates that the percentage contribution of renewable energy in fulfilling the total load demand is the maximum with respect to other cases.

Table 5. OOUI ranks of DSM.

Rank	Criteria for Ranking DSM				
	m_{pps}	m_{rei}	m_{dsmuf}^{s1}	m_{dsmuf}^{s2}	m_{dsmuf}^{s3}
1	Case 1	Case 2	Case 2	Case 2	Case 1
2	Case 5	Case 4	Case 1	Case 4	Case 5
3	Case 3	Case 1	Case 4	Case 1	Case 4
4	Case 4	Case 5	Case 5	Case 5	Case 2
5	Case 2	Case 3	Case 3	Case 3	Case 3

In case 2 total power requirements for the users is 793.92 kW and the renewable energy integration index value is 0.504, which indicates that the contribution of renewable energy in the total requirement is 50.4%. Cases 1–5 can be arranged according to renewable energy contribution in their total demand as follows: case 2 > case 4 > case 1 > case 5 > case 3.

Observation of Table 4, while focusing on the numeric values obtained for the overall usability index, giving equal weighting (0.5) to peak power shaving and the renewable energy integration index shows that it holds a maximum value for case 2 and a minimum for case 3.

The overall usability index indicates the relative effectiveness of DSM in reducing peak power and utilizing renewable energy simultaneously while achieving its objectives. Observation of Tables 4 and 5 for m_{dsmuf}^{s1} shows that case 2 has rank 1 as in this case the combined impact of peak power shaving and renewable contribution is the highest (Tables 3 and 4, verified this).

Table 5 shows that ranking of the overall usability index followed the pattern of m_{rei} ranking when high weighting ($w_{rei} = 0.9$) is associated with the renewable energy integration index (Table 4). The overall usability index tends to follow the pattern of m_{pps} ranking when weighting ($w_{pps} = 0.9$) shifted toward m_{pps}. Table 5 shows the ranking based on different criteria.

(B) Previous works have also tried to present a framework or factor for DSM. In a study, an impact analysis of DSM is presented through the Inverted Load Duration Curve (ILDC) [41]. This study remodels the power network by taking the Inverted Load Duration Curve. DSM is applied for the renewed model and the DSM impact is presented through a model, which is named as "VPI model". This presents the benefit of DSM in a quantified manner in corresponding to only the load, whereas this work presents the benefit quantification corresponding to both renewable generation and loads.

As discussed in the introduction section, similar indices are proposed in [30], wherein the economic and technical benefits are enumerated and quantified and two indices are proposed based on current reduction and cost of one kWh. However, the technical benefits take in account only the reduction in loads and do not consider the effect of different types of generation.

The VPI model as proposed in [41] is further extended and analyzed for a big power network in [42]. The influence of DSM is calculated and evaluated based on load reduction only. The impact of renewable energy generation and small DG's are not discussed. Additionally, the quantified value is not presented in terms of any general index.

In [43], a set of performance indicators are proposed to design a new energy management system. The focus of these indicators is consumption allocation to individual load, total consumption, cost allocation, total cost, maximum consumption, energy usage, and hourly distribution of energy. The systems under study in this paper were Supervisory Control and Data Acquisition (SCADA)-based connected loads and generation units. The formulation of DSM is not discussed, and the mathematical formulation of indicators is not presented through any simulation results. In another work, a decision support framework is presented for selection of the demand response method [44]. The proposed framework of [44] is based on cost minimization and load scheduling. The work, however, does not consider any role of renewable and focuses only on energy management.

Table 6 shows the uniqueness of the present work with respect to studies done by other authors [30,41–44] in a comparative manner.

Table 6. Comparative Study.

S.N.	Point of Comparison	Proposed	Dharme et al. [30] Year: 2006	Rahman et al. [41] Year: 1993	Khanh et al. [42] Year: 2011	Raihab et al. [43] Year: 2016	Dennis et al. [44] Year: 2018
1	Whether system under study is grid-tied microgrid	Yes	No	No	No	Yes	Yes
2	DSM Formulation	Multi-objective	Single objective	Single objective	Single objective	Single objective	Single objective
3	Optimization performed	Yes	No	Yes	No	No	Yes
4	Optimization method / solution strategies used	TLBO	No	Iterative	No	No	Greedy and multi-agent-system-based a decision support framework
5	Attributes for decision	OOUI	Index-based	VPI model-based	VPI model-based	Indicator-based	
6	Any proposed Indices	Yes	Yes	No	No	Yes	No
7	Any index for peak shaving	Yes	No	No	No	No	No
8	Any Index for renewable energy	Yes	No	No	No	No	No

7. Conclusions

The main finding of this paper is to provide the OOUI, namely, peak power shaving index, renewable energy integration index, and overall usability index. OOUI allows quantifying of the technical and economic benefits from DSM. The formulated indices are helpful for the operator in identifying the relatively more suitable operating condition, and relatively more beneficial, for DSM, which can help in policy-related decisions. The economic benefit of demand response can also be quantified using the proposed index mechanism.

The concluding numeric values of the indices obtained are listed in Table 4. The ranking of DSM, in Table 5, is done on the basis of these values. Observation of Table 5 gives clear information that the peak power shaving index value is the maximum in case 1. This indicates that the DSM application in case 1 results in the maximum peak shaving. Similarly, the renewable energy integration indicator value is the maximum in case 2. This indicates that the renewable energy contribution to fulfilling load demand is the maximum in case 2. Based on the two indices as described above, an overall usability index is designed. It is observed from results that the overall usability index has the maximum value in case 2 when other two indices are given equal weight.

The overall importance of this paper is that it provides a mathematical tool for OOUI. These indices provide additional aid to the operator in identifying the conditions in which DSM provides more benefit in terms of peak power shaving and renewable energy integration. Thus, it can help the operator in policy-related decisions. In other existing works, the quantification of the DSM benefit is made either through a load curve or through generation reduction. However, these existing works have not considered both load change and renewable energy integration simultaneously. The presented simulation-based analysis establishes that the proposed OOUI can successfully quantify the two important benefits, curtailments of peak load and harnessing the renewable energy. Additionally, the two benefits can be clubbed together and presented as a single overall usability index.

The proposed approach shows that OOUI have the potential to determine a DSM solution, which can provide relief to both the utility and customers by promoting higher renewable energy penetration and simultaneously shaving the peak load. A limitation of the current approach is that OOUI are obtained and analyzed for the DSM solutions obtained for multi-objective formulation, $F(X)$, using single objective optimization algorithm TLBO. In future work, this index-based approach can be analyzed for different multiobjective DSM models incorporating a multi-objective optimization algorithm. These DSM models can be based on different demand response programs, such as an emergency demand response program, and critical peak pricing.

Author Contributions: Conceptualization: M.S.; formal analysis: M.S.; methodology: M.S.; resources: R.C.J.; supervision: R.C.J.; writing—original draft: M.S.

Funding: This research received no external funding.

Conflicts of Interest: The authors declare no conflict of interest.

References

1. Jabir, H.J.; Teh, J.; Ishak, D.; Abunima, H. Impacts of Demand-Side Management on Electrical Power Systems: A Review. *Energies* **2018**, *11*, 1050. [CrossRef]
2. Amoasi Acquah, M.; Kodaira, D.; Han, S. Real-Time Demand Side Management Algorithm Using Stochastic Optimization. *Energies* **2018**, *11*, 1166. [CrossRef]
3. Naz, M.; Iqbal, Z.; Javaid, N.; Khan, Z.A.; Abdul, W.; Almogren, A.; Alamri, A. Efficient Power Scheduling in Smart Homes Using Hybrid Grey Wolf Differential Evolution Optimization Technique with Real Time and Critical Peak Pricing Schemes. *Energies* **2018**, *11*, 384. [CrossRef]
4. Jabir, H.J.; Teh, J.; Ishak, D.; Abunima, H. Impact of Demand-Side Management on the Reliability of Generation Systems. *Energies* **2018**, *11*, 2155. [CrossRef]
5. Nguyen, A.-D.; Bui, V.-H.; Hussain, A.; Nguyen, D.-H.; Kim, H.-M. Impact of Demand Response Programs on Optimal Operation of Multi-Microgrid System. *Energies* **2018**, *11*, 1452. [CrossRef]

6. Liu, Z.; Zheng, W.; Qi, F.; Wang, L.; Zou, B.; Wen, F.; Xue, Y. Optimal Dispatch of a Virtual Power Plant Considering Demand Response and Carbon Trading. *Energies* **2018**, *11*, 1488. [CrossRef]

7. Al-Alawi, A.; Islam, S.M. Demand side management for remote area power supply systems incorporating solar irradiance model. *Renew. Energy* **2004**, *29*, 2027–2036. [CrossRef]

8. Alham, M.H.; Elshahed, M.; Ibrahim, D.K.; El Zahab, E.E. A dynamic economic emission dispatch considering wind power uncertainty incorporating energy storage system and demand side management. *Renew. Energy* **2016**, *96*, 800–811. [CrossRef]

9. Kallel, R.; Boukettaya, G.; Krichen, L. Demand side management of household appliances in stand-alone hybrid photovoltaic system. *Renew. Energy* **2015**, *81*, 123–135. [CrossRef]

10. Kotur, D.; Đurišić, Ž. Optimal spatial and temporal demand side management in a power system comprising renewable energy sources. *Renew. Energy* **2017**, *108*, 533–547. [CrossRef]

11. Neves, D.; Brito, M.C.; Silva, C.A. Impact of solar and wind forecast uncertainties on demand response of isolated microgrids. *Renew. Energy* **2016**, *87*, 1003–1015. [CrossRef]

12. Nwulu, N.I.; Xia, X. Optimal dispatch for a microgrid incorporating renewable and demand response. *Renew. Energy* **2017**, *101*, 16–28. [CrossRef]

13. Rajanna, S.; Saini, R.P. Employing demand side management for selection of suitable scenario-wise isolated integrated renewal energy models in an Indian remote rural area. *Renew. Energy* **2016**, *99*, 1161–1180. [CrossRef]

14. Fan, S.; Ai, Q.; Piao, L. Hierarchical Energy Management of Microgrids including Storage and Demand Response. *Energies* **2018**, *11*, 1111. [CrossRef]

15. Khodaei, A.; Shahidehpour, M.; Choi, J. Optimal Hourly Scheduling of Community-Aggregated Electricity Consumption. *J. Electr. Eng. Technol.* **2013**, *8*. [CrossRef]

16. Oprea, S.-V.; Bâra, A.; Reveiu, A. Informatics Solution for Energy Efficiency Improvement and Consumption Management of Householders. *Energies* **2018**, *11*, 138. [CrossRef]

17. Cha, H.-J.; Won, D.-J.; Kim, S.-H.; Chung, I.-Y.; Han, B.-M. Multi-Agent System-Based Microgrid Operation Strategy for Demand Response. *Energies* **2015**, *8*, 14272–14286. [CrossRef]

18. Oprea, S.V.; Bâra, A.; Ifrim, G. Flattening the electricity consumption peak and reducing the electricity payment for residential consumers in the context of smart grid by means of shifting optimization algorithm. *Comput. Ind. Eng.* **2018**, *122*. [CrossRef]

19. Lizondo, D.; Rodriguez, S.; Will, A.; Jimenez, V.; Gotay, J. An Artificial Immune Network for Distributed Demand-Side Management in Smart Grids. *Inf. Sci.* **2018**, *438*, 32–45. [CrossRef]

20. Di Santo, K.G.; Di Santo, S.G.; Monaro, R.M.; Saidel, M.A. Active demand side management for households in smart grids using optimization and artificial intelligence. *Measurement* **2018**, *115*, 152–161. [CrossRef]

21. Mellouk, L.; Boulmalf, M.; Aaroud, A.; Zine-Dine, K.; Benhaddou, D. Genetic algorithm to Solve Demand Side Management and Economic Dispatch Problem. *Procedia Comput. Sci.* **2018**, *130*, 611–618. [CrossRef]

22. Ng, K.-H.; Sheblé, G.B. Direct load control-A profit-based load management using linear programming. *IEEE Trans. Power Syst.* **1998**, *13*, 688–694. [CrossRef]

23. Kurucz, C.N.; Brandt, D.; Sim, S. A linear programming model for reducing system peak through customer load control programs. *IEEE Trans. Power Syst.* **1996**, *11*, 1817–1824. [CrossRef]

24. Hsu, Y.Y.; Su, C.C. Dispatch of direct load control using dynamic programming. *IEEE Trans. Power Syst.* **1991**, *6*, 1056–1061.

25. Logenthiran, T.; Srinivasan, D.; Shun, T.Z. Demand Side Management in Smart Grid Using Heuristic Optimization. *IEEE Trans. Smart Grid* **2012**, *3*, 1244–1252. [CrossRef]

26. Samuel, G.G.; Rajan, C.C. Hybrid: Particle Swarm Optimization–Genetic algorithm and Particle Swarm Optimization–Shuffled Frog Leaping Algorithm for long-term generator maintenance scheduling. *Int. J. Electr. Power Energy Syst.* **2015**, *65*, 432–442. [CrossRef]

27. Rao, R.V.; Savsani, V.J.; Vakharia, D.P. Teaching-learning-based optimization: A novel method for constrained mechanical design optimization problems. *Comput.-Aided Des.* **2011**, *43*, 303–315. [CrossRef]

28. Rao, R.V.; Savsani, V.J.; Vakharia, D.P. Teaching-Learning-Based Optimization: An optimization method for continuous non-linear large scale problems. *Inf. Sci.* **2012**, *183*, 1–15. [CrossRef]

29. Khan, B.; Singh, P. Selecting a Meta-Heuristic Technique for Smart Micro-Grid Optimization Problem: A Comprehensive Analysis. *IEEE Access* **2017**, *5*, 13951–13977. [CrossRef]

30. Dharme, A.; Ghatol, A. Demand Side Management Quality Index for Assessment of DSM Programs. In Proceedings of the 2006 IEEE PES Power Systems Conference and Exposition, Atlanta, GA, USA, 29 October–2 November 2006; pp. 1718–1721. [CrossRef]

31. Tazvinga, H.; Zhu, B.; Xia, X. Energy dispatch strategy for a photovoltaic wind diesel battery hybrid power system. *Sol. Energy* **2014**, *108*, 412–420. [CrossRef]

32. Moghaddam, A.A.; Seifi, A.; Niknam, T.; Pahlavani, M.R. Multi-objective operation management of a renewable MG (microgrid) with back-up micro-turbine/fuel cell/battery hybrid power source. *Energy* **2011**, *36*, 6490–6507. [CrossRef]

33. Nwulu, N.I.; Fahrioglu, M. A soft computing approach to projecting Locational marginal price. *Neural Comput. Appl.* **2012**, *22*, 1115–1124. [CrossRef]

34. Albadi, M.H.; El-Saadany, E.F. Demand Response in Electricity Markets: An Overview. In Proceedings of the 2007 IEEE Power Engineering Society General Meeting, Tampa, FL, USA, 24–28 June 2007; pp. 1–5. [CrossRef]

35. FERC. Staff Report. Assessment of Demand Response and Advanced Metering. Available online: www. FERC.gov (accessed on 7 August 2006).

36. Fahrioglu, M.; Alvarado, F.L. Designing incentive compatible contracts for effective demand management. *IEEE Trans. Power Syst.* **2000**, *15*, 1255–1260. [CrossRef]

37. Nwulu, N.I.; Fahrioglu, M. A neural network model for optimal demand management contract design. In Proceedings of the 2011 10th International Conference on Environment and Electrical Engineering (EEEIC), Rome, Italy, 8–11 May 2011; pp. 1–4. [CrossRef]

38. Nwulu, N.I.; Fahrioglu, M. Power system demand management contract design: A comparison between game theory and artificial neural networks. *Int. Rev. Model. Simul.* **2011**, *4*, 106–112.

39. Fahrioglu, M.; Alvarado, F.L. Using utility information to calibrate customer demand management behavior models. *IEEE Trans. Power Syst.* **2001**, *16*, 317–322. [CrossRef]

40. Deb, K. *Optimization for Engineering Design: Algorithms and Examples*; Prentice Hall of India: New Delhi, India, 2012; ISBN 81-203-0943-X.

41. Rahman, S.; Rinaldy. An efficient load model for analyzing demand-side management impacts. *IEEE Trans. Power Syst.* **1993**, *8*, 1219–1226. [CrossRef]

42. Khanh, B.Q. Analysis of DSM's impacts on electric energy loss in distribution system using VPI model. In Proceedings of the Power and Energy Society General Meeting, Detroit, MI, USA, 24–29 July2011; pp. 1–8.

43. Khelifa, R.F.; Jelassi, K. An energy monitoring and management system based on key performance indicators. In Proceedings of the 2016 IEEE 21st International Conference on Emerging Technologies and Factory Automation (ETFA), Berlin, Germany, 6–9 September 2016; pp. 1–6. [CrossRef]

44. Behrens, D.; Schoormann, T.; Bräuer, S.; Knackstedt, R. Empowering the selection of demand response methods in smart homes: Development of a decision support framework. *Energy Inform.* **2018**, *1*, 53. [CrossRef]

![energies logo] *energies*

MDPI

Article

Decentralised Active Power Control Strategy for Real-Time Power Balance in an Isolated Microgrid with an Energy Storage System and Diesel Generators

Hyeon-Jin Moon [1], Young Jin Kim [2,*], Jae Won Chang [1] and Seung-Il Moon [1]

[1] Department of Electrical and Computer Engineering, Seoul National University, 1 Gwanak-ro, Seoul 08826, Korea; mhj7527@gmail.com (H.-J.M.); jwchang91@gmail.com (J.W.C.); moonsi@snu.ac.kr (S.-I.M.)
[2] Department of Electrical Engineering, Pohang University of Science and Technology (POSTECH), Pohang, Gyungbuk 37673, Korea
* Correspondence: powersys@postech.ac.kr; Tel.: +82-54-279-2368

Received: 8 December 2018; Accepted: 2 February 2019; Published: 6 February 2019

Abstract: Remote microgrids with battery energy storage systems (BESSs), diesel generators, and renewable energy sources (RESs) have recently received significant attention because of their improved power quality and remarkable capability of continuous power supply to loads. In this paper, a new proportional control method is proposed using frequency-bus-signaling to achieve real-time power balance continuously under an abnormal condition of short-term power shortage in a remote microgrid. Specifically, in the proposed method, the frequency generated by the grid-forming BESS is used as a global signal and, based on the signal, a diesel generator is then controlled indirectly. The frequency is controlled to be proportional to the AC voltage deviation of the grid-forming BESS to detect sudden power shortages and share active power with other generators. Unlike a conventional constant-voltage constant-frequency (CVCF) control method, the proposed method can be widely applied to optimise the use of distributed energy resources (DERs), while maintaining microgrid voltages within an allowable range, particularly when active power balance cannot be achieved only using CVCF control. For case studies, a comprehensive model of an isolated microgrid is developed using real data. Simulation results are obtained using MATLAB/Simulink to verify the effectiveness of the proposed method in improving primary active power control in the microgrid.

Keywords: microgrid; energy storage system; distributed generator; frequency control; active power control; autonomous control; droop control; frequency bus-signaling

1. Introduction

The penetration of renewable energy sources (RESs) such as photovoltaics (PVs) and wind turbines (WTs) is steadily increasing due to enhanced price competitiveness and the requirement of sustainable energy mixes. However, intermittent power outputs of RESs are detrimental to frequency and voltage stabilities, particularly in a small power system: e.g., a microgrid possessing a low moment of inertia [1,2]. A microgrid is an integrated platform that consists of power generation units, energy storage systems (ESSs), and demand response (DR) resources, whose operations are managed in a localised manner. The platform can be equipped with various functions for local power and energy management [3] under both grid-connected and islanded operating modes. In islanded mode, power demand and supply should be always balanced by itself using such distributed energy resources (DERs) in the system.

Electric utilities have technical difficulties in connecting remote microgrids on small oceanic islands to a bulk power grid on the mainland. It is also cost-ineffective and, therefore, electric power supply in these remote microgrids relies primarily on diesel generators located in the area.

Power outputs from diesel generators can be adjusted in real time to achieve power balance and hence maintain frequency and voltage stabilities [4]. However, heavy reliance on conventional diesel generators can lead to the following problems:

- Degradation of power quality and reliability due to the low inertia moment and constant-speed control scheme of diesel engines [2,5].
- High generation cost and volatility due to fuel import and transportation [6,7].
- Environmental issues such as carbon emission and air/noise pollution [6–9].

The development of ESS technologies has enabled RESs to become an attractive option for supplying electricity in such remote microgrids [4,6]. This is mainly because inverter-interfaced ESSs can alleviate the effects of intermittency of RESs on the operating stability of an isolated microgrid [10], given large penetration of RESs. Specifically, a battery energy storage system (BESS) can operate as a grid-forming unit that primarily regulates microgrid voltage and frequency [11–13]. A microgrid operator can access the BESS inverter and control the charging or discharging power of the battery within a short period of time [13], while maintaining state of charge (SOC) within an acceptable range for device protection and continuous use for other microgrid operating schemes (e.g., economic dispatch) [11–15]. Other DERs can be considered as grid-feeding units that follow the reference signals of active and reactive power.

In previous studies [14–39], grid-forming BESSs were widely used for the operation of isolated microgrids. In [16], a constant-voltage constant-frequency (CVCF) control scheme was adopted for BESSs to maintain frequency and voltages at their rated levels and consequently achieve real-time power balance in an isolated microgrid. In [17–26], the control scheme was practically applied to isolated microgrids in remote islands (including several islands in South Korea) mainly due to simple implementation and low fluctuations in grid frequency and voltages. However, the scheme requires large capacities of battery and inverters to achieve the real-time power balance under the condition on large variations in load demand and RES power outputs. To mitigate the operational burden of the BESS, centralised control methods were discussed in [15,27–30] for secondary active power regulation using real-time communications systems. Although such centralised methods are effective in optimal and robust operation of microgrids, they are vulnerable to a single point of failure (i.e., a part of a system that prevents the normal operation of the entire system if it fails). This degrades the reliability of isolated microgrids [31,32].

This challenge was considered in [33–37]; decentralised and autonomous control schemes were adopted as alternative options in which device-level controllers of individual units operate using local measurements of microgrid frequency. Unlike the centralised control schemes, the decentralised schemes allow the microgrid to operate continuously with the plug-and-play feature [31] (even under the condition of the single-point failure) and hence improve the operational reliability of the microgrid. In the decentralised schemes, droop controllers were commonly used to achieve power sharing among DERs. For example, power-and-frequency droop controllers were discussed to coordinate RESs and ESSs in [33,34] or adjust terminal voltage of the battery bank in [14]. In [34], the decentralised active power management scheme of multiple BESSs and distributed generators (DGs) was developed using multi-segment power-frequency characteristic curves. However, droop control has an inherent limitation that causes continuous fluctuations in microgrid frequency during the process of real-time power balancing [35]. Therefore, in [25,36–38], bus-signaling methods (rather than droop control methods) were exploited to induce the mode-change of DERs using the frequency and voltage control signals of BESSs while maintaining the SOC levels within allowable ranges. In [37], both droop control and bus-signaling methods were integrated with the CVCF controller mainly to maintain the SOC value of the BESS at a pre-determined level. However, these methods mainly focused on the battery energy management rather than the real-time power management in the microgrid.

Based on these observations, this paper proposes a new decentralised scheme integrated with frequency-bus-signaling and droop control based on a single-master and multiple-slaves configuration mainly to improve power sharing and real-time power balancing under the condition where serious disturbances occur in an isolated microgrid. In coordination with the CVCF controller, the proposed scheme enables the BESS to control the microgrid frequency in response to the change in AC voltages, given that unexpected voltage changes at the grid-forming converter bus indicate the instantaneous active power imbalance in the microgrid [39]. The proposed decentralised scheme enables autonomous power management in the microgrid via the coordination of the BESS with other DERs (particularly for the condition of lack of power generation), minimising the frequency and voltage deviations. Simulation case studies are performed using detailed models of an isolated microgrid in South Korea implemented with real parameters. Case study results demonstrate the effectiveness of the proposed control scheme in [40] mainly from the aspects of the following key advantages:

- Given extreme disturbances (such as a trip of the DGs), real-time power balance in an isolated microgrid is achieved without using communication systems.
- It has a simple structure and hence can be easily implemented in the outer control loop of the grid-forming BESS while ensuring the normal operations of inner control loops and, consequently, the device-level stability.
- Only the CVCF control is activated under normal operating conditions, minimising the fluctuation of microgrid frequency and active power of other DGs.

It needs to be noted that this paper is an extended version of our previous paper [40], including further improvements over [40] that can be summarized as follows:

- Performing additional literature review and further clarification between the proposed control scheme and conventional methods discussed in [11–42].
- Supplementing detailed explanations on the proposed control method and its simulation results, as well as the test bed with respect to load models and diesel generators
- Performing simulation case studies with consideration of practical microgrid components such as dead-bands and maximum/minimum limiters.

2. System Description

2.1. Geocha Island Microgrid

Figure 1 shows Geocha Island, located on the southwestern coast of South Korea. It consists of the West- and East-Geocha Islands which are 3.23 km^2 and 2.29 km^2 in area, respectively.

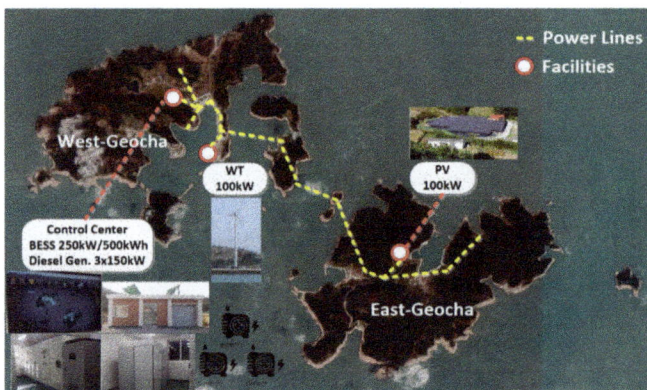

Figure 1. Isolated microgrid on Geocha Island, South Korea.

In 2017, a remote, isolated microgrid was established by the Korea Electric Power Corporation (KEPCO), mainly because of the difficulty in laying electrical power lines between the island and the Korean mainland. As shown in Figure 1, the main electrical facilities and power lines have no connection to the main grid. The west and east islands are connected via a 6.9-kV distribution line, which is characterised by a floating delta connection [26].

Figure 2 shows a schematic diagram of the microgrid configuration. The upper and lower parts represent the distribution lines in the west and east islands, respectively. The microgrid includes a 250-kW, 500-kWh BESS, a 100-kW PV, a 100-kW WT, and three 150-kW diesel generators. In the isolated microgrid, an energy management system (EMS) has been implemented to monitor the real-time operation of the DERs. Specifically, the PV and WT operate using maximum-power point tracking (MPPT) algorithms to maximise their power outputs for variations in solar insolation and wind speed, respectively [43,44]. In addition to the PV and WT, the diesel generators supply power as a grid-feeding unit to meet the load demand and maintain the SOC levels of the BESS within an acceptable range. The BESS operates as a grid-forming unit in charge of providing the primary reserve to the microgrid. While acting as a grid-forming unit, the BESS regulates the voltage and frequency of the isolated microgrid at the rated values using a CVCF control scheme to maintain the active and reactive power balance in a primary control level.

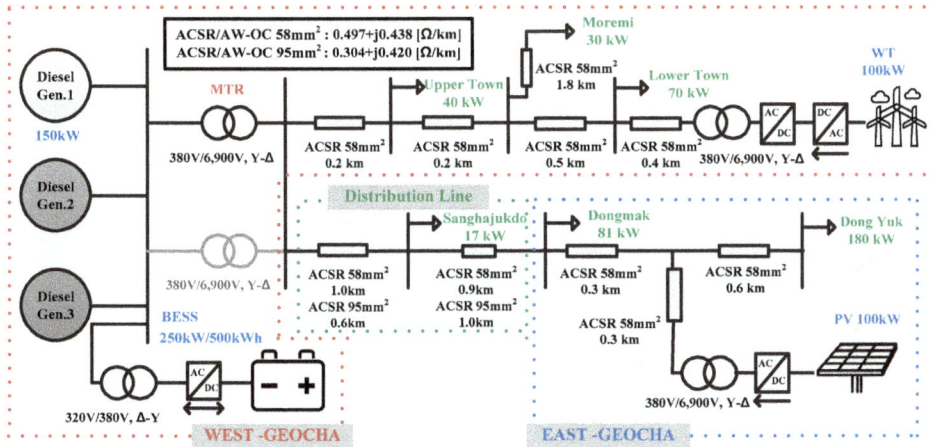

Figure 2. Schematic diagram of Geocha Island microgrid.

2.2. Grid-Forming BESS

A grid-forming BESS regulates the AC bus voltage and frequency by balancing power supply and demand in an isolated microgrid. The BESS consists of a battery pack, an LC filter, an inverter, and a transformer (see Figure 3). It operates as an AC voltage source and determines the levels of microgrid frequency and voltage by using conventional nested voltage and current control loops that operate on the dq reference frame. The BESS detects the instantaneous power imbalance by measuring the capacitor voltage V_c and recovers it to the reference value with the internal voltage and current controller. In the conventional CVCF control scheme, the dq voltage and frequency references are set to their rated values: i.e., $V_d^* = 1$ pu, $V_q^* = 0$, and $f^* = 1$ pu. The active and reactive power outputs of the BESS are indirectly controlled to maintain the bus voltage to the rated value. In this study, the frequency reference is calculated to share active power with other DERs by the proposed voltage-frequency proportional controller (VFPC), based on the level of voltage deviation, as explained in Section 3 in detail.

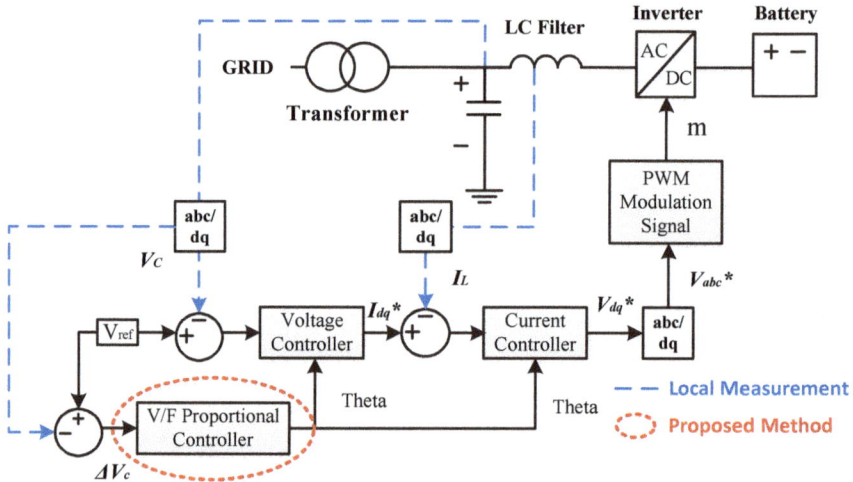

Figure 3. A schematic diagram of the grid-forming BESS and its device-level controllers.

2.3. Diesel Generator

Droop control is widely applied to improve the grid operational performance, stability, and reliability. It is based only on local measurements and allows generators in the system to operate autonomously. A droop control scheme is adopted such that the diesel generators can share active power with the BESS for real-time power balance. The droop control of active power can be expressed as:

$$f_{ref} = f_{set} - K_P(P_m - P_{set}),\tag{1}$$

where f_{ref} is the reference frequency of a generator and f_{set} and P_{set} are the presets of microgrid frequency and active power determined by the microgrid operator, respectively. P_m is a measured value of active output power of a generator and K_p is a droop coefficient that can be determined considering the operating frequency range in a microgrid as:

$$K_P = \frac{a(f_{max} - f_{min})}{P_{nom}},\tag{2}$$

where f_{max} and f_{min} are the maximum and minimum values of the grid frequency, respectively. Moreover, P_{nom} is the nominal active power of a generator and a is a constant for determining the droop coefficient.

Figure 4 shows a schematic diagram of the active power controller of the diesel generator. It includes an active power droop controller and a PI controller for time-delay dynamic models of a valve actuator, a diesel engine, and a synchronous machine. Note that the synchronous machine was modelled using the SI fundamental block in the MATLAB/Simulink. The parameters related to the controller are determined using those provided in [23]. The droop controller generates ω_{ref_di} and the PI controller is used to track the reference signal by comparing ω_{ref_di} and ω_{m_di}. Note that ω_{m_di} is the measured angular frequency of a diesel generator.

Similarly, the droop controller is widely used for reactive power control and consequently AC voltage control, as shown in Figure 5. It consists mainly of a reactive power droop controller, a PI controller, and a transfer function model of an exciter. The reactive power control can be represented using the reactive droop coefficient K_q as:

$$V_{ref} = V_{set} - K_q(Q_m - Q_{set})\tag{3}$$

$$K_q = \frac{b(V_{max} - V_{min})}{Q_{nom}} \qquad (4)$$

In Equation (3), V_{ref} is the reference value of the bus voltage of a diesel generator. The values of V_{set} and Q_{set} correspond to the preset values of bus voltage and reactive power, respectively. In addition, Q_m and Q_{nom} are the measured and nominal reactive powers of a generator. Moreover, V_{max} and V_{min} denote the maximum and minimum voltages of the system and b is a constant for determining the reactive power droop coefficient.

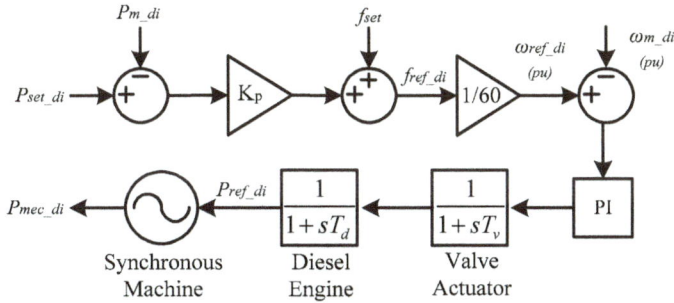

Figure 4. A schematic diagram for the active power controller of the diesel generator.

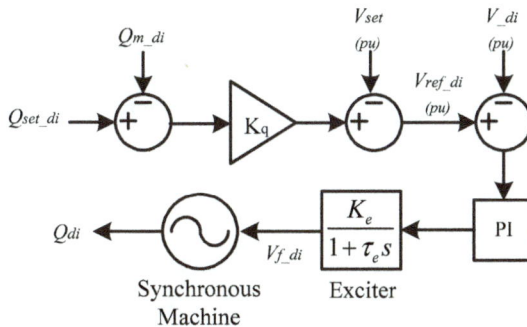

Figure 5. A schematic diagram for the reactive power controller of the diesel generator.

2.4. Basic Load Model

Load demand can be represented in terms of bus voltages [41] as:

$$P = P_0 \left(\frac{V}{V_0}\right)^{n_p} \qquad (5)$$

$$Q = Q_0 \left(\frac{V}{V_0}\right)^{n_q} \qquad (6)$$

where P_0 and Q_0 are active and reactive power demands at the rated operating voltage V_0. In addition, P and Q are active and reactive power demands for actual bus voltage V. Furthermore, n_p and n_q are the exponents that vary depending on the inherent characteristics of the load devices. These exponents essentially represent the sensitivities of the load demands with respect to the bus voltage V: i.e., $\partial P/\partial V$ and $\partial Q/\partial V$ at $V = V_0$. Equations (1) and (2) can then be represented equivalently using a ZIP model [41] that has been commonly adopted in power system analysis as:

$$P = P_0 \left[Z_p \left(\frac{V}{V_0}\right)^2 + I_p \left(\frac{V}{V_0}\right) + P_p \right] \qquad (7)$$

$$Z_p + I_p + P_p = 1 \tag{8}$$

$$Q = Q_0 \left[Z_q \left(\frac{V}{V_0} \right)^2 + I_q \left(\frac{V}{V_0} \right) + P_q \right] \tag{9}$$

$$Z_q + I_q + P_q = 1 \tag{10}$$

where Z_p, I_p, and P_p are the constant impedance, constant current, and constant power coefficients of active power demand, respectively. Similarly, Z_q, I_q, and Q_q are the ZIP coefficients of reactive power demand. The sum of the three coefficients must be equal to one, as shown in Equations (8) and (10), to meet the rated operating condition. Note that in this paper, we focus on the active power management in an isolated microgrid. In [42], the average value of n_p was known to be between 1.1 and 1.7. In isolated microgrids, n_p is expected to be larger due to the high proportion of resistive loads such as heating and lighting [42]. Therefore, n_p has been set to 2. This assumed that the isolated microgrid only includes constant impedance loads (i.e., $Z_p = 1$, $I_p = 0$, and $P_p = 0$).

2.5. Active Power Balance Equation in an Isolated Microgrid with CVCF Control

Under the normal operating condition where the grid-forming BESS has sufficient power reserve to cover the active power change of a system, the active power balance can be satisfied as:

$$\Delta P_{BESS} = \Delta P_{Load} + \Delta P_{Loss} - \Delta P_{DG} \tag{11}$$

where ΔP_{BESS} and ΔP_{DG} are the variations in the active power outputs of the grid-forming BESS and the DGs, respectively. ΔP_{Load} is the variation in the rated load demand and ΔP_{Loss} is the variation in active power losses in the microgrid. In contrast, the active power output of the BESS reaches its limit when the BESS does not have a sufficient power reserve. This limit can be estimated as:

$$\Delta P_{BESS\_max} = V_c (I_{d\_max} - I_{d\_set}) \tag{12}$$

where ΔP_{BESS_max} is the maximum variation in the active power output of the BESS. In Equation (12), I_{d_max} and I_{d_set} are the maximum and preset values, respectively, of d-axis current. In this situation, the BESS cannot recover the voltage completely and, consequently, the load shedding is initiated by the bus voltage reduction to achieve the active power balance in the microgrid, as shown in Equations (13) and (14):

$$\Delta P_{BESS\_max} = \Delta P_{Load} + \Delta P_{Loss} - \Delta P_{DG} - \Delta P_{Load\_VR} \tag{13}$$

$$\Delta P_{Load\_VR} = \sum \left[P_{Load\_i0} \left\{ 1 - \left(\frac{V_i}{V_0} \right)^{n_{pi}} \right\} \right] \tag{14}$$

In Equation (14), ΔP_{Load_VR} is a decrease in the total load demand at under-voltage buses. A shown in Equation (14), the value of n_{pi} significantly affects the value of V_i at which the power balance in Equation (13) is satisfied. The smaller n_{pi}, the smaller V_i that is required to induce the enough reduction of the load demand, causing the degradation of voltage stability and even voltage collapse. This implies that for the microgrid with less voltage-dependent loads, the proposed VFPC becomes more effective in alleviating the power shortage and consequently mitigate the voltage reduction. Moreover, with less voltage-dependent loads, the proposed VFPC is capable of adjusting the frequency more successfully by inducing active participations of other diesel generators in the real-time frequency regulation via their P-f droop controllers.

3. Proposed Control Method

The IEEE standard recommends that the bus voltages should be regulated within 88–110% of the rated value for the normal operating condition [45]. The CVCF controller, discussed in Section 2, is effective in maintaining the voltages within the acceptable range particularly when the grid-forming

BESS operates with sufficient primary reserve. Otherwise, a cooperative control scheme of the CVCF controller of the BESS and the local controllers of other DGs needs to be implemented to exploit the additional primary reserve capacities of the DGs for the reliable operation of the isolated microgrid. The proposed cooperative control scheme aims at the active power sharing in an isolated microgrid when active power balance cannot be achieved solely by the CVCF controller of the grid-forming BESS. Figure 6 shows the flow chart of the proposed method for the decentralised active power control. In the proposed method, the BESS operating with the CVCF controller detects insufficient active power supply in the microgrid by measuring the input voltage of the AC bus where the BESS is connected. It then controls the frequency proportional to the bus voltage deviation from the nominal voltage. Further details on each module in Figure 6 will be discussed in Sections 3.1 and 3.2.

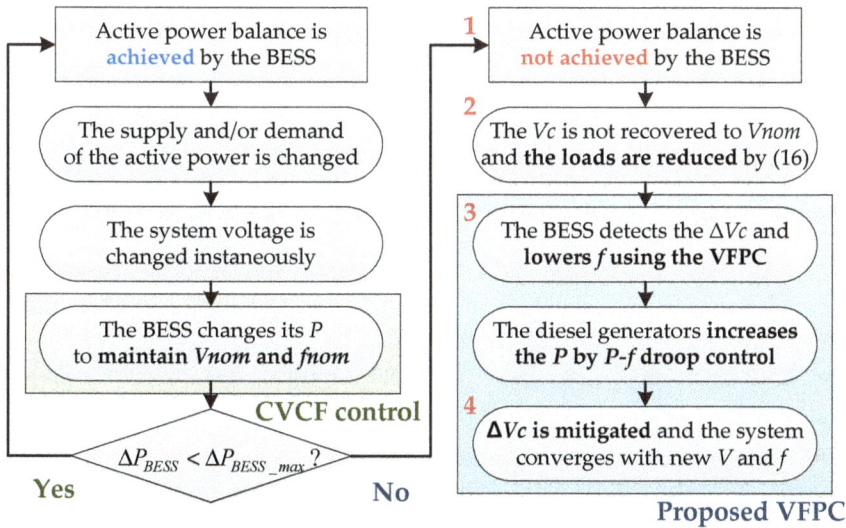

Figure 6. Flow chart of the proposed VFPC for the autonomous active power management at a primary control level.

In this paper, we focus on the coordinated, autonomous control scheme of the grid-forming BESS and the diesel generators via frequency-bus-signaling and droop control based on the single-master and multiple-slaves configuration. Note that for simplicity, the PV system and WT operate as independent current sources equipped with MPPT controllers without reactive power control. The proposed scheme also can be applied to a general microgrid including multiple BESS and diesel generators. In such microgrid, one BESS operates as a single-master unit that performs grid-forming control. The remaining BESSs and diesel generators operate as slave units using P-f droop control for power sharing. The operational burden on the master BESS will increase, causing the lack of active power reserve and hence frequent power shortages. This issue can be resolved using a multi-master droop control scheme (e.g., [13]). However, the scheme would cause continuous frequency deviations under the normal operating condition. Therefore, the microgrid operator needs to choose an appropriate configuration: i.e., the single-master and multiple-slaves or the multiple-masters and multiple-slaves. We leave the coordinated control of grid-forming master units in future work, while focusing on the development, analysis, and verification of the proposed control strategy based on the single-master-and-multiple-slaves configuration.

3.1. Frequency Control of BESS with VFPC

In this paper, a VFPC has been proposed to control the reference frequency of the grid-forming BESS, so that the primary active power reserve with the diesel generators is effectively exploited and shared in the microgrid under a power shortage condition. Specifically, in Figure 7, f_{nom} is the nominal frequency and f_{ref} is the reference frequency of the grid-forming BESS. The reference frequency is determined as:

$$f_{ref} = f_{nom} + K_v \Delta V_c \tag{15}$$

where ΔV_c is the variation in the AC voltage estimated by subtracting the actual AC voltage from the reference voltage of the BESS. The coefficient K_v denotes the proportional gain of the VFPC, which can be expressed as:

$$K_v = c \frac{f_{nom} - f_{min}}{V_{nom} - V_{min}} \tag{16}$$

where c is a constant to determine the V-f droop coefficient.

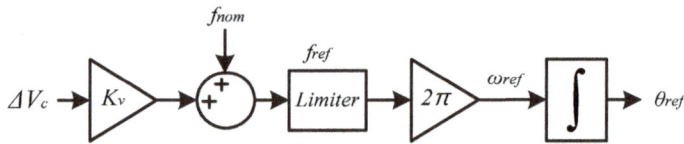

Figure 7. A schematic diagram of the *V-f* proportional controller for the grid-forming BESS.

In Figure 7, the limiter consists of dead-band, rate-limiter, and saturation blocks to operate only when the primary reserve of the grid-forming BESS is smaller than the magnitude of the load demand variation. The VFPC enables the BESS inverter to determine the system frequency for a decrease in the AC voltage, as shown in Equations (13) and (14), which is measured at the input port of the inverter. The diesel generators then measure the reduced system frequency and increase their active power outputs by conventional *P-f* droop controllers. Since the frequency is the same throughout the microgrid, the proposed frequency-bus-signaling method, discussed in Section 3.2, can achieve active power sharing in an autonomous manner.

3.2. Proposed Autonomous Active Power Management

The proposed VFPC of the grid-forming BESS is activated only if the active power balance cannot be achieved because of lack of the primary reserve of the BESS. Specifically, the capacitor voltage V_c is not fully recovered to V_{nom} when the BESS has insufficient the active power reserve and hence cannot compensate for all power imbalance in the microgrid. The BESS controls the frequency (i.e., f_{set} to f_{ref}) proportional to ΔV_c, as shown in Figure 8(Left), and then controls the diesel generators with active power droop curves (i.e., P_{set} to P_{ref}), as illustrated in Figure 8(Right). Considering the VFPC operation, a new power balance equation is represented as follows:

$$\Delta P_{BESS\_max} = \Delta P_{Load} + \Delta P_{Loss} - \Delta P_{DG} - \Delta P_{Load\_VR} - \Delta P_{Load\_FR} - \Delta P_{DG\_Droop} \tag{17}$$

$$\Delta P_{Load\_FR} = K_{pf} \Delta f * P_{Load\_0} \tag{18}$$

$$\Delta P_{DG\_Droop} = P_m - P_{set} = -\frac{1}{K_p}(f_{ref} - f_{set}) = -\frac{K_v}{K_p} \Delta V_c \tag{19}$$

where ΔP_{Load_FR} is a variation in the load demand for the change in the microgrid frequency and ΔP_{DG_Droop} is the variation in the active power generation by droop control. In large-scale power systems, the load demand variation with respect to the frequency deviation can be characterized using the sensitivity coefficient K_{pf} in Equation (18), which varies for the range from 0 to 3.0 [41]. However, in this paper, the load demand is assumed to remain unchanged for the frequency deviation for simplicity.

This assumption is also valid because the reference frequency of the BESS is controlled to vary within a very small range, resulting in a slight variation in the microgrid frequency. For the VFPC, Equation (19) can then be derived from Equations (1) and (15). Equation (17) also can be expressed as:

$$\Delta P_{BESS\_max} = \Delta P_{Load} + \Delta P_{Loss} - \Delta P_{DG} - \sum \left[P_{Load\_i0} \left\{ 1 - \left(\frac{V_i}{V_0} \right)^{n_{pi}} \right\} \right] - K_{pf} \Delta f \times P_{Load\_0} + \frac{K_v}{K_p} \Delta V_c \quad (20)$$

The integration of the VFPC with the CVCF controller of the grid-forming BESS enables additional primary reserve from the diesel generators to be exploited for real-time power balance in the microgrid via their active power droop control loops. In this way, autonomous active power management can be achieved for mitigating the active power shortage and the under-voltage, which often occur owing to the insufficient primary reserve of the BESS.

The proposed active power management strategy allows the grid-forming BESS to operate with the CVCF control under the normal condition where the BESS has sufficient primary reserve, so that the BESS can efficiently take charge of the primary reserve supply using its fast and accurate response characteristics. On the other hand, the frequency-bus-signaling method using the proposed VFPC is adopted for active power sharing with the diesel generators only under the abnormal condition where the BESS has the limited primary reserve. Consequently, the power sharing issue in the conventional CVCF control can be effectively resolved in the proposed strategy, minimising the microgrid frequency variations and the active power fluctuations of other droop-based DGs. The sequential operations of the VFPC can be summarised as follows:

1. The situation in which the remaining primary reserve of the BESS (ΔP_{BESS_max}) is not enough to cover the active power balance occurs due to a rapid increase in the net demand (e.g., sudden disconnection of a DG).
2. The overall bus voltage in the microgrid is reduced, inducing load reduction (ΔP_{load_VR}). This leads to the reduction of variations in the maximum power output of the BESS (see Equations (12) and (13)).
3. The BESS recognises the power shortage based on ΔV_c and reduces f_{ref} (see Equation (15)) by the VFPC. The diesel generator increases its active power with the *P-f* droop controller (see Equation (1)).
4. The participation of diesel generator, acting as a slave unit, enables the power shortage to be compensated for and consequently the microgrid voltages and load demand to be recovered. The reserve of the BESS is also procured and the microgrid starts operating with new levels of V and f (see Equation (20)).

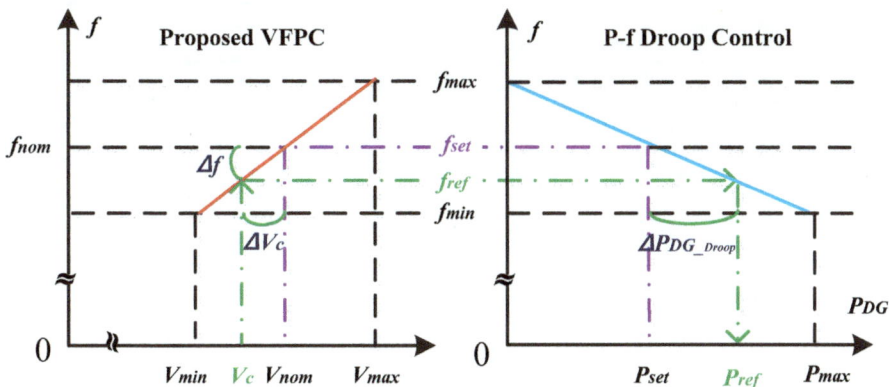

Figure 8. Comparison between (**Left**) the proposed control strategy (i.e., VFPC) of the grid-forming BESS) and (**Right**) the conventional control strategy (i.e., droop controller) of the diesel generators.

4. Case Studies and Simulation Results

Case studies have been performed using the model of the isolated microgrid on Geocha Island. As shown in Figure 9, the microgrid model has been implemented using MATLAB/Simulink with the parameters listed in Table 1. For simplicity, it was assumed in the case studies that the power factors of all loads are maintained at unity and a single generator is assumed to participate in the real-time frequency regulation. The wind speed is assumed to change continuously from 11 m/s to 14 m/s. The proposed control scheme has been implemented in the microgrid for primary active power sharing under the condition where the low voltages occur due to an increase in the load demand. The proposed method can be similarly tested under the high voltage condition.

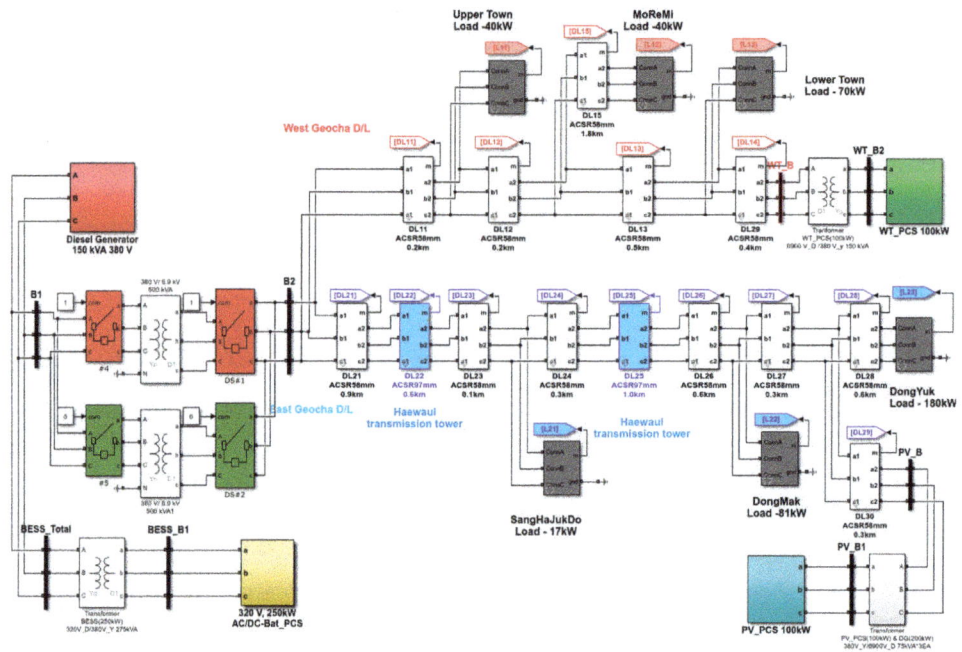

Figure 9. Simulation model of Geocha Island microgrid using MATLAB/Simulink.

Table 1. Parameters used for the simulation model of the isolated microgrid on Geocha Island.

Parameter Name	Symbol	Value	Units
Rated voltage of the system	V_{nom}	1	pu
System nominal frequency	f_{nom}	60	Hz
Minimum reference frequency	f_{min}	59.4	Hz
Dead-band of the VFPC of the BESS	-	±0.01	
Rate limit of the reference frequency	-	0.3	Per s
V-f proportional gain	K_v	6	-
Maximum limit of d-axis current in the BESS	I_{d_max}	1	pu
d-axis voltage reference of the BESS	V_d^*	1	pu
q-axis voltage reference of the BESS	V_q^*	0	pu
Sample time of the simulation	T_s	5×10^{-5}	s

The microgrid is assumed to experience three successive events: (i) the 180-kW load is connected at the Dong-yuk bus of the East Geocha distribution feeder at $t = 3$ s, (ii) the WT at the end bus of the West Geocha feeder is tripped at $t = 6$ s, and (iii) the PV system on the East Geocha feeder is tripped at $t = 9$ s. At $t = 0$ s, the total load demand and the power outputs of the PV system and diesel generator are set to 238-kW, 90-kW, and 100-kW, respectively. The reactive power output of the diesel generator is set to 10-kVAr, and the reactive power reference of PV and WT are set to 0.

4.1. Simulation Results under the Normal Condition (t < 9 s)

Figure 10 shows the active power profiles of the grid-forming BESS, WT, PV, and diesel generator for Cases 1 and 2, respectively: (i) where only the conventional CVCF controller was adopted and (ii) where the proposed VFPC was integrated with the CVCF controller. The proposed VFPC was not activated by its limiter (see Figure 7) under the normal condition where the active power imbalance can be compensated for by the BESS. Therefore, Cases 1 and 2 have the same profiles of the active powers, voltages, frequencies, and load demand until $t < 9$ s, as shown in Figures 10–15. In both cases, the BESS maintained the active power balance and, consequently, the V_{BESS} and f_{BESS} at the rated values in the microgrid during 0 s $\leq t \leq 9$ s. Specifically, during 0 s $\leq t \leq 3$ s, the bus voltages at the WT and PV system increased to levels slightly higher than 1 pu. The voltages at the BESS were maintained at almost 1 pu, as shown in Figure 11. Figures 12–15 compare the profiles of the microgrid frequency at the main transformer, the reference frequency of the BESS, the total load demand, and the reactive power outputs in the microgrid for Cases 1 and 2. As shown in Figure 14, the actual load demand was observed at approximately 243 kW, which was greater than 238 kW at the rated bus voltages; the voltages at several load buses were higher than 1 pu, increasing the load demand due to the load-voltage dependency, as shown in Equation (14). In Figure 15, the reactive power outputs of the BESS and the diesel generator were maintained at an almost constant level due to the small variations in the reactive power loss and the feeder voltage.

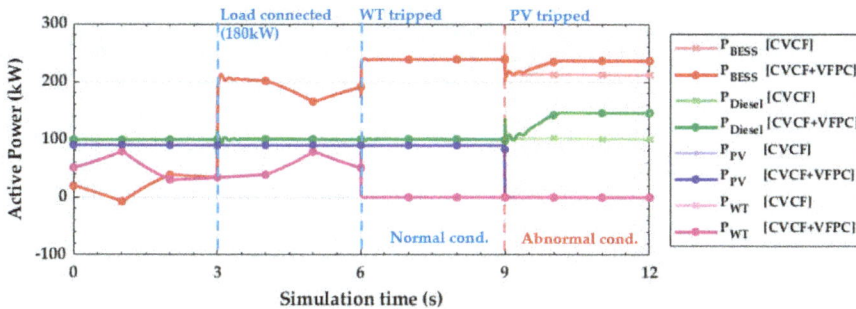

Figure 10. Profiles of the active power outputs of the BESS, WT, PV, and diesel generator.

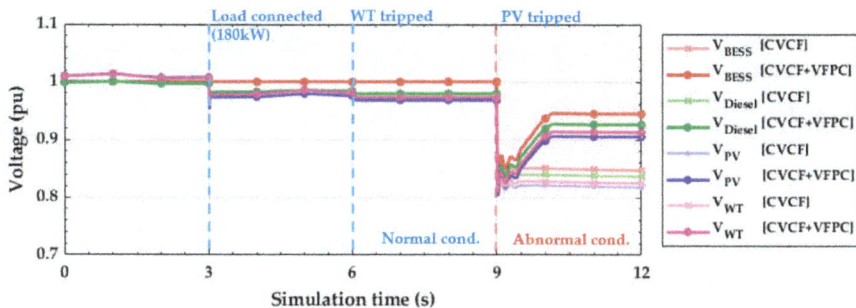

Figure 11. Profiles of the input voltages of the BESS, WT, PV, and diesel generator.

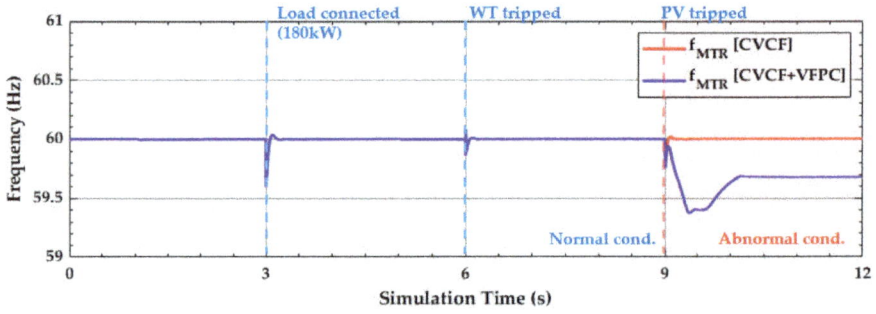

Figure 12. Comparison of the grid frequency profiles in the Geocha microgrid.

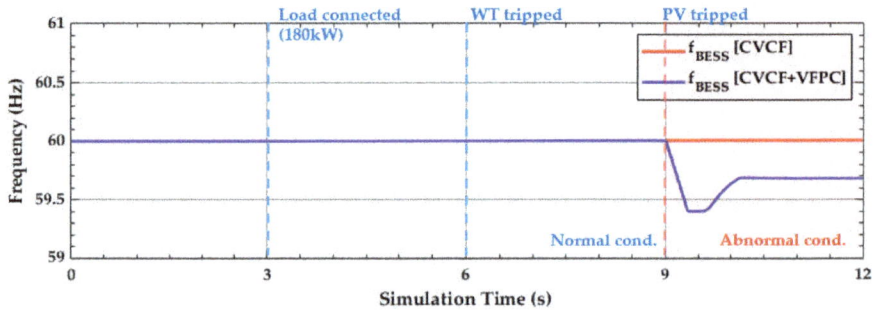

Figure 13. Comparison of the reference frequency profiles of the BESS in the Geocha microgrid.

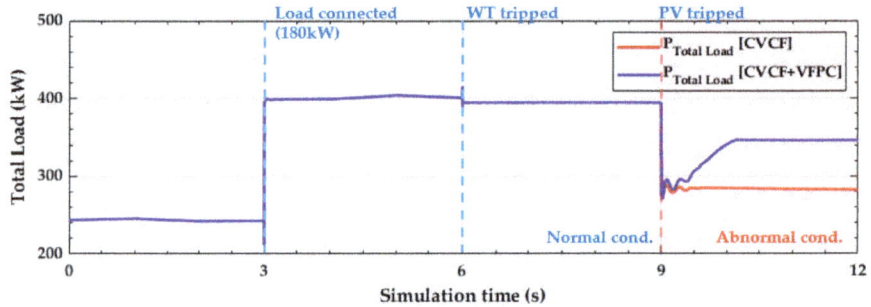

Figure 14. Comparison of the total load demand profiles in the Geocha microgrid.

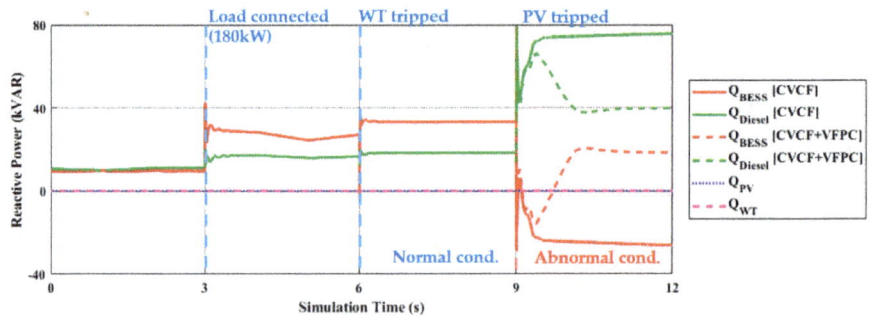

Figure 15. Profiles of the reactive power outputs of the BESS, WT, PV, and diesel generator.

When the 180-kW load was connected to the Dong-Yuk bus of the microgrid at $t = 3$ s, the nominal load demand increased from 238 kW to 418 kW in total. The microgrid voltage and frequency were then reduced instantaneously due to the large step increase in the load demand. The CVCF controller detected the voltage reduction and adjusted the output power of the BESS quickly from 34 kW to 213 kW, so as to maintain the power balance and the input voltage at the rated value. Meanwhile, the active power of the diesel generator fluctuated due to the frequency deviation during a short period of time. Figure 10 shows that the CVCF controller enabled the battery to respond to a large step-wise increase in the load demand within a short time period and maintain the real-time power balance in the microgrid. As shown in Figure 11, the small voltage drops were detected for the WT, PV system, and the diesel generator because of the increase in the power flowing along the distribution feeders connected from the west island to the east island. This increased active and reactive power losses on the substation transformers and in the distribution lines. The reactive power output of the BESS increased due to the increase in the reactive power loss in the microgrid, and the diesel generator then increased its reactive power output via the Q-V droop controller. During the period of 3 s $\leq t \leq 6$ s, as shown in Figures 11 and 14, the actual load demand increased about from 398 kW to 403 kW and the bus voltage was reduced about from 0.967 pu to 0.986 pu.

At $t = 6$ s, the WT was tripped from the distribution feeder. The active power of the BESS then increased to maintain the power balance via the CVCF control, causing a further drop in the voltage at the buses of other DERs owing to the increase in the power flow from the substation. The total load demand then decreased owing to the additional voltage drops at the load buses. Meanwhile, other DERs retain their active power because the BESS was solely responsible for maintaining the active power balance at the primary control level. The active power output of the BESS increased from 190 kW to approximately 238 kW, as shown in Figure 10, which is similar to the rated active power output of 250 kW. The power balance could be achieved within a short period of time. However, the primary active power reserve was significantly reduced in the microgrid, affecting the power system reliability. The reactive power outputs of the BESS and the diesel generator slightly increased for the same reason as aforementioned.

4.2. Simulation Results under the Abnormal Conditions ($t \geq 9$ s)

When the PV system was tripped at $t = 9$ s, the active power reserve of the BESS could not afford the power balance. As a result, the bus voltages and hence the load demand decreased significantly, as shown in Figures 11 and 14. In the CVCF-only case, the diesel generator maintained its active power output as constants, because the microgrid frequency was maintained at f_{set} and the diesel generator could not detect the power imbalance with only its local measurement, as shown in Figures 10, 12 and 13. Consequently, the total load demand was significantly reduced to 282 kW, which is about 67% of the nominal demand, owing to the severe voltage reductions beyond the lower voltage limit of 0.9 pu: i.e., $V_{aBESS} = 0.847$ pu, $V_{aDiesel} = 0.836$ pu, $V_{aWT} = 0.824$ pu, and $V_{aPV} = 0.817$, as shown in Figures 11 and 14. Moreover, the reduction of the bus voltage at which the BESS is located also affected the active power output of the BESS, as shown in Equation (12). This is illustrated in Figure 10 where the active power output of the BESS is about 212 kW; it is lower than the rated active power. This further decreased the bus voltage. The large reduction of the feeder voltage caused the excessive compensation of the Q-V droop controller for the reactive power. Since the BESS is located close to the diesel generator in the remote microgrid, the excessive compensation could be immediately balanced using the BESS. It can be seen that the total reactive power supply and the reactive power loss increased mainly because of a further increase in the power flowing from the west island to the east island.

In contrast, the reference frequency of the BESS was reduced to 59.4 Hz due to the voltage drop (as shown in Figure 13) for Case 2 where the proposed VFPC was applied to the CVCF control in the BESS. The microgrid frequency was then reduced to 59.37 Hz and the diesel generator increased the active power to 145.9 kW via the droop control. The active power output of the BESS was also increased to 236.1 kW, as shown in Equations (12) and (17), enabling the further recovery of the voltage drop. This coordinated control between the BESS and the diesel generator caused the microgrid voltage to rise by acquiring additional primary active power reserve in the microgrid. The compensation of the diesel generator for the reactive power was mitigated as the feeder voltage was gradually recovered. The microgrid frequency measured by a PLL was gradually restored to 59.68 Hz by increased power outputs of the diesel generator and the BESS. As shown in Figures 11 and 14, the proposed VFPC successfully mitigated the voltage drop (V_{aBESS} = 0.944 pu, $V_{aDiesel}$ = 0.926 pu, V_{aWT} = 0.913 pu, and V_{aPV} = 0.905 pu) and the load reduction (about 72 kW, about 17% of the nominal demand).

4.3. Simulation Results with Less Voltage-Dependent Loads

In Case 3, the constant current load with the rated power of 81-kW was taken into consideration at the Dong-Mak bus (see Figure 2); n_p was reduced to approximately 1.65 at t = 0 s. After the constant impedance load of 180-kW was connected at the Dong-Yuk bus at t = 3 s, n_p increased from 1.65 to 1.81. The power shortage then led to larger decreases in the feeder voltages and consequently in the microgrid frequency during $t \geq 9$ s, compared to the original condition (i.e., Case 2) where only the constant impedance loads were considered. Figures 16–19 and Table 2 show that the lower n_p, the lower active power output of the BESS, further reducing the bus voltage, the frequency, and actual total load demand, particularly when the diesel generator failed to completely follow the command of the master BESS owing to the insufficient reserve capacity. In Figures 16–19, the full and dotted lines represent the cases of n_p = 2 and 1.81, respectively.

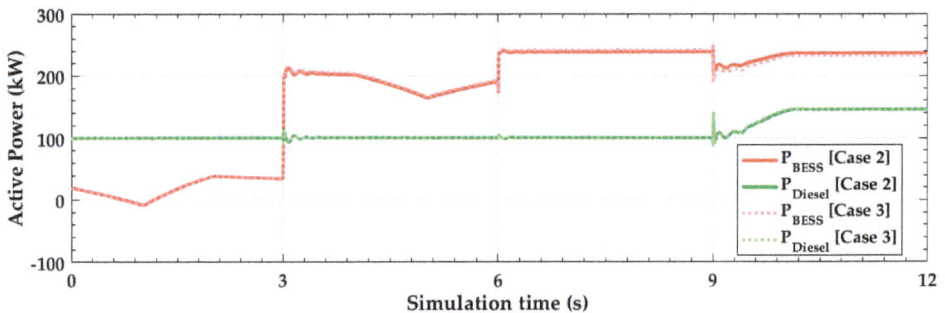

Figure 16. Profiles of the active power outputs of the BESS and diesel generator in Cases 2 and 3.

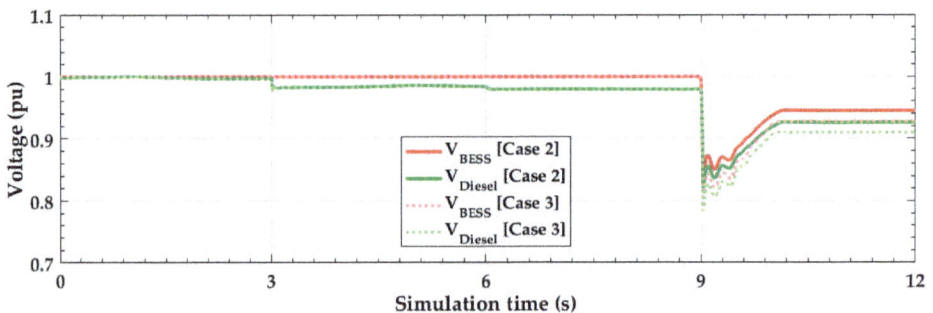

Figure 17. Profiles of the input voltages of the BESS and diesel generator in Cases 2 and 3.

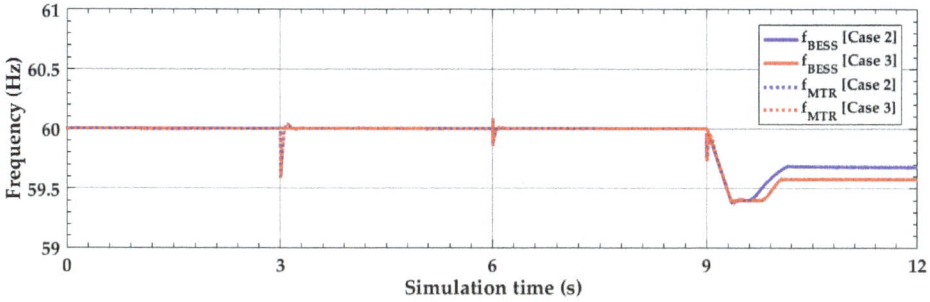

Figure 18. Comparisons of the frequency profiles at the main transformer and BESS in Cases 2 and 3.

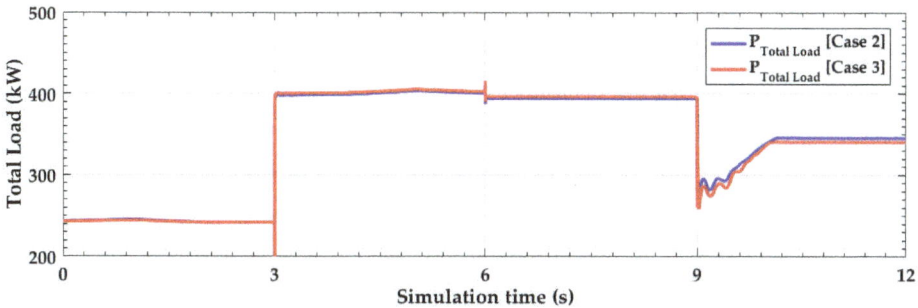

Figure 19. Comparisons of the total load demand profiles in the Geocha microgrid.

Table 2. Comparisons between the simulation results acquired for different load demand and compositions at $t = 12$ s.

	n_p	P_{load_0} (kW)	V_{aBESS} (pu)	$V_{aDiesel}$ (pu)	P_{BESS} (kW)	P_{Diesel} (kW)	P_{load} (kW)	f_{MTR} (Hz)
Case 2	2	418	0.944	0.926	236.1	145.9	345.6	59.68
Case 3	1.81	418	0.927	0.910	231.8	145.9	340.8	59.58
Case 2*	2	388	0.975	0.955	243.8	135.4	343.7	59.86
Case 4	1.56	388	0.969	0.949	242.3	143.7	350	59.83

In Cases 2* and 4, to simulate the less severe power shortage condition, the original condition was slightly modified to reduce Dong-Mak load from 81-kW to 51-kW. In addition, in Case 4, the 51-kW Dong-Mak load and the 40-kW Upper-Town load were modeled as constant power loads, reducing n_p to approximately 1.24 at $t = 0$ s. After the constant impedance load of 180-kW was connected at the Dong-Yuk bus at $t = 3$ s, increasing n_p from about 1.24 to 1.56. The power shortage occurred at $t = 9$ s in the microgrid, resulting in $n_p = 1.56$. It led to the bigger drop in the voltage and consequently caused the larger decrease in the frequency, in comparison to Case 2*. The maximum variation in the power output of the BESS was also reduced owing to the voltage drop, as shown in Equation (12). The diesel generator measured the frequency, which was further reduced, and increased its output power larger than those in Case 2*. This allowed the power shortage to be better compensated for and, consequently, the microgrid voltages and total load demand to be more recovered. Figures 20–23 then show that the lower n_p, the higher active power supply when the diesel generators had the sufficient reserve capacities and succeeded in following completely the command of the master BESS. This mitigated the reduction of actual load demand. Note the full and dotted lines represent the cases of $n_p = 2$ and 1.56, respectively. Table 2 shows that although the bus voltage, microgrid frequency, and BESS power output were reduced at $t = 12$ s for $n_p = 1.56$, the output power of the diesel generator and the total load demand were higher in Case 4 than those in Case 2*.

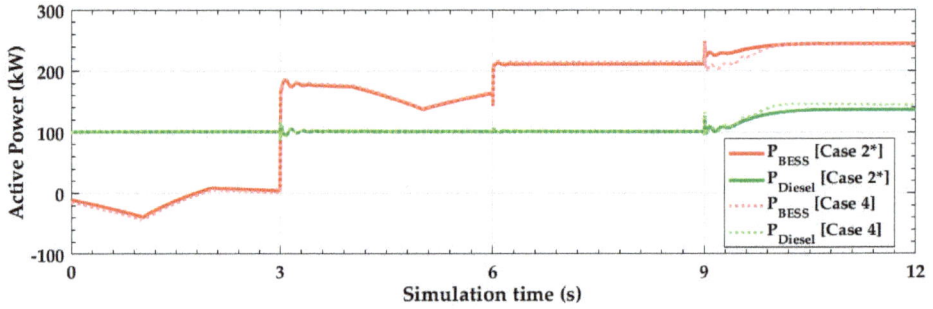

Figure 20. Profiles of the active power outputs of the BESS and diesel generator in Cases 2* and 4.

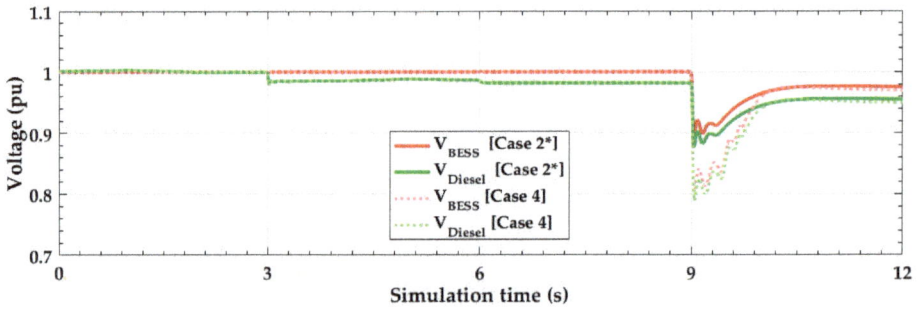

Figure 21. Profiles of the input voltages of the BESS and diesel generator in Cases 2* and 4.

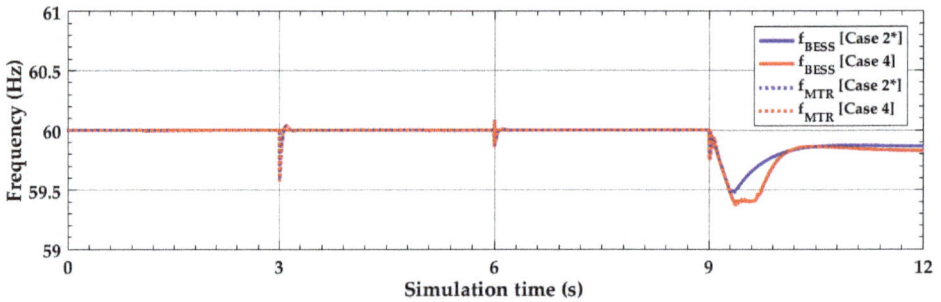

Figure 22. Comparisons of the frequency profiles at the MTR and BESS in Cases 2* and 4.

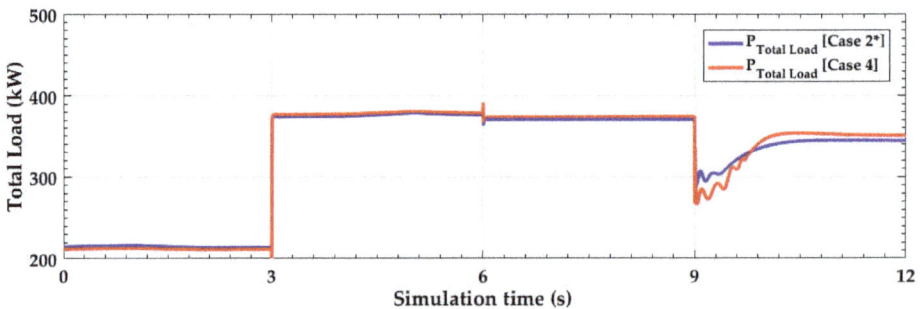

Figure 23. Comparisons of the total load demand profiles in Cases 2* and 4.

As shown in Cases 2–3 and 2*–4, the voltage drop becomes larger as n_p is reduced, particularly, under the power shortage condition. This implies that the proposed controller is more effective in alleviating the power shortage problem and voltage stability issue for a remote microgrid with less voltage-dependent load. The effect of the proposed controller becomes more evident when the slave units have sufficient reserve capacities.

4.4. Discussion

The CVCF control enables the grid-forming BESS to maintain the active power balance of the microgrid immediately by utilising the fast time response of the inverter. The grid-forming BESS is solely responsible for the active power balance in the microgrid at the primary level. Therefore, a cooperative method to induce the reserve provision of other DGs is required particularly when the limit of the BESS is reached. The simulation results in Section 4.2 show that when the primary active power reserve of the BESS is insufficient, and the load demand exceeds the power supply (after $t = 9$ s), the grid voltage is then greatly reduced overall, which reduces the load demand by the voltage-dependent characteristics of the loads. Furthermore, the active power reserve of the BESS is also reduced by the voltage drop at the BESS, exacerbating the active power imbalance in the microgrid.

The proposed VFPC operates only when the primary active power reserve of the microgrid is not enough by measuring the AC voltage of the BESS under the low-voltage condition. Under the normal condition (until $t = 9$ s in Section 4.1), the proposed method is able to reduce the microgrid frequency variation and the active power fluctuation, compared to the conventional *f-P* droop control method of the grid-forming BESS. In addition, this control method adjusts the frequency, which is proportional to the bus voltage of the BESS by the VFPC, to react to the diesel generator with active power droop control in the microgrid. Therefore, case studies results show that this method does not require any communication and can be easily applied with other conventional control methods such as CVCF and conventional droop control of other DERs.

Simulation results suggest that the conventional *f-P* droop control of the grid-forming BESS cannot operate properly when a large change in active power, such as a generator trip and step-load change, instantaneously occurs in the microgrid. These changes induce an instantaneous voltage drop, which reduces the maximum active power output of the grid-forming BESS. In this case, the conventional *f-P* controller receives the reduced active power of the BESS due to the voltage drop as an input signal, outputting the less reduced frequency. Then other DGs using *P-f* droop controllers cannot increase their active power outputs enough due to the frequency not being sufficiently reduced in the transient state.

It needs to be noted that in practice, it is likely for the microgrid operator to limit the reactive power supply to ensure the stable operation of the microgrid without significantly affecting the overall performance of the proposed active power management. In addition, due to the independent control of active and reactive power, the proposed method does not prevent the normal operation of the conventional reactive power controllers, mitigating the excessive compensation even for the case where the active power reserve is not sufficient.

5. Conclusions

In this paper, the VFPC has been proposed as an additional controller that can be easily integrated with the conventional CVCF controller of the BESS in the isolated microgrid. This effectively assists the microgrid operator to resolve the under-voltage problem owing to the limited reserve of the grid-forming BESS. This method uses the microgrid frequency as a global bus signal, enabling the indirect control of other DERs without communications systems. Moreover, the detailed simulation model was implemented in MATLAB/Simulink using actual system parameters to verify the effectiveness of the proposed active power management strategy. The simulation results show that the proposed VFPC method can mitigate the active power shortage and the bus voltage reduction using the frequency-based operation of the DERs in the microgrid. From the simulation results, the main advantages of the proposed controller can be summarised as follows:

- The proposed VFPC can be easily applied to the existing CVCF controller of the grid-forming BESS and enables the coordinated control with other DERs that operate with conventional *P-f* droop controllers.
- The proposed VFPC can be activated based on the local measurement of its bus voltage, not active power, even when sudden and severe imbalance of active power takes place in the microgrid.
- The proposed controller is activated only during the period of active power imbalance Unlike the conventional *f-P* droop method, the CVCF controller can still reduce the fluctuation of frequency and active power under the normal microgrid condition.

Author Contributions: Conceptualization, H.-J.M.; methodology, H.-J.M.; software, H.-J.M and J.W.C.; validation, H.-J.M., Y.J.K., J.W.C., and S.-I.M.; formal analysis, H.-J.M.; investigation, H.-J.M., Y.J.K.; resources, H.-J.M.; data curation, H.-J.M., J.W.C.; writing—original draft preparation, H.-J.M., Y.J.K., J.W.C., and S.-I.M.; writing—review and editing, H.-J.M., Y.J.K., J.W.C., and S.-I.M.; visualization, H.-J.M.; supervision, Y.J.K., S.-I.M.;

Acknowledgments: This work was supported by the Human Resources Development program of Korea Institute of Energy Technology Evaluation and Planning (KETEP) grant funded by Korea government Ministry of Trade, Industry and Energy (No. 20174030201540). This work was partly supported by the Korea Institute of Energy Technology Evaluation and Planning (KETEP) grant funded by the Korea government (MOTIE) (No. 70300037). This article is an extension of the work presented at ICEER2018 and published in Energy Procedia.

Conflicts of Interest: The authors declare no conflict of interest.

Abbreviations

Parameters of synchronous machine

Parameters	Symbols	Values	Units
Inertia coefficient	J	3.35	$kg \cdot m^2$
Friction factor	F	0	$N \cdot m \cdot s$
Pole pairs	p	2	-
Stator resistance per phase	R_s	1.66×10^{-2}	Ω
Stator leakage inductance	L_l	1.68×10^{-4}	H
d-axis magnetizing inductance viewed from stator	L_{md}	5.86×10^{-3}	H
q-axis magnetizing inductance viewed from stator	L_{mq}	5.05×10^{-3}	H
Field resistance	R_f	5.25×10^{-3}	Ω
Field leakage inductance	L_{lfd}	6.82×10^{-4}	H
d-axis resistance of Damper	R_{kd}	1.53×10^{-1}	Ω
d-axis leakage inductance of Damper	L_{lkd}	3.40×10^{-3}	H
q-axis resistance of Damper	R_{kq1}	4.06×10^{-2}	H
q-axis leakage inductance of Damper	L_{lkq1}	6.08×10^{-4}	H
P gain of PI controller for active power control	K_{pp}	20	-
I gain of PI controller for active power control	K_{ip}	60	-
P gain of PI controller for reactive power control	K_{pq}	5	-
I gain of PI controller for reactive power control	K_{iq}	13	-

Parameters of synchronous machine controller

Parameters	Symbols	Values	Units
Time constant of diesel engine	T_d	0.5	s
Time constant of valve actuator	T_v	0.05	s
P-f droop coefficients of the diesel generator	K_p	4.0×10^{-6}	-
Amplification gain of the exciter	K_e	70	-
Time constant of the exciter	τ_e	2.0×10^{-3}	-
Q-V droop coefficient of the diesel generator	K_q	2.5×10^{-6}	-

P gain of PI controller for active power control	K_{pp}	20	-
I gain of PI controller for active power control	K_{ip}	60	-
P gain of PI controller for reactive power control	K_{pq}	5	-
I gain of PI controller for reactive power control	K_{iq}	13	-

References

1. Teodorescu, R.; Liserre, M.; Rodriguez, P. *Grid Converters for Photovoltaic and Wind Power Systems*; John Wiley & Sons: West Sussex, UK, 2011; Volume 29.
2. Jing, L.M.; Son, D.H.; Kang, S.H.; Nam, S.R. A Novel Protection Method for Single Line-to-Ground Faults in Ungrounded Low-Inertia Microgrids. *Energies* **2016**, *9*, 459. [CrossRef]
3. Hatziargyriou, N. *Microgrids Architectures and Control*; Wiley, IEEE-Press: Chichester, UK, 2014; pp. XXI, 317. [CrossRef]
4. Lu, X.N.; Wang, J.H. A Game Changer Electrifying remote communities by using isolated microgrids. IEEE Electrif. Mag. **2017**, *5*, 56–63. [CrossRef]
5. Lee, J.H.; Lee, S.H.; Sul, S.K. Variable-Speed Engine Generator with Supercapacitor: Isolated Power Generation System and Fuel Efficiency. *IEEE Trans. Ind. Appl.* **2009**, *45*, 2130–2135. [CrossRef]
6. Neto, M.R.B.; Carvalho, P.C.M.; Carioca, J.O.B.; Canafistula, F.J.F. Biogas/photovoltaic hybrid power system for decentralized energy supply of rural areas. *Energy Policy* **2010**, *38*, 4497–4506. [CrossRef]
7. Arent, D.; Barnett, J.; Mosey, G.; Wise, A. The potential of renewable energy to reduce the dependence of the state of Hawaii on oil. In Proceedings of the 42nd HICSS'09 Hawaii International Conference on System Sciences, Waikoloa, HI, USA, 5–8 January 2009; pp. 1–11.
8. Tong, Z.M.; Zhang, K.M. The near-source impacts of diesel backup generators in urban environments. *Atmos. Environ.* **2015**, *109*, 262–271. [CrossRef]
9. Uddin, M.N.; Rahman, M.A.; Sir, M. Reduce Generators Noise with Better Performance of a Diesel Generator Set using Modified Absorption Silencer. *Glob. J. Res. Eng.* **2016**, *XVI*. Available online: https://globaljournals.org/GJRE_Volume16/5-Reduce-Generators-Noise.pdf (accessed on 6 February 2019).
10. Prull, D.S. *Design and Integration of an Isolated Microgrid with a High Penetration of Renewable Generation*; University of California: Berkeley, CA, USA, 2008.
11. Green, T.C.; Prodanovic, M. Control of inverter-based micro-grids. *Electr. Power Syst. Res.* **2007**, *77*, 1204–1213. [CrossRef]
12. Katiraei, F.; Iravani, R.; Hatziargyriou, N.; Dimeas, A. Microgrids management. *IEEE Power Energy Mag.* **2008**, *6*, 54–65. [CrossRef]
13. Wang, X.F.; Guerrero, J.M.; Blaabjerg, F.; Chen, Z. A Review of Power Electronics Based Microgrids. *J. Power Electron.* **2012**, *12*, 181–192. [CrossRef]
14. de Matos, J.; Silva, F.; Ribeiro, L. Power Control in AC Isolated Microgrids with Renewable Energy Sources and Energy Storage Systems. *IEEE Trans. Ind. Electron.* **2014**, *62*. [CrossRef]
15. Miao, Z.X.; Xu, L.; Disfani, V.R.; Fan, L.L. An SOC-Based Battery Management System for Microgrids. *IEEE Trans. Smart Grid* **2014**, *5*, 966–973. [CrossRef]
16. Bevrani, H.; François, B.; Ise, T. *Microgrid Dynamics and Control*; John Wiley & Sons: Hoboken, NJ, USA, 2017.
17. Kim, H.; Bae, J.; Baek, S.; Nam, D.; Cho, H.; Chang, H.J. Comparative Analysis between the Government Micro-Grid Plan and Computer Simulation Results Based on Real Data: The Practical Case for a South Korean Island. *Sustainability* **2017**, *9*, 197. [CrossRef]
18. Kim, S.-Y.; Mathews, J.A. Korea's Greening Strategy: The role of smart microgrids. *Asia-Pac. J.* **2016**, *14*, 6.
19. Hwang, W.H.; Kim, S.K.; Lee, J.H.; Chae, W.K.; Lee, J.H.; Lee, H.J.; Kim, J.E. Autonomous Micro-grid Design for Supplying Electricity in Carbon-Free Island. *J. Electr. Eng. Technol.* **2014**, *9*, 1112–1118. [CrossRef]
20. Chae, W.K.; Lee, H.J.; Won, J.N.; Park, J.S.; Kim, J.E. Design and Field Tests of an Inverted Based Remote MicroGrid on a Korean Island. *Energies* **2015**, *8*, 8193–8210. [CrossRef]
21. Kim, S.-H.; Chung, I.-Y.; Lee, H.-J.; Chae, W.-K. Voltage and Frequency Control Method Using Battery Energy Storage System for a Stand-alone Microgrid. *Trans. Korean Inst. Electr. Eng.* **2015**, *64*, 1168–1179. [CrossRef]

22. Chae, W.K.; Won, J.N.; Lee, H.J.; Kim, J.E.; Kim, J. Comparative Analysis of Voltage Control in Battery Power Converters for Inverter-Based AC Microgrids. *Energies* **2016**, *9*, 596. [CrossRef]

23. Kim, Y.S.; Kim, E.S.; Moon, S.I. Frequency and Voltage Control Strategy of Standalone Microgrids with High Penetration of Intermittent Renewable Generation Systems. *IEEE Trans. Power Syst.* **2016**, *31*, 718–728. [CrossRef]

24. Kim, H.; Baek, S.; Choi, K.H.; Kim, D.; Lee, S.; Kim, D.; Chang, H.J. Comparative Analysis of On- and Off-Grid Electrification: The Case of Two South Korean Islands. *Sustainability-Basel* **2016**, *8*, 350. [CrossRef]

25. Moon, H.-J.; Chang, J.W.; Kim, E.-S.; Moon, S.-I. Frequency-based wireless control of distributed generators in an isolated microgrid: A case of Geocha Island in South Korea. In Proceedings of the 2017 52nd International Universities Power Engineering Conference (UPEC), Crete, Greece, 28–31 August 2017; pp. 1–6.

26. Won, J.; Chae, W.; Lee, H.; Park, J.; Sim, J.; Shin, C. Demonstration of remote microgrid system in Korean island. *CIRED-Open Access Proc. J.* **2017**, *2017*, 2212–2214. [CrossRef]

27. Belvedere, B.; Bianchi, M.; Borghetti, A.; Nucci, C.A.; Paolone, M.; Peretto, A. A Microcontroller-Based Power Management System for Standalone Microgrids with Hybrid Power Supply. *IEEE Trans. Sustain. Energy* **2012**, *3*, 422–431. [CrossRef]

28. Tan, K.T.; So, P.L.; Chu, Y.C.; Chen, M.Z.Q. Coordinated Control and Energy Management of Distributed Generation Inverters in a Microgrid. *IEEE Trans. Power Deliv.* **2013**, *28*, 704–713. [CrossRef]

29. Olivares, D.E.; Canizares, C.A.; Kazerani, M. A Centralized Energy Management System for Isolated Microgrids. *IEEE Trans. Smart Grid* **2014**, *5*, 1864–1875. [CrossRef]

30. Kaur, A.; Kaushal, J.; Basak, P. A review on microgrid central controller. *Renew. Sustain. Energy Rev.* **2016**, *55*, 338–345. [CrossRef]

31. Yazdanian, M.; Mehrizi-Sani, A. Distributed Control Techniques in Microgrids. *IEEE Trans. Smart Grid* **2014**, *5*, 2901–2918. [CrossRef]

32. Abdelaziz, M.M.A.; Farag, H.E. An Enhanced Supervisory Control for Islanded Microgrid Systems. *IEEE Trans. Smart Grid* **2016**, *7*, 1941–1943. [CrossRef]

33. Wu, D.; Tang, F.; Dragicevic, T.; Vasquez, J.C.; Guerrero, J.M. A Control Architecture to Coordinate Renewable Energy Sources and Energy Storage Systems in Islanded Microgrids. *IEEE Trans. Smart Grid* **2015**, *6*, 1156–1166. [CrossRef]

34. Mahmood, H.; Jiang, J. Decentralized Power Management of Multiple PV, Battery, and Droop Units in an Islanded Microgrid. *IEEE Trans. Smart Grid* **2017**. [CrossRef]

35. Solanki, A.; Nasiri, A.; Bhavaraju, V.; Familiant, Y.L.; Fu, Q. A New Framework for Microgrid Management: Virtual Droop Control. *IEEE Trans. Smart Grid* **2016**, *7*, 554–566. [CrossRef]

36. Schonbergerschonberger, J.; Duke, R.; Round, S.D. DC-bus signaling: A distributed control strategy for a hybrid renewable nanogrid. *IEEE Trans. Ind. Electron.* **2006**, *53*, 1453–1460. [CrossRef]

37. Wu, D.; Tang, F.; Dragicevic, T.; Vasquez, J.C.; Guerrero, J.M. Autonomous Active Power Control for Islanded AC Microgrids with Photovoltaic Generation and Energy Storage System. *IEEE Trans. Energy Conver.* **2014**, *29*, 882–892. [CrossRef]

38. Peyghami, S.; Mokhtari, H.; Blaabjerg, F. Autonomous Power Management in LVDC Microgrids Based on a Superimposed Frequency Droop. *IEEE Trans. Power Electr.* **2018**, *33*, 5341–5350. [CrossRef]

39. Mastromauro, R.A. Voltage control of a grid-forming converter for an AC microgrid: A real case study. In Proceedings of the 3rd Renewable Power Generation Conference (RPG 2014), Naples, Italy, 24–25 September 2014; pp. 1–6.

40. Moon, H.-J.; Chang, J.W.; Lee, S.-Y.; Moon, S.-I. Autonomous active power management in isolated microgrid based on proportional and droop control. *Energy Procedia* **2018**, *153*, 48–55. [CrossRef]

41. Kundur, P.; Balu, N.J.; Lauby, M.G. *Power System Stability and Control*; McGraw-Hill: New York, NY, USA, 1994; Volume XXIII, p. 1176.

42. Farrokhabadi, M.; Canizares, C.A.; Bhattacharya, K. Frequency Control in Isolated/Islanded Microgrids Through Voltage Regulation. *IEEE Trans. Smart Grid* **2017**, *8*, 1185–1194. [CrossRef]

43. Femia, N.; Petrone, G.; Spagnuolo, G.; Vitelli, M. Optimization of perturb and observe maximum power point tracking method. *IEEE Trans. Power Electr.* **2005**, *20*, 963–973. [CrossRef]

44. Abdullah, M.A.; Yatim, A.H.M.; Tan, C.W.A.; Saidur, R. A review of maximum power point tracking algorithms for wind energy systems. *Renew. Sustain. Energy Rev.* **2012**, *16*, 3220–3227. [CrossRef]
45. IEEE Guide for Design, Operation, and Integration of Distributed Resource Island Systems with Electric Power Systems. *IEEE Access* **2011**, 1–54. [CrossRef]

Article

Optimal Scheduling of Energy Storage Using A New Priority-Based Smart Grid Control Method

Luis Galván *, Juan M. Navarro *, Eduardo Galván *, Juan M. Carrasco * and Andrés Alcántara *

Electronical Engineering Department, University of Seville, 41092 Seville, Spain
* Correspondence: lgalvan@gte.esi.us.es (L.G.); jmnavarro@gte.esi.us.es (J.M.N.); egalvan@us.es (E.G.); jmcarrasco@us.es (J.M.C.); aalcantara@gte.esi.us.es (A.A.)

Received: 16 January 2019; Accepted: 12 February 2019; Published: 13 February 2019

Abstract: This paper presents a method to optimally use an energy storage system (such as a battery) on a microgrid with load and photovoltaic generation. The purpose of the method is to employ the photovoltaic generation and energy storage systems to reduce the main grid bill, which includes an energy cost and a power peak cost. The method predicts the loads and generation power of each day, and then searches for an optimal storage behavior plan for the energy storage system according to these predictions. However, this plan is not followed in an open-loop control structure as in previous publications, but provided to a real-time decision algorithm, which also considers real power measures. This algorithm considers a series of device priorities in addition to the storage plan, which makes it robust enough to comply with unpredicted situations. The whole proposed method is implemented on a real-hardware test bench, with its different steps being distributed between a personal computer and a programmable logic controller according to their time scale. When compared to a different state-of-the-art method, the proposed method is concluded to better adjust the energy storage system usage to the photovoltaic generation and general consumption.

Keywords: batteries; energy storage; microgrids; optimal scheduling; particle swarm optimization; power system management; smart grid; supply and demand; trade agreements

---

1. Introduction

Energy storage systems' quick development, consumers' interest in playing a more active role in the energy market, and the increasing penetration of noncontrollable renewable energy sources, which is also leading to stability concern, underline the need of a new model for the electrical system. This means we are currently in need of both management and operation methods which can be applied to the electrical system in the near future.

Several methods have been presented for this very purpose. Some of these methods attempt to control real time power market while others try to choose the optimal schedule to dispatch or receive energy. Most of the former methods rely on multiagent systems (MAS) like the so-called Power Matcher, which finds an equilibrium point on a real-time price-regulated market. According to [1,2], a Power-Matcher-driven smart city would be capable of acting as a virtual power plant. However, consumers require some dedicated appliances, such as programmable dishwashers and washing machine, in order to use power matcher. Another MAS-based method is described in [3], where two sets of priorities are used instead of prices to include more possibilities. Once again users are required to specify their power profile into dedicated intelligent devices, which act as agents. A different method is proposed in [4], involving multiple time scale markets to increase the flexibility of the auctions. This method, which is especially appropriate for prosumers, also proposes a strategy to place power biddings. Although revolutionary, all these methods would be hard to fully implement in the near future. A method that can be applied to the current market and help it evolve toward this smart market concept would be preferable.

MASs are also useful in residential areas with shared energy sources or energy storage systems (ESS). These areas can easily be considered as microgrids connected to the main grid. Agents can internally agree on the amount of power that users are to exchange, possibly creating a local market for the microgrid. In general, these methods attempt to minimize the total cost of the energy. This can be done collaboratively, as in [5], where the whole microgrid attempts to receive as much energy as possible from their own photovoltaic generation. It can also be done competitively, as in [6], where each agent represents a different user. In general, MASs help reaching agreements quickly. However, the typical consumer might be reluctant to let a machine act on their behalf on a market.

Optimal schedule methods, on the other hand, focus on finding the best possible use of ESS. A typical application of these methods is the search of the best bid for the day-ahead auction using batteries as prosumers. Appropriate optimization methods are exposed in [7] and [8], among others. Some researchers have gone one step beyond by modeling batteries and deciding their behavior according to their predicted state of charge and state of health [9]. Similar strategies have also been proposed for generation that combines controllable and noncontrollable energy sources [10–12].

Other optimal schedule methods aim to reduce electrical bills and operational costs by finding the best schedule of ESS charges and discharges along the day. Regardless of whether these methods are intended for a grid [13–17] or for a particular facility [18–22], these methods use heuristic minimization methods. It is typical to add constraints to ensure power balance and to keep the power of the ESSs (and other devices) within a realistic range. It is most common to employ either the Particle Swarm Optimization (PSO) method [15] or to design a variant of it [19–21].

The optimization requires a prediction of the generation and consumption. Some of these methods consider the uncertainty of these predictions in their objective function [14,20]. Other methods reassess the situation and repeat the optimization every certain time (usually every hour) to adjust their plan [13,15,22]. In [17], authors combine these strategies in a two-step method: a day-ahead optimization, which considers uncertainty, and several hourly reoptimizations to correct the ESS power schedule.

The main drawback of these methods is that they ultimately apply the optimized power schedule blindly. In most cases, there is no real-time control that considers the feedback from noncontrollable devices every few seconds. There are some exceptions where deviation from the optimized ESS is allowed in order to adapt "optimal" plan to contingencies:

- In [18], the ESS state of charge is optimized instead of its power. This adds some feedback regarding the ESS, but is still blind to the unpredicted behavior of other devices.
- In [13], authors consider a series of priority rules for the power generators, but do not let the optimization play with these rules to improve the found solution.
- The method from [22] does consider a real-time control, although it is limited to grid power peak prevention.

Methods have also been presented for smart buildings, attempting to minimize power consumption while maintaining the comfort. Strategies include scheduling appliances [23], prioritizing them [24,25], or allowing a small deviation of comfort variables: temperature, humidity, light, or CO_2 concentration [26,27]. The disadvantage with these methods is how subjective it is to define comfort. Constraining the usage of appliances is also inconvenient for users.

The method proposed in the present paper attempts to combine the best of both approaches: real-time power markets and optimal scheduling. It achieves better results than previous state of the art while at the same time lowering the computational cost, increasing the robustness of the strategy and, above all, achieving significant savings in the price of the electric bill. To do so, a variant of the algorithm exposed in [3], hereinafter referred to as "E-Broker", is used with optimal parameters. To prove its capabilities, the proposed method is applied to a laboratory building comprising its load, a battery, and a photovoltaic (PV) facility. These tests are run on real hardware, thus demonstrating that the method can be implemented right away. Since the original algorithm from [3] was considered for microgrids, this step brings us nearer to the smart market without requiring users to change their appliances or behavior.

2. Materials and Methods

Due to the complexity of the method, Section 2.1 describes the method in a theoretical way, without considering its embodiment. Section 2.2, on the other hand, describes the method implementation and the employed test bench.

2.1. Proposed Method

The proposed method is designed to optimally use an energy storage system on a microgrid with a load and generation system connected to the distribution grid. For the rest of this document, a battery and a PV facility will be assumed to play the energy storage and generation systems, as those were employed during the tests. Figure 1 shows the model of the microgrid.

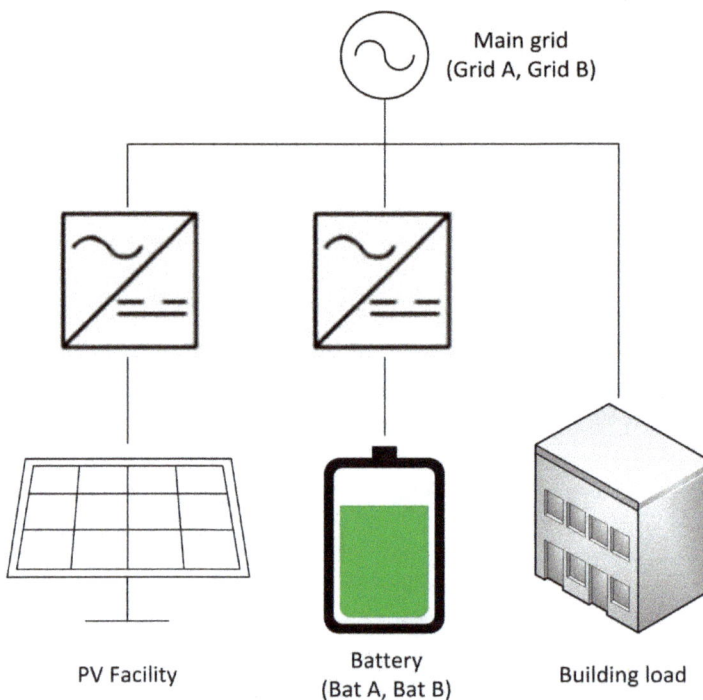

Figure 1. Microgrid model.

The method comprises four key operations: a prediction, an optimization, several power requests definitions, and an auction. The prediction and the optimization are applied once a day, while the power requests and the auction are executed continuously as a real time (RT) decision cycle.

The power requests (R), which are made on behalf of the devices of the microgrid, represent how much power each device is capable provide as a supplier (R_S) or absorb as a demander (R_D). For example, a power request is associated with the PV facility and indicates how much power it can provide. Similarly, a power request is associated with the load indicating how much power it will consume. Power requests are made for all devices, including the main grid. The auction algorithm, or E-Broker, then calculates how much power must be transferred between the different devices according to these power requests and a priority system. The power requests corresponding to the battery and main grid are split, as if they were respectively modeled as two batteries (Bat A and Bat B) and two main grids (Grid A and Grid B). The E-Broker treats these power requests with different priorities, so the whole system will behave differently depending on how these power request divisions are made. These power requests divisions (between Grid A and B, and between Bat A and B) depend on two variables: P^b and SOC^b.

The values of P^b and SOC^b are obtained through the aforementioned optimization process. At the beginning of each day, a heuristic optimization function chooses the best evolution of these variables along the day to reduce the expected electricity bill as much as possible. To do so, the optimization process considers the initial state of the battery and the electricity prices. Additionally, since the load consumption and the PV generation cannot be controlled (except for generation curtailment), the optimization process considers a prediction for the power consumed $\hat{P}_{Load}$ or generated $\hat{P}_{Pv}$ by these devices. This prediction is made just before the optimization process.

Figure 2 shows the inputs and outputs of each operation. These operations are further described below in its own subsection. The whole method has been programed in Python.

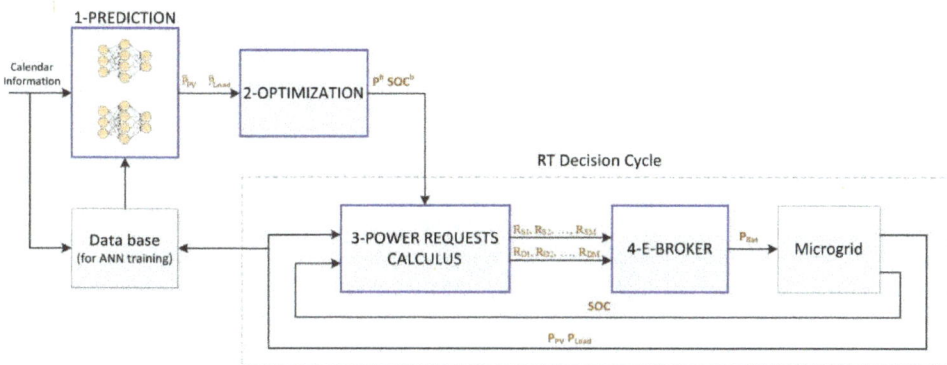

Figure 2. Proposed method.

2.1.1. Prediction

As previously stated, the optimization process needs some prediction of the uncontrolled devices power: the load and the PV facility. The more accurate the prediction is, the more reliable the optimization will be. Nevertheless, the strategy employed to obtain this prediction is not a fundamental part of the general method. The procedure described here was used to obtain the results of Section 3; other prediction methods, like the future work mentioned in Section 4, would work as well.

Predictions of loads and generation are obtained at the beginning of each day using artificial neural networks (ANNs). These predictions occur at midnight and consider only calendar information: the month, the day of the week, and the type of day (working day or holiday). For the tests, the load corresponded to a laboratory building in the University of Seville. Since the usage of the building follows a schedule, the calendar information is enough to obtain good predictions. The PV facility is harder to predict as it depends on the weather. However, calendar information is good enough for average and cloudless days. In the future, more weather-related inputs will be included to cover the possibility of cloudy days.

The ANNs used for the tests have been programmed in Python using the Keras package for Deep Learning, which runs on top of Theano (see [28] for more information about the Keras package). The activation function of the ANNs uses the rectifier activation function or Rectified Linear Unit (ReLU). The mean square error is minimized in the compilation. The "adam" efficient gradient descent algorithm is also used for its high efficiency. The number of iterations run for the training process (epochs) is 200. The number of evaluated instances before updating the weights (batch size) is 2.

Two ANNs are used for the predictions: one for the PV ($\hat{P}_{Pv}$) facility and another for the load ($\hat{P}_{Load}$). The ANNs are structured in three layers of 70, 50, and 96 neurons; with 3 inputs for calendar information and 96 outputs for the predicted quarter-hourly power values. These output power values are assumed to be the average power produced or consumed by the corresponding device during a 15-min window. All ANNs have been trained with 1-year historical data, using 70% of the data for training and the remaining 30% for evaluation.

2.1.2. Optimization

After the prediction, an optimization process is run to find the best suited evolution for P^b and SOC^b according to the prediction. These values will later serve as boundaries to determine the behavior of the grid and the battery regarding the power auction.

The selected algorithm for this optimization is the PSO, which is typical and efficient for this type of problem according to [15]. If the priorities of the E-Broker auction, which only consider relative order, needed to be optimized, then the genetic algorithm (GA) would be suitable. This is so because the GA works better with discrete variables. However, the proposed priorities are already optimal for this microgrid. Since PSO works better than GA for continuous variables, PSO is preferred.

The PSO method is imported from library pyswarm for Python. The swarm size is specified to be 2000 particles. The stop conditions are also specified: the maximum number of iterations is 200, the minimum step of the particles is 0.0001 and the minimum change of the objective function is 0.0001€/month. This way, once iterations have no real impact on the yearly electricity cost, the method stops. The default values are used for the remaining parameters, which define the movement of the particles.

Aside from the upper and lower limits of P^b and SOC^b, no constraints need to be imposed. This is so because the power requests and the auction algorithm always command the battery to exchange a feasible amount of power with another device capable to do so. The absence of constraints ensures a connected and convex space of candidate solutions, which simplifies the movement of the particles in the PSO.

The optimization variables are the necessary parameters to parametrize P^b and SOC^b. P^b is a boundary for the grid power that triggers a different behavior of the grid regarding the power requests. The grid power requests will have a higher priority as long as the grid power is below P^b. Tests indicate that the best evolution for P^b is to remain constant, so it can be defined with only one optimization variable x_p. P^b is chosen to be proportional to the contracted power P_{cont} and to the optimization variable x_p, which is bounded between 0 and 1. Although this is not expected, P^b is allowed to be greater than the contracted power.

$$P^b = 2 \cdot P_{cont} \cdot x_p \tag{1}$$

Similar to P^b, SOC^b is a boundary for the battery state of charge (SOC) that triggers different power requests. Several piecewise polynomial interpolations have been tested to parametrize the best behavior of SOC^b with the PSO selecting both the time and the SOC^b value for each point. Best results have been obtained with a lineal interpolation between 13 points for each day. The first is at midnight and its value corresponds to the actual SOC of the battery at the time of the optimization. The last point is 24 h later. Both coordinates of the remaining points, as well as the value of SOC^b for the last point, are optimization variables for the PSO algorithm. Variables that define the time coordinate of these points are bounded between 0 and 1, which correspond to the beginning and the end of the day respectively. Variables that define the SOC^b coordinates are bounded between 0.1 and 0.95, which correspond to 10% and 95% of the battery allowed SOC range.

The objective function is the total grid electricity cost for a month with 30 equal days according to Spanish tariff 3.0A. This tariff considers the time of use (TOU) for energy and power by dividing the day into three periods with different prices. On each period, the energy cost depends on the total amount of energy provided by the grid during each period while the power cost depends on the highest power peak P_{Peak} of each period. Details of these costs can be consulted on [29]. At the moment of writing this paper, the Spanish legislation does not allow consumers to sell electricity, so this situation is not considered. Nevertheless, it is possible to consider this case by adding power requests on behalf of the grid as a demander.

To evaluate the objective function, the PSO simulates the actions of the RT decision cycle (the Power requests calculus and the E-Broker auction algorithm) according to the values of P^b and SOC^b. Thus, each evaluation of the objective function requires simulating a day divided into 96 equal intervals of 15 min. Please note that the sample time of the actual RT decision cycle is 5 s. The 15-min sample time is just a simplification to be able to run so many simulations. On all these simulations, the load and the PV facility are assumed to consume or generate the previously predicted average power for each interval.

Each simulation obtains the total energy cost C_{ETotal} as

$$C_{E\ Total} = \sum_{k=1}^{96} C_{E\ k} \cdot P_{Grid\ k} \cdot 0.25h, \tag{2}$$

where $C_{E\ k}$ and $P_{Grid\ k}$ are the energy cost and the power provided by the grid during interval k. The total power cost C_{PTotal} is calculated as

$$C_{P\ Total} = \sum_{k=1}^{3} C_{P\ k} \cdot f(P_{Peak\ k}), \tag{3}$$

where $C_{P\ k}$ and $P_{Peak\ k}$ are the power cost and the maximum peak registered for period k, and function $f(P_{Peak\ k})$ is defined according to the Spanish legislation.

$$f(P_{Peak\ k}) = \begin{cases} 0.85P_{cont} \ \forall P_{Peak\ k} < 0.85P_{cont} \\ P_{Peak\ k} \ \forall P_{Peak\ k} \in [0.85P_{cont}, 1.05P_{cont}] \\ 1.05P_{cont} + 3(P_{Peak\ k} - 1.05P_{cont}) \ \forall P_{Peak\ k} > 1.05P_{cont} \end{cases} \tag{4}$$

P_{cont} is the contracted power, which can be changed once a year if the user so desires, but for the optimization it is considered to be a known constant. The value of the objective function F_{Obj} is calculated by adding the total energy cost of 30 equal days and the total power cost for the month:

$$F_{Obj} = C_{PTotal} + 30 \cdot C_{ETotal} \tag{5}$$

2.1.3. Power Requests Calculus

Power requests are calculated on real time according to the measures and optimization values. Afterwards, they are sent to the E-Broker auction algorithm, where they function similar to power bids. Several power requests may be associated with the same device. Figure 3 summarizes how the values of power requests are calculated for each device.

The actual power generated by the PV facility P_{PV} and the power consumed by the load P_{Load} are measured. A power request is made on behalf of each one (R_{SPV} and R_{DL}) to provide or receive the same power they are respectively generating or consuming. These two devices are not actually requesting permission to produce or consume such power; they will do it anyway as they are not controlled. Thus, to ensure the E-Broker auction algorithm always serves these requests, they will be given the highest priority.

The power grid is associated to two power requests: R_{SGA} and R_{SGB}. This can be understood as dividing the grid into two different suppliers (Grid A and Grid B) that, together, can provide up to the grid maximum power P_{MAX}. The value of R_{SGA} is P^b, the optimal power limit of the grid. The rest of the power the grid can provide is assigned to R_{SGB}. R_{SGA} will later have a higher priority than R_{SGB}, so an attempt will be made to limit the grid power to P_b. Nevertheless, thanks to R_{SGB}, it is possible to go beyond this limit, if absolutely necessary.

The battery is also symbolically divided into two sections: Bat A, whose charge capacity is only up to SOC^b, and Bat B, with the rest of the capacity. Each battery section will make a power request to provide power (R_{SBA} and R_{SBB}) and a power request to absorb power (R_{DBA} and R_{DBB}). These power requests depend on the actual SOC of the whole battery, which is measured by the battery management system.

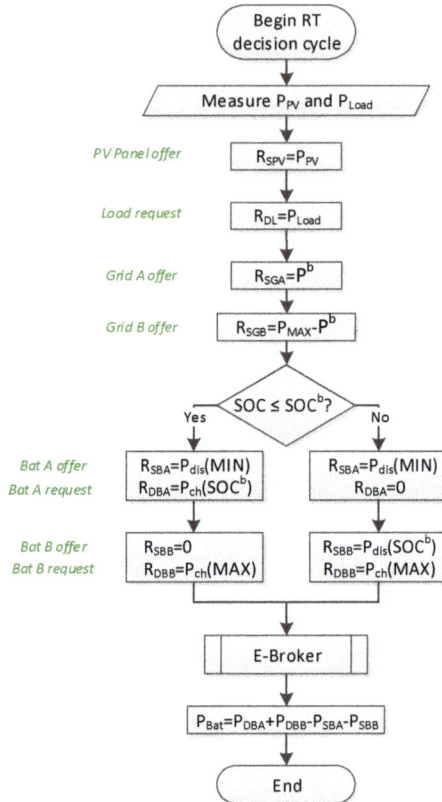

Figure 3. Flow chart of the real time (RT) decision cycle. Power requests are detailed.

When the actual SOC is lower than SOC^b, battery section A will request to absorb the necessary power $P_{ch}(SOC^b)$ to reach SOC^b as quickly as possible within the limits of the battery power capabilities. If it is possible to reach SOC^b in just one iteration of the RT decision cycle, then $P_{ch}(SOC^b)$ is the necessary power to do so; otherwise, $P_{ch}(SOC^b)$ is the battery nominal charge power. Battery section B will not request to provide any power while SOC is below SOC^b.

On the other hand, when the actual SOC is higher than SOC^b, battery section B will request to provide the necessary power $P_{dis}(SOC^b)$ to discharge back to SOC^b as quickly as possible within the battery capabilities. Again, if it is possible to reach SOC^b in one iteration, $P_{dis}(SOC^b)$ is the necessary power to do so; otherwise, it is the battery nominal discharge power. Battery section A will not request to charge any power.

Either way, with any remaining discharge power, battery section A will request to discharge to the battery minimum charge $P_{dis}(MIN)$. Similarly, battery section B will request to charge to the battery maximum charge with any available charge power left $P_{ch}(MAX)$.

Battery section A has a higher priority to charge and battery B has a higher priority to discharge. Hence, the battery SOC will tend to follow SOC^b when possible. Nevertheless, it is still possible to deviate from SOC^b, since the battery corresponding section always requests to charge or discharge to its limits.

Once all power requests have been calculated, the E-Broker auction algorithm is executed.

2.1.4. E-Broker Auction Algorithm

The algorithm used for the auction is a version of the one described in [3], where only the own priorities of the suppliers and the limit priorities of the demanders are used. It can be defined as a real time (RT) multi-agent system (MAS) that receives the power requests (R) made by M suppliers and N demanders and decides whether to address them or not according to the suppliers' priority values (O_{Si}) and demanders' priority values (O_{Dj}).

Any prosumer (such as a battery) is considered as both a supplier and a demander. In addition, as previously explained, optimized elements such as the battery and the network can be included in the auction as several participants each: Bat A, Bat B, Grid A, and Grid B.

Suppliers and demanders are sorted according to their priority values. Higher priorities are represented with greater priority values for demanders and with smaller values for suppliers. In addition, for an exchange between supplier i and demander j to occur, the priority value of the demander (O_{Dj}) must be greater than or equal to that of the suppliers (O_{Si}). This is similar to a market auction where priority values act as prices. Table 1 shows the sorted suppliers and demanders, their priority values and whether a combination is allowed (A) or forbidden (X).

Table 1. Suppliers and demanders priorities and allowed combinations.

	Supplier	**1: PV**	**2: Bat B**	**3: Grid A**	**4: Bat A**	**5: Grid B**
Demander	O_{Si} or O_{Dj}	1	3	4	6	7
1: Load	8	A	A	A	A	A
2: Bat A	5	A	A	A	X	X
3: Bat B	2	A	X	X	X	X

Devices are checked in priority order and each demander is searched for a supplier with equal or lower priority that has power available to feed it. This way, the algorithm calculates the power each supplier must provide $P_{S1} \ldots P_{SM}$ and each demander must receive $P_{D1} \ldots P_{DN}$, having a balance between the total power delivered and received. Figure 4 summarizes the E-Broker Auction Algorithm.

Since the power requests for the PV facility and the loads were based on the actual power of these devices, and the distribution grid is not controlled from the microgrid, only the battery power needs to be applied. To do so, the inverter connected to the battery is commanded to charge (or discharge) the battery at a certain power rate. The net power the battery must absorb is P_{Bat}, which is calculated from the power values selected by the E-Broker algorithm as shown in Figure 3.

$$P_{Bat} = P_{DBA} + P_{DBB} - P_{SBA} - P_{SBB} \qquad (6)$$

$$P_{Bat} = P_{D2} + P_{D3} - P_{S4} - P_{S5} \qquad (7)$$

A detailed example is provided for better clarity. Table 2 shows the power requests (R), as well as the final powers (P) each device must provide or receive.

For this example, the battery can charge or discharge at 5 kW and at the point of the shown execution of the RT cycle SOC is lower than SOC^b. Therefore, Bat A makes a request as a demander to get to SOC^b: 3 kW, which is enough to get to SOC^b by the next cycle. Bat B requests to keep charging with the remaining charge power (2 kW) in case there is an excess of PV power. Bat A also requests to discharge at up to its maximum rate (5 kW) if needed.

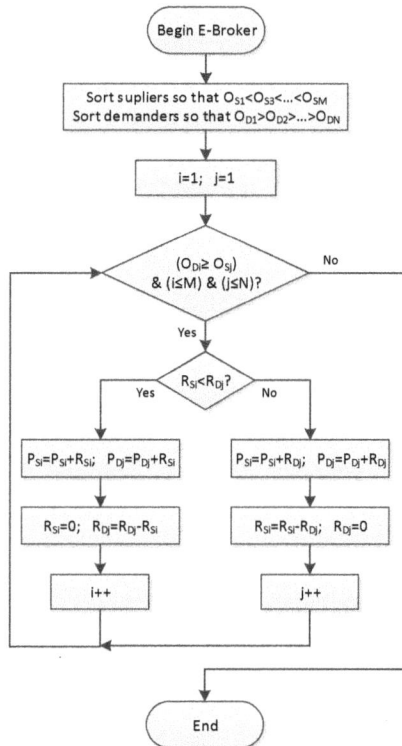

Figure 4. E-Broker Auction Algorithm flow chart.

Table 2. Algorithm example.

	Device	O_{Si}/O_{Dj}	R (kW)	P (kW)
S1	PV facility	1	4	4
S2	Bat B	3	0	0
S3	Grid A	4	8.5	8.5
S4	Bat A	6	5	0
S5	Grid B	7	10	0
D1	Load	8	12	12
D2	Bat A	5	3	0.5
D3	Bat B	2	2	0

The power request of Grid A is given by P^b which, in this example, is 8.5 kW. Due its priority, the power of Grid B will only be used if the power required by the load is greater than the sum of the powers of the PV facility, Grid A, and the battery. This will help preventing grid power peaks greater than P^b.

The E-Broker algorithm starts by checking the first demander (the Load) and the first supplier (the PV facility). It verifies that the priority of the demander is greater than the priority of the supplier, so the exchange is allowed. The load request is greater than the PV facility request, so all the power of the PV facility is assigned to the load.

The PV facility has exchanged all its power, but the Loads still request to receive 8 kW more. The E-Broker algorithm searches for the next supplier with a power request greater than 0: Grid A. It is verified that the priority of the loads is still greater than that of Grid A. Thus, the remaining power that was missing from the load is exchanged, 8 kW, leaving Grid A with 0.5 kW available to assign.

The algorithm confirmed that the priority of the next demander (Bat A) is greater than that of Grid A. The exchange is possible, so the remaining 0.5 kW of Grid A are assigned to Bat A. The remaining 2.5 kW of Bat A must stay unassigned since all other suppliers have greater priority values. There are no other demanders whose priority is greater than those of the remaining suppliers, so no more power can be assigned.

In the end, the PV facility provides 4 kW (as originally measured), the load consumes 12 kW (as measured), the battery is commanded to charge at 0.5 kW and the grid provides the difference: 8.5 kW. Please note that even though the E-Broker algorithm considers power transferring in pairs of devices (a supplier and a demander) the actual route of the power may be different. For example, the 0.5 kW used to charge the battery may actually be coming from the PV facility. The E-Broker only decides how much each device exchanges, not which device it is exchanged with.

2.2. Implementation and Test Description

A test bench has been built to validate the proposed method, as shown in Figure 5. Here, a 20-kW PV system and a 10-kWh-energy and 5-kW-power battery are employed. The load is emulated using a revertible power source and an inverter controlled by a raspberry pi (model 3B). This load emulation system is programed to consume power according to historic data from previous days.

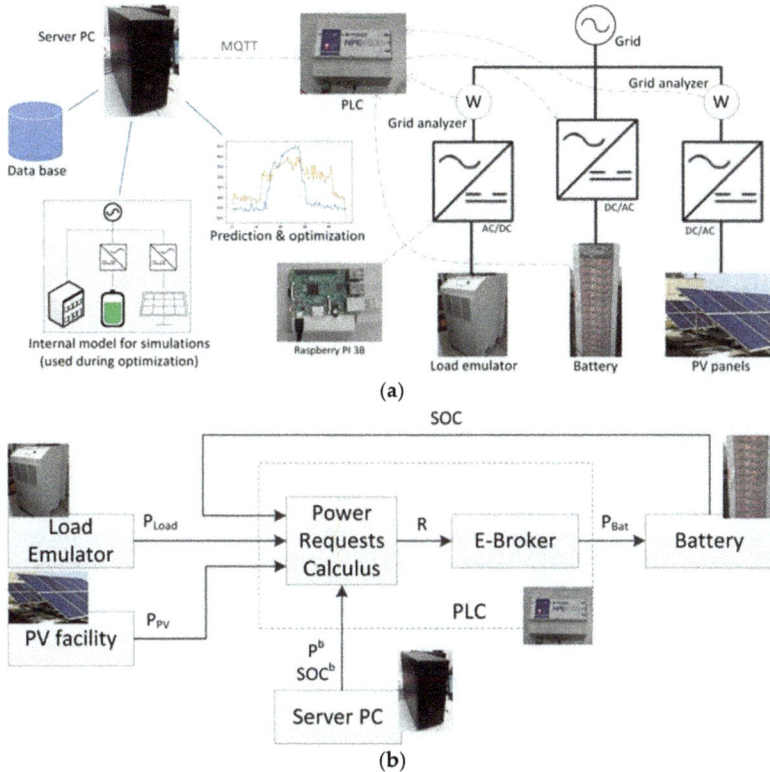

Figure 5. Schemes of the test bench. (**a**) Devices and connections scheme. Black bold solid lines represent electrical connections, gray dashed lines represent communications and blue thin solid lines represent PC functions (**b**) Control scheme of the RT decision cycle.

The proposed method is distributed among a programmable logic controller (PLC) and a personal computer (PC) acting as a server, which communicate with one another using the message queue telemetry transport (MQTT) protocol.

As shown in Figure 6, the server PC runs the aforementioned prediction and optimization at the beginning of each day. After each optimization, the PC server sends to the PLC the necessary parameters so that the latter can reconstruct P^b and SOC^b through linear interpolation.

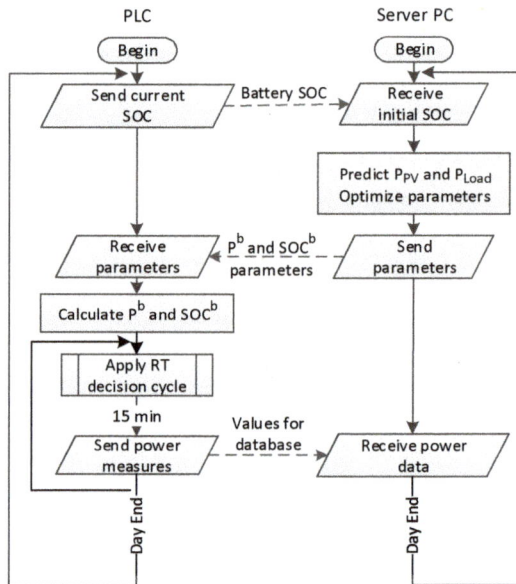

Figure 6. Combined flow chart of the programmable logic controller (PLC) and Server PC (personal computer) duties.

Using this information, the PLC receives the measured power values of the load and PV systems, runs the RT decision cycle shown in Figures 3 and 4, and commands the battery inverter to charge or discharge the battery accordingly. This RT decision cycle is executed every 5 s.

Every 15 min, the PLC reports the average PV power production, load consumption, and battery SOC for that time period. This information is stored in a data base to train the ANNs, as well as to evaluate the results of the proposed method.

The tests have been performed considering days independently. Since the power cost can only be evaluated for periods of at least one month, the grid electricity bill has been calculated considering 30 equal days. The prices used and their corresponding time periods are shown in Table 3. The objective function has been calculated considering the Contracted Power is 10 kW.

Table 3. Prices by day period.

Price	18:00–22:00	8:00–18:00 & 22:00–24:00	0:00–8:00
C_{Ek} (€/kWh)	0.018762	0.012575	0.004670
C_{Pk} (€/kWmonth)	3.384797	2.030890	1.353907

Among the referenced publications for optimal scheduling, [17,22] are the newest ones. The method from [22] aims for the same purpose and considers some real-time control. The method from [17], on the other hand, intends to control a whole distribution grid and would be harder to adapt. Thus, the proposed method is compared to [22], which can be considered the state of the art. Since the method from [22] did not originally consider PV generation, an additional adjustment has been made to it: whenever the PV generation is greater than the load power, the battery will attempt to absorb the power excess in the same way it would attempt to prevent power peaks. This can be understood as the battery avoiding a negative power peak. This adjustment always produces a better result as otherwise this extra power would be unused. The simulation is run for the same day, with the same power values for the PV generation and load consumption, and the method from [22] is provided the same prediction as the proposed method had at the beginning of the day, so that they can be compared under the same conditions.

3. Results

Figure 7 shows the results of a real test performed on 26 September 2018 in the Laboratory of Electronic Engineering Department of the University of Seville.

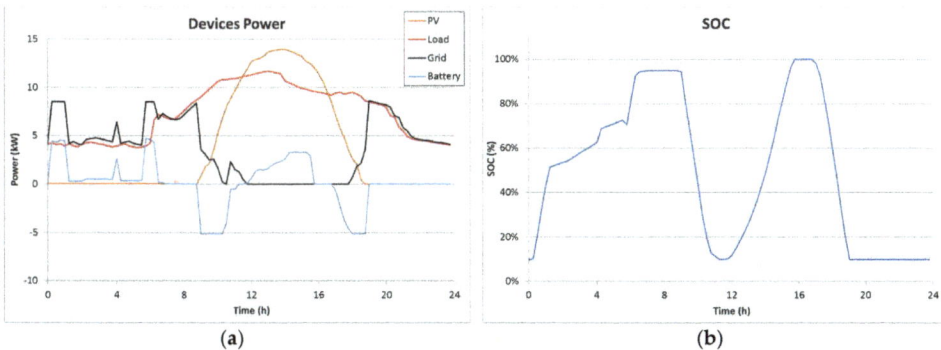

Figure 7. Real hardware tests. (**a**) Devices power. Battery and load power are considered positive when received. Grid and PV power are considered positive when delivered. (**b**) Battery state of charge (SOC).

The battery is used mainly for energy shifting. It first charges during the cheapest period, before dawn, and uses that energy at the beginning of the second period. This, along with the PV power keeps the grid power to a minimum. Afterwards, when the PV generation exceeds the load power, the battery absorbs the difference, which is used in the evening, during the most expensive period. In addition, the grid power never exceeds the 85% of the contracted power, which keeps the power cost at the minimum.

It is worth noting that the battery power follows the PV generation peaks near midday, maintaining the grid power low and stable. This is possible because the RT decision cycle acts according to the measured load and PV power. If an optimal power plan obtained from imperfect predictions had been applied on an open loop, the grid power would have oscillated around zero. Sometimes, power would have flown to the grid (which is not allowed). This exemplifies why the RT decision cycle is important.

Figure 8 shows the results of the optimization for SOC^b and compares them with the evolution of the actual SOC.

Two important details should be appreciated in this figure. The first one is how the battery SOC tends to follow SOC^b during the first half of the day only. As previously explained, the priorities are chosen so that SOC follows SOC^b when possible, so the first half of the day demonstrates that the optimization has an impact on the trend of the SOC. However, after midday, there is a surplus of PV generation, which the battery must absorb due to the specified priorities. Again, the RT decision cycle overruns the optimization plan so that the PV power surplus is not wasted.

The second detail to note is how all the points that describe SOC^b are before midday. As explained, the optimization process selects both coordinates of these points. When the predictions are considered, the optimization finds no benefit in adding points to the second half of the day. After all, according to the predictions, the battery behavior during the evening will not depend on SOC^b, but on the PV power surplus. Consequently, the PSO uses all the points to produce and parametrize the optimal trend of SOC^b during the morning.

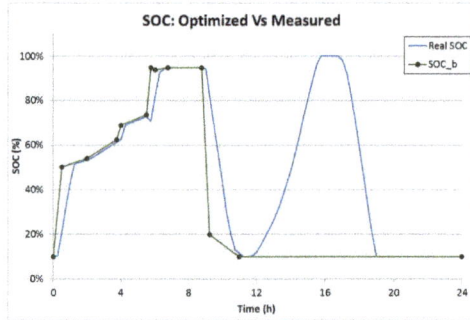

Figure 8. Optimized SOC^b and actual SOC evolution through the day.

Figure 9 shows the predicted and real power values for the PV facility and load respectively. As previously stated, these predictions were obtained using artificial neural networks whose input solely consists of calendar information.

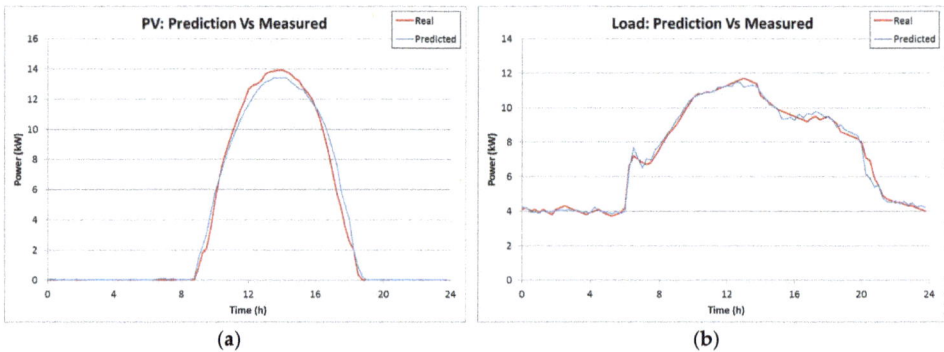

Figure 9. Predictions and actual values for (**a**) the PV generation and (**b**) the loads.

It is important to note that the predictions are similar to the reality, but not equal. The fact that the general shape of each curve is correct is enough to let the PSO find the optimal plan for SOC^b. The small differences will be corrected during by the RT decision cycle. Nevertheless, the predictions were quite accurate the day of the test.

Figure 10 shows the simulation results for the method from [22] applied to the same day.

Although the power peaks are still prevented and the battery power follows the PV generation, the energy shifting is much poorer. The battery even charges and discharges on the same tariff period with no real benefit (and with the consequent losses due to its efficiency). Since the battery is empty when the load power rises, there is an unavoidable peak which slightly increases the power cost. The filling of the battery after midday occurs due to the prevention of a (negative) power peak on the grid, and would also have occurred regardless of the optimization result.

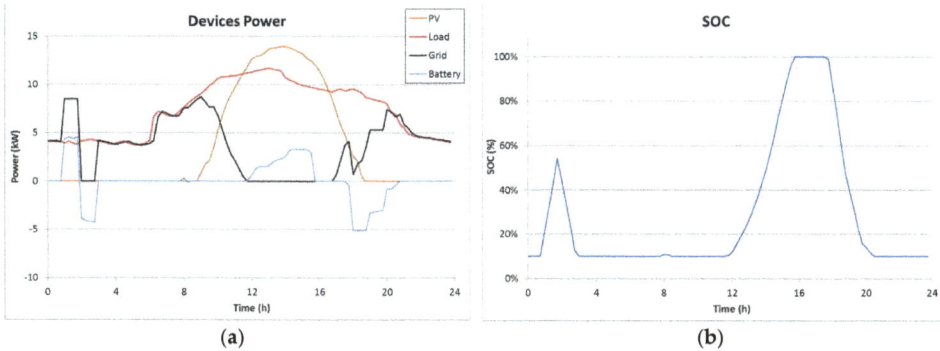

Figure 10. Simulations results for the state-of-the-art method: (**a**) Devices power; (**b**) Battery SOC.

In general, the optimization in this method is not as good as in the presented one. It might have done better with a higher number of particles or longer time. Since both methods used the same parameters to define the PSO, it can be concluded that the proposed method converges better. Considering the proposed method only requires one optimization at the beginning of the day, while method from [22] requires an optimization each hour, this is an important advantage.

Table 4 shows the values of the electrical bill for both methods so that they can be compared with ease. Additionally, the case where no battery is present has been added to the table so that both methods can be compared with the absence of any method. As shown, the proposed method saves 4.41% over the state-of-the-art method, and 8.12% over the original cost. It is also worth noting that the method from [22] cannot save any power cost with peak shaving in this situation.

Table 4. Cost comparison.

Method	Energy Cost (€/month)	Power Cost (€/month)	Total Cost (€/month)
Proposed method	23.3472	57.54155	80.8887
Method from [22]	26.8039	57.81233	84.6162
Without battery	30.2297	57.81233	88.0420

4. Discussion

The most evident result is the fact that the proposed method can reduce the grid power and energy cost more than the previous method. The strategy of the proposed method is shown to be more consistent, while the successive optimizations of the method from Ref. [22]'s alternative tend to produce an erratic behavior, with the battery charging and discharging on the same price period for no apparent reason. This occurs because the optimization problem is badly conditioned. The grid peak limit, which is recalculated on every hourly optimization, conflicts with the remaining optimization variables. All variables represent power (whether from the battery or from the grid) and the RT control applies only the limiting one. Consequently, some of the variables have no impact on the objective function while others compete with one another to produce any impact. In contrast, the method proposed in the present paper is capable of reallocating the points that define SOC^b, so that all variables have some impact on the objective function. This helps particles of the PSO find a unique solution rather than iterating along a family of very similar solutions.

In addition, the PV generation and load behavior allows for several quasi optimal alternatives and the hourly optimization keeps changing the strategy constantly. While the optimal battery power plan (obtained in [22]) is heavily dependent on the PV generation and load consumption, the optimal SOC boundary (obtained according to the proposed method) is much more robust and stable. This is so because the SOC, being the integral of the battery power, acts as a low pass filter. Methods that optimize the power of the battery (instead of the SOC) will find similar difficulties, particularly if they must optimize periodically to reassess the plan.

It is possible to add constraints to the PSO that would prevent the battery from discharging and charging in the same tariff period. For example, the total number of battery power changing signs could be limited. However, adding many constraints produces non-convex search spaces for the particles, which makes it harder to find the solution. An excess of constraints could even isolated particles from one another if the search space becomes non-connected. The proposed method does not require any constraint except for the limits of each variable (typically normalized between 0 and 1). This helps the convergence speed.

The lack of constraints is possible due to the way the power requests are calculated. Regardless of the optimization results, no device in the microgrid is ever forced to produce or absorb an impossible amount of power. For example, if for any reason the load consumption was cut unexpectedly, the battery would not provide more power. In fact, even if the prediction or the optimization is badly performed, the battery will not be commanded to provide or exchange power if there is no other device to exchange it with.

Optimizing only the boundaries of the battery behavior, as proposed is another advantage as it allows the battery to absorb or provide unexpected power to prevent peaks or avoid power losses. The battery can act as a dynamic reserve both for peak shaving and energy shifting. This, combined with the priority system from the E-Broker Auction Algorithm is responsible for the robustness of the method. This is important especially because the predictions only consider calendar information (and not the weather forecast) and thus may not be perfectly accurate. This proves that the proposed method can cope with some unpredicted situations.

Nevertheless, since the predictions must be accurate enough for the PSO to find a general plan for the SOC^b, future versions of the proposed method will consider weather forecast information. This information is expected to include humidity, clarity of the sky, and temperature of every hour. Currently, the proposed method does not require hourly optimizations, so the strategy is maintained coherent for longer periods of time and computational costs are lowered. However, authors consider the possibility of optimize again midway through the day if predictions are found to deviate too much from measured power values. This has not been implemented yet but will be tried in future versions.

Another aspect of the proposed method that shares with the method from [22] is the fact that no division needs to be imposed on the battery to distinguish peak shaving from energy shifting. The proposed method optimizes the appropriate variables (SOC^b for energy shifting and P^b for peak shaving) so that no capacity needs to be wasted.

In the future, other applications for the proposed method are expected, such as demand control by price or frequency response. Although this will require predicting and optimizing additional magnitudes, the general concept of the method will remain intact: long-term optimization and real time decision making. These new applications, just like the presented one, will save storage costs because the method will provide the best possible use for the batteries. As for the exposed application, since the power and energy cost reflects how much power the main grid is moving at any given hour, reducing the energy bill helps decongesting the grid during the peak hours.

Finally, an important aspect of the proposed method is the fact that the E-Broker auction algorithm is based on a method originally designed for microgrids operation with distributed resources. Consequently, the proposed method is an evolutionary step towards the smart grids and distributed paradigm that can be applied in the present time. This serves as an intermediate state between the current energy system situation and the utopic approaches from other publications (see [1–4] for examples). We hope the proposed method helps us reach a new and better future.

5. Patents

International patent application WO2015/113637A1 results from part of the work reported in this manuscript. The described E-Broker auction algorithm is based on a simplification of the method disclosed in the patent. Please note that the rest of the proposed method is original and helps improving the performance of the patented method.

Author Contributions: Conceptualization, L.G., E.G., and A.A.; Software and data curation, J.M.N.; Investigation, L.G., J.M.N., and A.A.; Resources, E.G. and J.M.C.; Methodology, Validation and formal analysis, all authors; Writing—original draft preparation, L.G. and J.M.N.; Writing—review and editing, E.G., J.M.C., and A.A.; Visualization, all authors.; Supervision, E.G., J.M.C., and A.A.; Project administration and funding acquisition, E.G. and J.M.C.

Funding: This research was funded partially by the European Commission for the Seventh Framework Programme under grant agreement ID 771066 and partially by the Spanish Ministry of Economy and Competitiveness under, grants ENE2016-80025-R and RTC-2016-5634-3.

Acknowledgments: The starting point of this research were the results of the projects Pollux and IoE, which were funded by the European Commission for the Seventh Framework Programme under grant agreement IDs 100205 and 269374. Green Power Technologies SL is also acknowledged for supporting the experiments with previous demonstrators.

Conflicts of Interest: The authors declare no conflict of interest. The funders had no role in the design of the study; in the collection, analyses, or interpretation of data; in the writing of the manuscript, or in the decision to publish the results.

References

1. Bliek, F.; Noort, A.; Roossien, B.; Kamphuis, R.; Wit, J.; Velde, J.; Eijgelaar, M. PowerMatching City, a living lab smart grid demonstration. In Proceedings of the 2010 IEEE PES Innovative Smart Grid Technologies Conference Europe (ISGT Europe), Gothenberg, Sweden, 11–13 October 2010. [CrossRef]
2. Kamphuis, I.G.; Eijgelaar, B.R.M.; de Heer, H.; van de Velde, J.; van den Noort, A. Real-time trade dispatch of a commercial VPP with residential customers in the PowerMatching City SmartGrid living lab. In Proceedings of the 22nd International Conference and Exhibition on Electricity Distribution, Stockholm, Sweden, 10–13 June 2013. [CrossRef]
3. Martín, P.; Galván, L.; Galván, E.; Carrasco, J.M. System and Method for the Distributed Control and Management of a Microgrid. Patent No. WO2015113637, 6 August 2015.
4. Flikkema, P.G. A Multi-Round Double Auction Mechanism for Local Energy Grids with Distributed and Centralized Resources. In Proceedings of the 2016 IEEE 25th International Symposium on Industrial Electronics (ISIE), Santa Clara, CA, USA, 8–10 June 2016.
5. Raju, L.; Rajkumar, I.; Appaswamy, K. Integrated Energy Management of Micro-Grid using Multi Agent System. In Proceedings of the 2016 International Conference on Emerging Trends in Engineering, Technology and Science (ICETETS), Pudukkottai, India, 24–26 February 2016. [CrossRef]
6. Nunna, H.K. An Agent based Energy Market Model for Microgrids with Distributed Energy Storage Systems. In Proceedings of the 2016 IEEE International Conference on Power Electronics, Drives and Energy Systems (PEDES), Trivandrum, India, 14–17 December 2016. [CrossRef]
7. Mohsenian-Rad, H. Optimal Bidding, Scheduling, and Deployment of Battery Systems in California Day-Ahead Energy Market. *IEEE Trans. Power Syst.* **2016**, *31*, 442–453. [CrossRef]
8. de la Nieta, A.A.S.; Tavares, T.A.M.; Martins, R.F.M.; Matias, J.C.O.; Catalão, J.P.S.; Contreras, J. Optimal Generic Energy Storage System Offering in Day-Ahead Electricity Markets. In Proceedings of the 2015 IEEE Eindhoven PowerTech, Eindhoven, The Netherlands, 29 June–2 July 2015. [CrossRef]
9. Zhang, Z.; Wang, J.; Wang, X. An improved charging/discharging strategy of lithium batteries considering depreciation cost in day-ahead microgrid scheduling. *Energy Convers. Manag.* **2015**, *105*, 675–684. [CrossRef]
10. Wang, R.; Wang, P.; Xiao, G. A robust optimization approach for energy generation scheduling in microgrids. *Energy Convers. Manag.* **2015**, *106*, 597–607. [CrossRef]
11. Mohamed, F.A.; Koivo, H.N. Multiobjective optimization using Mesh Adaptive Direct Search for power dispatch problem of microgrid. *Int. J. Electr. Power Energy Syst.* **2012**, *42*, 728–735. [CrossRef]
12. Hosseinnezhad, V.; Rafiee, M.; Ahmadian, M.; Siano, P. Optimal day-ahead operational planning of microgrids. *Energy Convers. Manag.* **2016**, *126*, 142–157. [CrossRef]
13. Morais, H.; Kádár, P.; Faria, P.; Vale, Z.A.; Khodr, H.M. Optimal scheduling of a renewable micro-grid in an isolated load area using mixed-integer linear programming. *Renew. Energy* **2010**, *35*, 151–156. [CrossRef]

14. Sousa, T.; Morais, H.; Vale, Z.; Faria, P.; Soares, J. Intelligent energy resource management considering vehicle-to-grid: A simulated annealing approach. *IEEE Trans. Smart Grid* **2012**, *3*, 535–542. [CrossRef]

15. Yang, Y.; Zhang, W.; Jiang, J.; Huang, M.; Niu, L. Optimal Scheduling of a Battery Energy Storage System with Electric Vehicles' Auxiliary for a Distribution Network with Renewable Energy Integration. *Energies* **2015**, *8*, 10718–10735. [CrossRef]

16. Faria, P.; Soares, J.; Vale, Z.; Morais, H.; Sousa, T. Modified Particle Swarm Optimization Applied to Integrated Demand Response and DG Resources Scheduling. *IEEE Trans. Smart Grid* **2013**, *4*, 606–616. [CrossRef]

17. Wang, Z.; Negash, A.; Kirschen, D.S. Optimal scheduling of energy storage under forecast uncertainties. *IET Gener. Transm. Distrib.* **2017**, *11*. [CrossRef]

18. Yoon, Y.; Kim, Y.-H. Charge Scheduling of an Energy Storage System under Time-of-Use Pricing and a Demand Charge. *Sci. World J.* **2014**, *2014*, 937329. [CrossRef] [PubMed]

19. Jia, X.; Li, X.; Yang, T.; Hui, D.; Qi, L. A Schedule Method of Battery Energy Storage System (BESS) to Track Day-Ahead Photovoltaic Output Power Schedule Based on Short-Term Photovoltaic Power Prediction. In Proceedings of the International Conference on Renewable Power Generation (RPG 2015), Beijing, China, 17–18 October 2015. [CrossRef]

20. Lee, T.-Y. Operating Schedule of Battery Energy Storage System in a Time-of-Use Rate Industrial User with Wind Turbine Generators: A Multipass Iteration Particle Swarm Optimization Approach. *IEEE Trans. Energy Convers.* **2007**, *22*, 774–782. [CrossRef]

21. Choi, Y.; Kim, H. Optimal Scheduling of Energy Storage System for Self-Sustainable Base Station Operation Considering Battery Wear-Out Cost. *Energies* **2016**, *9*, 462. [CrossRef]

22. Kim, S.-K.; Kim, J.-Y.; Cho, K.-H.; Byeon, G. Optimal Operation Control for Multiple BESSs of a Large-Scale Customer Under Time-Based Pricing. *IEEE Trans. Power Syst.* **2018**, *33*, 803–816. [CrossRef]

23. Celik, B.; Roche, R.; Bouquain, D.; Miraoui, A. Decentralized neighborhood energy management with coordinated smart home energy sharing. *IEEE Trans. Smart Grid.* **2017**, 1–10. [CrossRef]

24. Shah, S.; Khalid, R.; Zafar, A.; Hussain, S.M.; Rahim, H.; Javaid, N. An optimized priority enabled energy management system for smart homes. In Proceedings of the 2017 IEEE 31st International Conference on Advanced Information Networking and Applications (AINA), Taipei, Taiwan, 27–29 March 2017. [CrossRef]

25. Li, V.; Logenthiran, W.T.; Woo, W.L. Intelligent multi-agent system for smart home energy management. In Proceedings of the 2015 IEEE Innovative Smart Grid Technologies-Asia (ISGT ASIA), Bangkok, Thailand, 3–6 November 2015; pp. 1–6. [CrossRef]

26. Wang, L.; Wang, Z.; Yang, R. Intelligent multiagent control system for energy and comfort management in smart and sustainable buildings. *IEEE Trans. Smart Grid* **2012**, *3*, 605–617. [CrossRef]

27. Hurtado, L.A.; Nguyen, P.H.; Kling, W.L. Smart grid and smart building interoperation using agent-based particle swarm optimization. *Sustain. Energy Grids Netw.* **2015**, *2*, 32–40. [CrossRef]

28. Keras: The Python Deep Learning Library. Available online: https://keras.io/ (accessed on 17 January 2018).

29. Spain Ministry of Economy. Real Decreto 1164/2001, de 26 de octubre, por el que se establecen tarifas de acceso a las redes de transporte y distribución de energía eléctrica; Boletín Oficial del Estado, BOE-A-2001-20850, No. 268. November 2001, pp. 40618–40629. Available online: https://www.boe.es/buscar/doc.php?id=BOE-A-2001-20850 (accessed on 17 January 2018).

Article

A Horizon Optimization Control Framework for the Coordinated Operation of Multiple Distributed Energy Resources in Low Voltage Distribution Networks

Konstantinos Kotsalos [1,*,†], **Ismael Miranda** [2], **Nuno Silva** [3] **and Helder Leite** [4]

[1] Efacec, Division of Smart Grids, 4471-907 Porto, Portugal
[2] Efacec, Division of Storage, 4471-907 Porto, Portugal; ismael.miranda@efacec.com
[3] Efacec, Division of T&I, 4466-952 Porto, Portugal; nuno.silva@efacec.com
[4] Faculty of Engineering (FEUP), University of Porto, 4200-465 Porto, Portugal; hleite@fe.up.pt
*   Correspondence: konstantinos.kotsalos@efacec.com; Tel.: +35-196-007-8114
†   Current address: Via de Francisco Sá Carneiro Apartado 3078, 4471-907 Moreira da Maia, Porto, Portugal.

Received: 27 February 2019; Accepted: 21 March 2019; Published: 26 March 2019

Abstract: In recent years, the installation of residential Distributed Energy Resources (DER) that produce (mainly rooftop photovoltaics usually bundled with battery system) or consume (electric heat pumps, controllable loads, electric vehicles) electric power is continuously increasing in Low Voltage (LV) distribution networks. Several technical challenges may arise through the massive integration of DER, which have to be addressed by the distribution grid operator. However, DER can provide certain degree of flexibility to the operation of distribution grids, which is generally performed with temporal shifting of energy to be consumed or injected. This work advances a horizon optimization control framework which aims to efficiently schedule the LV network's operation in day-ahead scale coordinating multiple DER. The main objectives of the proposed control is to ensure secure LV grid operation in the sense of admissible voltage bounds and rated loading conditions for the secondary transformer. The proposed methodology leans on a multi-period three-phase Optimal Power Flow (OPF) addressed as a nonlinear optimization problem. The resulting horizon control scheme is validated within an LV distribution network through multiple case scenarios with high microgeneration and electric vehicle integration providing admissible voltage limits and avoiding unnecessary active power curtailments.

Keywords: low voltage networks; multi-period optimal power flow; multi-temporal optimal power flow; active distribution networks; unbalanced networks

---

1. Introduction

Nowadays, an increasing number of small-scale units, typically referred to Distributed Energy Resources (DER), is connected along the Low Voltage (LV) distribution networks posing several technical challenges, whilst bringing novel and diverse opportunities. Most commonly there is already a large share of injected power at the distribution level by the renewable energy merely based on solar energy through Photovoltaic (PV). The connection of such resources at the LV grid and end-users' premises is foreseen to increase substantially in the close future with small rooftop installation usually coupled with Battery Storage System (BESS), controllable loads (e.g., Electric Water Heaters) and Electric Vehicles (EV). Therefore, it shall be critical to exploit the DER controllability and active participation through demand response schemes in order to support or even improve techno-economically the operation of the distribution networks operation [1,2]. For instance, DERs might be used to provide technical support to tackle voltage or congestion problems delivering profits to the end-consumers accordingly.

Traditionally, LV networks used to be the most passive circuits within the power systems, since power flows were solely headed from distribution transformers to consumers without the operation of automation elements [3]. In particular, the entire segment from the secondary substation and its downstream connected LV networks is very often not monitored nor controlled [4].

The Distribution System Operators (DSOs) address such technical challenges by increasing the observability and controllability of the grids, envisioning the active management of the DERs for ancillary services, throughout new operation stages [5,6]. Such attributes of advanced control and monitoring techniques do typically refer to Advanced Distribution Management Systems (A-DMS) [7]. The active management of distribution networks through the engagement of DERs in the operation of the grid is regarded to occur with the provision of flexibilities services such as active and reactive power control (i.e., inverter based control). The smart grid deployment, as an alternative paradigm for the operation of distribution networks, envisions the active management of DER taking advantage of advanced control infrastructures and communications through demand side management schemes. Advanced control methodologies need to be implemented to determine control actions related to controllable DER, which can techno-economically improve distribution networks' operation delivering benefits to residential users.

Virtual Power Plants and aggregation of flexibilities in higher levels for market participation (i.e., in balancing market) and provision of ancillary services through DER has been foreseen in [8,9]. In several European countries already, flexible loads are incentivized in order to defer network investments or decrease congestions by load shifting [1].

A particular concern in recent research works regards the potential flexibility provided by the DER connected along the LV distribution level to address voltage problems. In past years, research was focused on active power curtailment following local droop based rules or even combined with reactive control [10–12]. Moreover, self-consumption is a common practice adopted by the DSOs lately, in order to avoid voltage rise during the peak period of PV generation. In several European countries (e.g., Belgium, Denmark and the Netherlands), residential PV self-consumption measures based on net metering schemes target matching the endogenously generated power with local consumption [13]. On the contrary, Portugal and Germany promote lower remuneration for energy produced by microgeneration, thus attracting instantaneous consumption. Towards the path to maximizing renewable generation into distribution networks, the focus in research remains in controlling the microgeneration itself. A distributed scheme with more sophisticated rules to mitigate overvoltages due to a large integration of PVs was proposed in [14]. In addition, real-world LV four-wire distribution networks are in practice fairly unbalanced, since single-phase grid elements (e.g., end-users, micro-generation and EVs) do impact the voltage, not only of the connected phase, but also of the other two phases due to the neutral-point shifting phenomenon [15]. Consequently, local droop based controls via single-phase PVs might be insufficient for voltage regulation in unbalanced grids; hence, the deployment of centralized optimization schemes can tackle such issues since topological (i.e., making use of the most efficient controllable DER or asset) and phase coupling considerations among phases can be regarded [11,14].

Interest has also been attended for the efficient integration and exploitation of distributed BESS [16,17]. Recently, BESS has been introduced by electric utilities to accommodate the increased generated power by solar energy in LV grids, though the deployment of residential BESSs has been limited up to the previous years, due to the relatively high capital cost of such devices. Lately, the continuous reducing cost of batteries in addition to the rising electricity costs and incentives for investments in storage [18]. According to Directive 2009/72/EC [19], the utilization of energy storage systems by grid operators is very limited at present; meanwhile, unbundling requirements for DSOs under EU directives do not allow energy storage units to be directly owned, or controlled by them. Concurrently, the growing number of BESS owned by residential consumers is likely to undermine the current business model of the electric utilities [20]. The following trend aims to maximize the revenue brought to the consumer in particular when home energy management systems are utilized to optimize the local generation and consumption.

Other research works have proposed advanced controlling more DER types such as EV and Controllable Loads (CL) for the mitigation of overvoltages or line congestions by [21–26]. A centralized control scheme for the voltage regulation and the mitigation of high unbalance instances is proposed [27], the efficiency of which is compared with typical local control droops. An extension of the same authors provide a framework for the coordination of an On Load Tap Changer (OLTC), installed at the secondary substation, with BESS and controllable microgeneration [20].

Optimal Power Flow (OPF) is widely applied as a tool within DMS application for the planning and operation of the power systems. Clearly, OPF problems are deemed challenging since they require solving of non-convex problems. Nonconvexity stems from the nonlinear relationship between voltages and the complex powers consumed or injected at each node [28]. Further adaptations and assertions have to regarded for power flow equations in particular for LV grids as they present purely unbalanced loading conditions and mainly resistive line characteristics. The widely used DC power flow methodologies in transmission grid studies, but cannot be applied due to the higher R/X ratios [29]. The application of non-convex and nonlinear AC power flows in an OPF framework possibly leads to computational complexity according to [30]. Therefore, in literature, convex relaxations are settled, based on e.g., semidefinite relaxations [31]; such approaches explore solutions that are globally optimal for the original problem in many practical cases, leading though non-valid solutions in some cases [28]. Further computational complexity can be certainly added to the OPF formulation if it is settled for multiple periods leading to multi-period OPF.

Recent works have dealt with proposing efficient linearizations to resort tractable multi-period OPF [6,17]. For instance, authors in [17] take advantage of the linearization to reduce the convergence of the programming and utilize it for planning of the distribution network. In [6], a step forward advances the multiperiod OPF framework incorporating uncertainties brought by forecasts through chance constrained optimization. Nevertheless, both works do not address the three-phase nature and the subsequent unbalances may arise in LV distribution networks. In this work, a three-phase multi-period OPF based on the exact (i.e., nonlinear) AC power flows is proposed, incorporating multiple DER within the operation of the distribution grid.

The main contributions of this paper can be summarized as follows:

- A decision tool which provides support to the DSO for the minimization of the operational costs based on the coordinated operation of multiple DER. The tool is capable of mitigating the regulating of nodal voltages, minimizing curtailments of active power by the microgeneration, ensuring nominal rated power for the secondary for all time instances. Multiple active measures are posed based on different DER technologies.
- A three-phase multi-period OPF framework based on the exact formulation of the AC power flow equations. The overall problem is resolved through a nonlinear optimization problem addressed interior-point method where efficient explicit calculation for the gradients of the constraints and the Hessian of the Lagrangian are proposed leaning on sparsities.
- Analytical inter-temporal constraints (i.e., providing the limitations of each type of DER) and the counterpart inter-temporal cost dependencies are discussed with their subsequent burdens. In particular, a technique is proposed to address singularity of Jacobian matrix (i.e., of the nonlinear problem) induced by the inter-temporal constraints.

This paper is structured as follows: Section 2 describes the mathematical models of the power system as well as the temporal models deployed for the DER. Section 3 discusses the proposed coordinated control scheme, which is mathematically stated through a three-phase Multiperiod AC-OPF (MACOPF). An analytical discussion regarding the resolution of the nonlinear programming through the Interior-Point (IP) primal-dual algorithm takes places in the same section. Sections 4 and 5 present the study for the validation of the proposed control scheme. The final section provides the conclusions.

2. Proposed Models for Distributed Energy Resources and Distribution Network Models

2.1. Distribution Network Models (Lines and Transformer)

The LV distribution network is assumed to be comprised by a typical three-phase four wire unbalanced network with a multi-earthed neutral; this fact allows the application of the Kron's reduction [32]. All buses are considered to have three terminals, where each one represents the phase connection point: a, b, c. The voltage magnitude for bus j is given by the real vector $v_j \in \mathbb{R}^3$, $v_j = [v_{j,a}, v_{j,b}, v_{j,c}]^T$, and, accordingly, the voltage angles by the real vector $\vartheta_j \in \mathbb{R}^3$.

The connection between buses k and m is a square (e.g., Kron's reduction) symmetric matrix $z_{k,m} \in \mathbb{C}^{\Phi_{k,m} \times \Phi_{k,m}}$, where $\Phi_{k,m}$ the number of phases interconnected nodes k and m. The active conductors (e.g., the three-phases, neutral follows the reduction) present coupling, amongst them the $[z_{k,m}]$ has off-diagonal elements different from 0. The admittance matrix ($Y_{bus} \in \mathbb{C}^{3N_b \times 3N_b}$) determines the overall topological structure of the distribution network, where assuming no shunt admittances modeled for the distribution lines (i.e., negligible according to [33]), the element $Y_{k_{p_k}, m_{p_m}}$ of Y_{bus} which refers to the connection between phase p_k of bus k and phase p_m of m can be expressed as:

$$(Y_{k_{p_k}, m_{p_m}}) = \begin{cases} -\frac{1}{(z_{k,m})_{p_k,p_m}} & \text{if } k \neq m, \\ \sum_n \frac{1}{z_{k,n}} & \text{if } k = m. \end{cases}$$

The distribution transformer (MV/LV) according to its type most commonly delta–wye grounded, wye–delta, wye–wye, open-wye–open-delta, delta–delta can be represented and included in the Y_{bus} of the network by constant impedances (i.e., for steady state analysis). For each type of transformer configuration, the admittance matrices for the distribution transformer can be found in [34]. Building the admittance matrix of the transformer is commonly known that it is not always invertible other than wye-g–wye-g; hence, an addition of a fictitious small admittance from the isolated transformer sides to the ground remedies the issue [35].

The interconnection of the LV grid with the MV grid is represented with an ideal voltage source that is assigned with the voltage vector $V_{source} = [1, 1, 1]^T$, which is also considered as the reference bus with angles $\underline{/V_{source}} = \left[0, \frac{2\pi}{3}, -\frac{2\pi}{3}\right]^T$. The ideal voltage source is connected in series with proper impedances that can be assessed according to the MV grid's short circuit power (Figure 1).

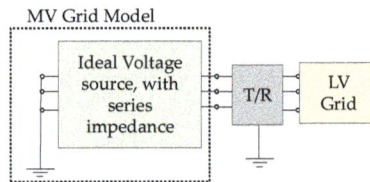

Figure 1. Configuration for interconnection of MV with the LV grid.

2.2. Battery Energy Storage System (BESS)

The BESS model is based on a first order battery model. Two distinct auxiliary variables are settled as power injections. The positive term refers to the discharging mode of operation $p_{dch} \geq 0$, $p_{dch} \in \mathbb{R}_+$, while the charging of the storage unit is negative $p_{ch} \leq 0$, $p_{ch} \in R_-$. This model captures the losses during charging and discharging modes, through the corresponding efficiencies (η_{ch}, η_{dch}). $\mathcal{E}_0$ is considered the initial (i.e., $\tau = 0$) stored energy of the BESS. The available energy capacity of a BESS at time step τ can calculated by Equation (1), which bundles the instant energy state with the former one:

$$\mathcal{E}(\tau) = \mathcal{E}(\tau - 1) - \Delta\tau \begin{bmatrix} \eta_{ch} & \eta_{dch} \end{bmatrix} p(\tau), \tag{1}$$

where

$$p(\tau) = \begin{bmatrix} p_{ch}(\tau) \\ p_{dch}(\tau) \end{bmatrix}.$$

The energy state for the BESS for the last step will be accordingly defined as a linear combination with the previous states of its energy. As it will be presented in Section 3.4, such inter-temporal couplings led to problematic conditions for the resolution of an optimal control, hence special treatment is proposed. In the proposed optimization framework, as a primary decision variable for each BESS is considered, its power injection P_{BESS}, which should be subjected to the equality constraint (2a) for each instant τ, followed by some operational constraints for both operation modes as follows:

$$P_{BESS}(\tau) = p_{ch}(\tau) + p_{ch}(\tau), \tag{2a}$$

$$\overline{P_{ch}} \leq p_{ch}(\tau) \leq 0, \tag{2b}$$

$$0 \leq p_{dch}(\tau) \leq \overline{p_{dch}}, \tag{2c}$$

$$\underline{SoC} \leq SoC(\tau) \leq \overline{SoC}, \tag{2d}$$

$$SOC(\tau) = \frac{\mathcal{E}(\tau)}{\mathcal{E}_{rated}}. \tag{2e}$$

The constraints (2a)–(2e) are posed $\forall \tau \in \mathcal{T}, \mathcal{T} := \{1, \ldots, H_\tau\}$ and H_τ the last instant defining the horizon of the optimization. The constraints (2a)–(2e) pose the technical constraints for the BESS charging and discharging power. Accordingly, the State-of-Charge (SoC)—defined in Equation (2e)—is constrained based on the BESS's characteristics. To avert simultaneous charging and discharging of the BESS, a penalty cost is assigned with each auxiliary decision variables p_{ch}, p_{dch}, both of which should greater—at least one order—than the use of the BESS (c_{BESS}) itself, i.e., P_{BESS}.

As an additional note, for any three-phase BESS the different phases are hereby considered to follow the same mode of operation; therefore, the mathematical expression is comprised of three single-batteries installed in the different phases, coupled with the equality constraints for their active and reactive power injections as in Equation (3):

$$\begin{array}{ccccc} P^A_{BESS}(\tau) &=& P^B_{BESS}(\tau) &=& P^C_{BESS}(\tau) \\ Q^A_{BESS}(\tau) &=& Q^B_{BESS}(\tau) &=& Q^C_{BESS}(\tau) \end{array} \forall \tau \in \mathcal{T}. \tag{3}$$

2.3. Electric Vehicles (EVs)

The EVs are structured following the same rationale as the BESS model. In this study, the EVs are considered as flexible DER according, certainly, to their availability each time. Their provided flexibility is essentially regarded to be the intervals when they are parked to their owner's house premises. Being in this state (i.e., parked), if there is need for charge, this will be decided by the proposed control following the coordinated smart charging scheme. When the owner of an EV desires to provide a signal of flexibility, the time interval when the estimated trip will occur together with the estimated consumed energy should be dispatched to the DSO. These two signals are captured for each controllable (i.e., willing to be charged in concordance to the smart charging scheme) with $[y_{trip}]$ that is added to discharge the EV and E_{tr}, where n_{tr} corresponds to the number of trips for an EV. The fictitious variable $[y_{trip}]_{n_{tr} \times H_\tau}$ is added to discharge the EVs during their trips.

The Vehicle-to-Grid (V2G), where the EV injects power to the grid, operation is also incorporated within the proposed EV model. Whenever the V2G mode of operation is not deemed to be followed, the p_{dch} is simply constrained to zero.

One can define the energy state for each instant for one EV given by the vector $\mathcal{E}_{EV} \in \mathbb{R}^{H_\tau}$ recasting Equation (1), which infers to a linear combination of preceding instances inherent to the controllability that its flexibility allows, and the initial stored energy $\mathcal{E}_0$. The energy storage for one EV at instant τ can be calculated by Equation (4):

$$\mathcal{E}(\tau) = \mathcal{E}_0 + \underbrace{\Delta\tau[\text{diag}\{n_{dch}\} \, \text{diag}\{1/n_{ch}\}]}_{\Lambda} \cdot \underbrace{\begin{bmatrix} p_{dch} \\ p_{ch} \end{bmatrix}}_{p_{EV}(\tau)} - y_{\text{trip}}(\tau) \cdot E_{tr}(\tau). \tag{4}$$

For one EV, let it j, connected along the distribution network, the energy state function (4) can be rewritten in a compact matrix format capturing both operating modes where the energy stored to each EV towards the time evolution H_τ can expressed as the vector $\mathbf{E}_{EV}^j = [E_{EV}^j(0), \ldots, E_{EV}^j(H_\tau)]^T$:

$$\mathbf{E}_{EV}^j = \begin{bmatrix} I \\ \vdots \\ I \end{bmatrix} \mathcal{E}_0^j + \begin{bmatrix} \Lambda & & 0 \\ \vdots & \ddots & \\ \Lambda & \cdots & \Lambda \end{bmatrix} \begin{bmatrix} p_{EV}^j(1) \\ \vdots \\ p_{EV}^j(H_\tau) \end{bmatrix} - y_{\text{trip}}^j \cdot \mathbf{E}_{tr}^j. \tag{5}$$

2.4. Microgeneration (μG)

The microgeneration in this work is considered to be single phase inverter based installations. In case the DSO desires to incorporate the control of microgeneration in the control, the type of control pertaining the active power through Active Power Curtailment (APC) and/or Reactive Power Control (QR) have to be opted for the centralized controller.

Regarding the APC, the following settings define the maximum possible curtailed power as a percentage of the instant injected power (i.e., maximum curtailment β=20%), given the following rule in Equation (6):

$$\overline{p_{pac}(\tau)} = \begin{cases} \beta \cdot p_{inj}(\tau), & \text{if } p_{inj}(\tau) \geq \xi \cdot p_{rated}, \\ 0, & \text{else.} \end{cases} \tag{6}$$

ξ (in this study ξ=0.5) stands for a parameter which leads to control PV with larger injected power in proportion to their installed power at the instant period. The reactive power control is defined in similar way, though allowing capacitive and inductive operation (i.e., injecting and absorbing reactive power accordingly). Nevertheless, if the option of controlling the PVs in both PAC and QR mode, to avoid the nonlinear constraint inherent to the operation of the inverter; a simpler linear constraint is posed to ensure that the microgenerator's inverter does not exceed its rated power:

$$\begin{aligned} \overline{Q_{QR}(\tau)} &= (p_{rated} - p_{inj}(\tau)) \cdot \tan(\theta_{\min}) \\ \underline{Q_{QR}(\tau)} &= -\overline{Q_{QR}(\tau)}, \end{aligned} \tag{7}$$

where $\theta_{\min}$ is given by the minimum power factor ($\text{PF}_{\min}$) applied; $\theta_{\min} = \cos^{-1}(\text{PF}_{\min})$.

2.5. Three-Phase Power Flow

A three-phase Power Flow (PF) is implemented following the main notions described in [36]. The PF tool is incorporated in the overall proposed scheme, as an algorithmic step, either for the calculation of the initial point of the optimization, or to validate of the set-points yielded by the control scheme.

Th three-phase power is based on a Backward-Forward Sweep (BFS) technique, where, in the Backward stage, the branch current calculations occur, whilst, in the Forward Sweep stage, the nodal voltage is calculated. This method, unlikely for classical power flow methods, copes with a branch-oriented technique rather than nodal relations.

Note that this power flow algorithm presents quick convergence, i.e., iterations do not exceed 4 for tolerance convergence $\epsilon_\tau = 1e - 4$. The performance can be further accelerated by the valid assertion that the angle displacement in LV distribution networks between adjacent nodes is fairly small [17], i.e., $\Delta\theta$ leads to zero which leads to the conception in Equation (8) for the voltage drop:

$$\Delta V_{abc}^{(\kappa+1)} = Re\{Z_\ell \cdot J_{abc}^{(\kappa)}\}, \tag{8}$$

where Z_ℓ is the corresponding impedance among the connected branches and $J_{abc}^{(\kappa)}$ is the vector for the line section currents at iteration κ. Regarding the power flow framework, it is hereby structured in such a way that each load might have a different load model among constant PQ and a constant I or constant Z model. Accordingly, the injection current at node j is given by Equation (9):

$$I_{abc}^{j} = (S_{abc} \cdot diag^{-1}(V_{L-L}))^* \left\| \frac{V_j}{V} \right\|^{\kappa}, \tag{9}$$

where S_{abc} stands for the apparent power consumed at node j, V_{L-L} the line-to-line voltage. The operator $diag(.)$ is settled as an operator that returns a diagonal vector and κ is considered the load model parameter, which is 0 for constant PQ load, 1 for constant current and 2 in case of constant impedance.

3. Multi-Period AC-OPF (MACOPF) Formulation-Coordinated Control Scheme

The centralized coordinated management of the DER is discussed in this section through a multi-period three-phase AC-Optimal Power Flow (MACOPF) where the different periods $\tau \in \mathcal{T}$ are coupled with inter-temporal costs and the DER are assigned with inter-temporal technical constraints accordingly.

The MACOPF is stated for a horizon of operational planning H_t. All subsequent time steps belong to the set $\mathcal{T} := \{1, \ldots, H_\tau\}$. The main objective (i.e., $\mathcal{O}_1$) of the scheme is to minimize the operating costs assigned with all the controllable assets providing their coordination given their availability. Penalty costs assigned to auxiliary variables described with the term Φ_p. Such penalty costs refer to relaxation of voltage bounds to ensure feasibility and thus convergence as well as penalties to avert simultaneous charging and discharging or even auxiliary variables described within Section 2.

Assume that the state vector $(x_{g,\tau})$ at the time instant τ is given by (10) and the set of decision variables τ corresponds to the vector u_τ comprised of active and reactive power of each controllable DER as shown in Equation (11) as well as auxiliary variables y_τ in Equation (12). The voltage angles can be omitted to reduce the scale of the optimization problem, since the angle displacement between adjacent nodes in LV grids is typically less than $10°$ [17]. Nonetheless, angles are included for completeness:

$$x_{g,\tau} = \begin{bmatrix} \Theta \\ \mathcal{V} \end{bmatrix}_\tau , \forall \tau \in \mathcal{T}, x_\tau \in \mathbb{R}^{(2*3N_b)}, \tag{10}$$

$$u_\tau = \begin{bmatrix} P_c \\ Q_c \end{bmatrix}_\tau , \forall \tau \in \mathcal{T}, u_\tau \in \mathbb{R}^{(2*n_c)}, \tag{11}$$

$$y_\tau = \begin{bmatrix} p_{ch} \\ p_{dch} \\ y_{\pi,ch} \\ y_{\pi,dch} \\ \varepsilon_V \\ \varepsilon_{sub} \\ y_{trip} \end{bmatrix}_\tau , \forall \tau \in \mathcal{T}, y_\tau \in \mathbb{R}^{\ell_y}, \tag{12}$$

where N_b refers to the number of buses and N_c the number controllable units and $\ell_y = (4*N_{BESS}) + 3*N_b + N_{ev} + 3$ with n_{BESS} the number of BESS and n_{EV} the number of EVs. The real vector

$\mathcal{V} = [v_1, v_2, \ldots, v_{N_b}]_{\tau}^T$ corresponds to the voltage magnitudes for each bus (each bus has three terminals) at each time instant τ, and respectively Θ to the voltage magnitudes. The sets $\mathcal{N}, \mathcal{J}, \mathcal{T}$, denote the buses ($N_b$), branches and the horizon of the multi-period scheme. Let for the optimization problem the state vector and the decision variables correspond to the respective matrices $X = [x_1, \ldots, x_{H_T}]^T$ and $\mathcal{U} = [u_1, \ldots, u_{H_T}]^T$, essentially defined as stacked vectors of each subsequent time period. For the sake of comprehension, Figure 2 presents the described structure of the optimization variables. The fact that the auxiliary variables (y_τ) are appended as last elements of vector X eases the extension of the stated problem. Additional objective terms might be assigned in the current formulation unless there is no dependence or conflicting interest upon the aforementioned objective.

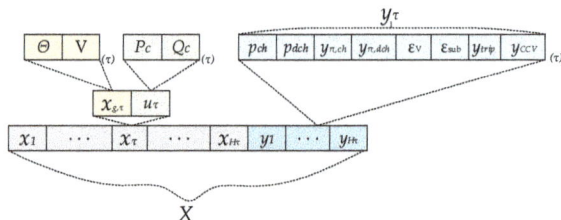

Figure 2. Structure of optimization variables; discriminated by state vector, control variables and the auxiliary variables per each time step.

The overall control scheme can be described by the set of Equations (13):

$$\min_{u} \sum_{\tau=1}^{H_t} \left\{ \underbrace{\sum_{k}^{N_b} \left([c_k(\tau)]^T \cdot u_{k,\tau} \right)}_{\mathcal{O}_1} \right\} \Delta \tau + \Phi_p, \tag{13}$$

subjected to

$$G_j(x_\tau, u_\tau, y_\tau) = 0 \qquad \forall j, \tau \in \mathcal{N}, \mathcal{T}, \tag{13a}$$

$$H_{Sub}(x_\tau, u_\tau, y_\tau) + \varepsilon_{sub} \le 0 \qquad \forall i, \tau \in \mathcal{J}, \mathcal{T}, \tag{13b}$$

$$V_{\min} - \varepsilon_{\mathcal{V}} \le v_j(x_\tau) \le V_{\max} + \varepsilon_{\mathcal{V}} \qquad \forall i, \tau \in \mathcal{N}, \mathcal{T}, \tag{13c}$$

$$h_\xi(x_\tau, u_\tau, y_\tau) = 0 \qquad \forall \xi, \tau \in \mathcal{U}, \mathcal{T}, \tag{13d}$$

$$g_\xi(x_\tau, u_\tau, y_\tau) \le 0 \qquad \forall \xi, \tau \in \mathcal{U}, \mathcal{T}, \tag{13e}$$

where vector $[c_k(\tau)]^T$ includes the price of each controllable unit per time step τ in €/kWh or €/kVArh. The constraints in (13a) set the nonlinear power flow equation at each bus of the network; inequality constraint (13b) poses the technical constraint for the MV/LV transformer; the boxed constraints in (13c) to respect all nodal voltages to range strictly within the admissible bounds. The additional positive variable $\varepsilon_{\mathcal{V}}$ is used to relax the voltage constraints and avoid infeasibility. The latter is applicable substantially when the active measures are not adequate for addressing voltage problems. Accordingly, since a transformer is capable of being operated in full load conditions or slightly higher for a certain short interval, an auxiliary variable is also applied to turn these constraints less tight. Th constraints (13d)–(13e) correspond to the operational limits of the controllable DER.

In the following subsection, the resolution of the optimal control problem is analytically discussed through the optimization techniques used to address it efficiently. The proposed control scheme evidently deals with a large number of decision variables -$X \in \mathbb{R}^{N_X} : N_X = (2 * 3N_b + 2 * N_c + 2 * N_{BESS} + 2 * N_{ev} + 3) \cdot H_T$-. Accordingly, the power flow through the nonlinear equations are accounted $N_{nonlin} = 2 * 3 * N_b$. Such large optimization problems, where the structure of the Jacobian of the nonlinear constraints present very sparse blocks, should possibly reveal particular framing of the problem re-structuring it and inherently leading to improved computational efficiency. In [37], the authors restructure the multi-period OPF to a tailored approach exploiting its particularities. Hereby, particular techniques are proposed to speedup the largeness of the nonlinear optimal control programming. The explicit calculation of the Jacobian and the Hessian is proposed, taking advantage of the sparsities, as well as slight pivotal elements to the Jacobian address any singularities met by the inter-temporal constraints. The initial point X_0 for the optimizer is either obtained through a local database which has stored previous occurrences or by running sequentially (H_τ) power flows. Additionally, if load and weather forecasts fail to be obtained, a worst case scenario can be interrogated by a local historical database.

3.1. Interior-Point Primal-Dual Method for the Proposed Control Scheme

The proposed control scheme based on the MACOPF is stated with the set of Equations (13). Hereby, a discussion will follow based on an interior-point method, which follows the above form of the previous section, and will involve lengthy and complicated notation. To ease the description, the control problem is reformulated in a more compact manner (i.e., grouping the equalities and inequality constraints) as proposed in the literature [38–40]. Note that, with bold notation, a vector form is implied. The decision variables and state vectors are simply represented by one vector $\mathbf{x}$:

$$\min_{\mathbf{x}} f(\mathbf{x}), \tag{14}$$

subject to

$$g_E(\mathbf{x}) = \mathbf{0}, \tag{14a}$$

$$h_I(\mathbf{x}) \leq \mathbf{0}, \tag{14b}$$

$$x_{\min} \leq \mathbf{x} \leq x_{\max}, \tag{14c}$$

where $g_E(\mathbf{x})$ contains any type of equality constraint (i.e., linear and nonlinear) and h_I any type of inequality constraint. The inequalities constraints can be introduced as equality constraints by the addition of slack variables s_j, such that $h_I(\mathbf{x}) - \mathbf{s} = \mathbf{0}$. Then, a *penalty function* introduces a new form to the initial objective function as follows:

$$\min_{\mathbf{x}} f_p = f(\mathbf{x}) - \mu^{(k)} \sum_{j=1}^{N_x} \ell n(x_j - x_{j,\min}) - \mu^{(k)} \sum_{j=1}^{N_x} \ell n(x_{j,\max} - x_j) - \mu^{(k)} \sum_{j=1}^{N_{ineq}} \ell n(s_j), \tag{15}$$

subject to

$$\begin{aligned} g_E(\mathbf{x}) &= \mathbf{0}, \\ h_I(\mathbf{x}) + \mathbf{s} &= \mathbf{0}, \\ \mathbf{x}, \mathbf{s} &\geq \mathbf{0}, \end{aligned} \tag{16}$$

where $\mu^{(k)}$ is the the logarithmic barrier parameter for iteration k that essentially reduces monotonically to 0 as iteration progresses. The non-negativity conditions at (16) are handled by incorporating them into logarithmic barrier terms. The Lagrangian ($\mathcal{L}_p$) of the f_p can be defined as:

$$\mathcal{L}_p(\mathbf{x}, \lambda, \sigma, \mathbf{s}) := f_p(\mathbf{x}) - \lambda^T g_E(\mathbf{x}) - \sigma^T(h_I(\mathbf{x}) + \mathbf{s}) \Leftrightarrow$$

$$\mathcal{L}_p(\mathbf{x}, \lambda, \sigma, \mathbf{s}) = f(\mathbf{x}) - \mu^{(k)} \sum_{j=1}^{N_x} \ell n(x_j - x_{j,\min}) - \mu^{(k)} \sum_{j=1}^{N_x} \ell n(x_{j,\max} - x_j) -$$

$$\mu^{(k)} \sum_{j=1}^{N_{ineq}} \ell n(s_j) - \lambda^T g_E(\mathbf{x}) - \sigma^T(h_I(\mathbf{x}) + \mathbf{s}), \tag{17}$$

where vectors λ, σ are the Lagrange multipliers for the corresponding equality and inequality constraints. The first-order Karush–Kuhn–Tucker (KKT) condition (**iff** g_E, h_I are first order differentiable) for a local optimum point $p^* = (\mathbf{x}^*, \lambda^*, \sigma^*, \mathbf{s}^*)$ are the following:

$$
\begin{aligned}
\nabla_x \mathcal{L}_p(\mathbf{x}^*, \lambda^*, \sigma^*, \mathbf{s}^*) &= \nabla_x f_p(x) - \lambda \nabla_x g_E(\mathbf{x})^T - \sigma \nabla_x h_I(\mathbf{x})^T - \mathcal{D} &= \mathbf{0}, \\
\nabla_\mu \mathcal{L}_p &= \mu[\mathbf{s}]^{-1} e - \lambda &= 0, \\
\nabla_\lambda \mathcal{L}_p &= g_E(\mathbf{x}^*) &= 0, \quad (18) \\
\nabla_\sigma \mathcal{L}_p &= h_I(\mathbf{x}^*) - \mathbf{s}^* &= 0, \\
\end{aligned}
$$
$$
\mathcal{D} = \mu[\mathbf{x} - \mathbf{x}_{\min}]^{-1} e + \mu[\mathbf{x}_{\max} - \mathbf{x}]^{-1} e,
$$

where e is the appropriate size vector with all entries equal to one.

The first-order KKT conditions are necessary and sufficient for global optimality for a convex problem when the Linear Constraint Qualification (LICQ) holds [41]. In the proposed control scheme, the non-convex nature of the—exact—nonlinear power flow necessitates the verification second-order KKT conditions to certify the local optimality of p^*. The second-order conditions can be found analytically in the literature [38], since here the first-order will be analytically discussed due to the examination of linear dependence of the inter-temporal constraints introduced by DER.

The IP algorithm at iteration k requires the solution of the following nonlinear system:

$$
\begin{bmatrix}
\nabla_x^2 \mathcal{L}_p & \nabla_x g_E(\mathbf{x}) & \nabla_x h_I(\mathbf{x}) & 0 \\
\nabla_x g_E(\mathbf{x})^T & 0 & 0 & 0 \\
\nabla_x h_I(\mathbf{x})^T & 0 & 0 & I \\
0 & 0 & I & \nabla_s^2 \mathcal{L}_p
\end{bmatrix}
\begin{bmatrix}
\Delta \mathbf{x} \\
\Delta \lambda \\
\Delta \sigma \\
\Delta \mathbf{s}
\end{bmatrix}
= -
\begin{bmatrix}
\nabla_x f_p(x) - \lambda \nabla_x g_E(\mathbf{x})^T - \sigma \nabla_x h_I(\mathbf{x})^T \\
\mu[\mathbf{s}]^{-1} e - \lambda \\
g_E(\mathbf{x}) \\
h_I(\mathbf{x}) - \mathbf{s} \\
[\mathbf{z}] \mathbf{x} - \mu e
\end{bmatrix}. \quad (19)
$$

The KKT system in Equation (19) is nonlinear and its solution most commonly in the literature [42] is addressed by applying the Newton's method. In the proposed control scheme, the gradients for the nonlinear constraints and the Hessian of the Lagrangian are explicitly calculated. In case where these derivations are not provided, approximations based on finite differences are typically applied [43,44].

3.2. Gradients of Nonlinear Constraints and Hessian of Lagrangian

The gradient (see also Appendix B) of the objective function, the Jacobian of nonlinear constraints and Hessian of the Lagrangian are implicitly provided to the optimizer, by expanding the calculations presented in [45]. On this point, it is important to be stated that the voltage vectors are expressed using polar coordinates: expressing voltage in rectangular coordinates eliminates trigonometric functions from the PF equations [46]. Nevertheless, in [42,47], a benchmarked comparison of both types of coordinates present the same order of computational performance as well as an equivalent number of iterations for convergence for a typical OPF. On one hand, rectangular coordinates can ease the process of first and second order gradients leading to quadratic and constant forms; both types cannot avoid the nonlinear equalities (and inequality) constraints and form a convex region. Concurrently, rectangular coordinates may provide slightly faster evaluation of particular gradients and Hessian, but the voltage bounds are handled as functional bounds in many OPF problems.

Therefore, the complex voltage vector might be denoted by $V \in \mathbb{R}^{3N_b}$. The element at bus j at phase a is $v_{j,a} = |v_{j,a}| e^{j\theta_{j,a}}$. The derivation with the state vectors for instant period τ followed, obviously, for the other periods are 0 entries-:

$$
V_{\Theta_\tau} = \frac{\partial V_\tau}{\partial \Theta_\tau} = j[V_\tau] \quad V_{V_\tau} = \frac{\partial V_\tau}{\partial V_\tau} = [V_\tau][V_\tau]^{-1} := [E]. \quad (20)
$$

The analytical AC power flow equations over all periods of the horizon window can be derived by the resolution of Equation (21). The operator $\odot$ is used for element-wise matrix product. Complex number equations are not addressed by state-of-the art optimizers and only in a few cases yield faster solutions [48]. Thus, a segregation is proposed—as shown in Equation (22)—which introduces the power flows through the vector $G(X)$, due to their complex nature, which essentially cannot be posed at the optimization stage:

$$G_c(X) = S_{bus}(V) + S_d - C_g S_g \quad , G_c : \mathbb{C}^{n_b} \to \mathbb{C}^{n_b}$$
$$S_{bus}^\tau(V_\tau) = V_\tau \odot (I_{inj}^\tau)^* = V_\tau \odot (Y_{bus} \cdot V_\tau)^*, \tag{21}$$

$$G(X) = \begin{bmatrix} \Re\{G_c(X)\} \\ \Im\{G_c(X)\} \end{bmatrix} \quad , G : \mathbb{R}^{2n_b} \to \mathbb{R}^{2n_b}. \tag{22}$$

The current bus injection I_{inj} appears in the power flow Equations (21). It would be useful for the power flow expressions derivation to present the corresponding for the current injections:

$$\frac{\partial I_{inj}^\tau}{\partial x_\tau} = \begin{bmatrix} \frac{\partial I_{inj}^\tau}{\partial \Theta_\tau} & \frac{\partial I_{inj}^\tau}{\partial V_\tau} & 0 & 0 \end{bmatrix}, \tag{23}$$

$$\frac{\partial I_{inj}^\tau}{\partial \Theta_\tau} = Y_{bus} \frac{\partial V_\tau}{\partial \Theta_\tau} \overset{(20)}{=} jY_{bus}[V_\tau], \quad \frac{\partial I_{inj}^\tau}{\partial V_\tau} = Y_{bus} \frac{\partial V_\tau}{\partial V_\tau} \overset{(20)}{=} Y_{bus}[E].$$

The first and second derivatives for the nonlinear constraints, which substantially refer to the power flow equation, will be based on the introduced vector $G_c(X)$:

$$G_X = \frac{\partial G}{\partial X} = \begin{bmatrix} \frac{\partial \Re\{G_c(X)\}}{\partial X} \\ \frac{\partial \Im\{G_c(X)\}}{\partial X} \end{bmatrix} = \begin{bmatrix} \frac{\partial G}{\partial x_1} & \frac{\partial G}{\partial x_2} & \cdots & \boxed{\frac{\partial G}{\partial x_{H_t}}} & \cdots & \frac{\partial G}{\partial x_{H_t}} & \frac{\partial G}{\partial y_1} & \cdots & \frac{\partial G}{\partial y_{H_t}} \end{bmatrix}, \tag{24}$$

$$\frac{\partial G}{\partial x_\tau} = \begin{bmatrix} G_{\Theta_\tau} & G_{v_\tau} & G_{P_{g_\tau}} & G_{Q_{g_\tau}} \end{bmatrix}. \tag{25}$$

The first order partial derivatives are presented for the defined G_c, which can thereafter appended in Equation (25):

$$G_{c,\Theta_\tau} = \frac{\partial S_{bus}(V_\tau)}{\partial \Theta_\tau} = [I_{inj}^\tau] \frac{\partial V_\tau}{\partial \Theta_\tau} + [V_\tau] \frac{\partial (I_{inj}^\tau)^*}{\partial \Theta_\tau} \overset{(20)}{=} j[V_\tau] \left([I_{inj}^\tau] - Y_{bus}[V_\tau]\right)^*, \tag{26}$$

$$G_{c,V_\tau} = \frac{\partial S_{bus}(V_\tau)}{\partial V_\tau} = [I_{inj}^\tau] \frac{\partial V_\tau}{\partial V_\tau} + [V_\tau] \frac{\partial (I_{inj}^\tau)^*}{\partial V_\tau} \overset{(20)}{=} [V_\tau] \left([I_{inj}^\tau] + Y_{bus}[V_\tau]\right)^* [V_\tau]^{-1}, \tag{27}$$

$$G_{c,P_g^\tau} = -C_g, \quad G_{c,Q_g^\tau} = -jC_g, \tag{28}$$

where $(C_g)_{N_b \times N_c}$ stands for the injection connectivity matrix that each element (i,j) is one if at bus i_{th} controllable asset j_{th} is connected, else the element is zero. It is obvious that the partial derivatives of $G_c(x_\tau)$ with respect to other $x_{g,\tau_2}, u_{g,\tau_2}$—with $\tau_2 \neq \tau$—results in zero entries. The G_X is therefore formed by two stacked matrices, which present sparse block diagonalities. The corresponding partial derivatives with respect to the auxiliary variables will be zero entries as well.

The second derivatives for the the complex power injections are necessary only for the assessment of the Hessian of the Lagrangian, which appears in iteration of the KKT-system—Equation (18). As it can be observed, the Hessian matrix of the Lagrangian can be given by Equation (1):

$$\mathcal{H}_p(\mathbf{x},) = \nabla_{XX} L_p = \nabla_{XX} f(X) + \nabla_{XX} g_E(\lambda) + \nabla_{XX} H(X). \tag{29}$$

The second order derivatives for the complex power flows are calculated in proportion to the instant λ. The derivative is provided discretized in two parts:

$$\frac{\partial^2 G_c \lambda}{\partial x_\tau^2} = \frac{\partial}{\partial x_\tau} \left(\lambda G_{c,x_\tau}^T (X) \right)$$

$$= \begin{bmatrix} G_{c,\Theta_\tau \Theta_\tau} & G_{c,\Theta_\tau V_\tau} & 0 & 0 \\ G_{c,V_\tau \Theta_\tau} & G_{c,V_\tau V_\tau} & 0 & 0 \\ 0 & 0 & 0 & 0 \\ 0 & 0 & 0 & 0 \end{bmatrix}. \tag{30}$$

A brief presentation of all the subsequent expressions follows:

$$G_{c,\Theta_\tau \Theta_\tau} = \frac{\partial}{\partial \Theta_\tau} \left(j \left([I_{inj}^\tau] - [V_\tau] Y_{bus}^T \right)^* [V_\tau] \lambda_\tau \right)$$

$$= \underbrace{[V_\tau]^* \left((Y_{bus}^*)^T [V_\tau][\lambda_\tau - [(Y_{bus})^*[V_\tau]\lambda_\tau]] \right)}_{A_1} + \underbrace{C_2 \left(Y_{bus}^* [V_\tau]^* - [I_{bus}]^* \right)}_{A_2}. \tag{31}$$

Accordingly, the $G_{c,V_\tau \Theta_\tau}$ can be calculated as:

$$G_{c,V_\tau \Theta_\tau} = j\mathcal{B}(A_1 - A_2) = (G_{c,\Theta_\tau V_\tau})^T, \tag{32}$$

$$G_{c,VV_\tau} = \mathcal{B}(C + C^T)\mathcal{B}, \tag{33}$$

where

$$\begin{aligned} \mathcal{B} &= [V_\tau]^{-1}, \\ C_2 &= [\lambda_\tau][V_\tau], \\ C &= C_2(Y_{bus}[V_\tau])^*. \end{aligned}$$

The overall assessment of the second order gradient can be induced to a simple routine reducing computational effort and memory allocation by saving certain matrices presented above.

The subsequent gradients of the objective functions can be assessed for the proposed scheme since the $f(X) = \sum_{\tau \in \mathcal{T}} f(x_\tau)$ substantially corresponds to a linear combination of convex functions. The first-order gradient of the objective will be comprised of constant and null entries since the costs are linear functions, as described in Section 3.5. By extension, the second-order derivatives of the objective function will be a null matrix.

If a solution of the barrier problem satisfies the primal-dual Equations (19) of the nonlinear KKT system, its solution may be approximated by an iteration of Newton's method. The search direction can be obtained as a solution of the linearization of the KKT system, which is presented in Section 3.3.

3.3. Solution of Karush–Kuhn–Tucker Equations

Hereby, the barrier solution is presented for the optimal control problem (14), assuming that the decision variables x are positive (i.e., to avoid lengthy notation). The dual variables can be introduced as:

$$z_i := \frac{\mu}{x_i} \tag{34}$$

$$\begin{bmatrix} \mathcal{H} & 0 & -J_E^T & -J_I^T & -I \\ 0 & [\lambda] & 0 & [s] & 0 \\ J_E & 0 & 0 & 0 & 0 \\ J_I & -I & 0 & 0 & 0 \\ [z] & 0 & 0 & 0 & [x], \end{bmatrix}^{(k)} \begin{bmatrix} \Delta x \\ \Delta s \\ \Delta \lambda \\ \Delta \sigma \\ \Delta z \end{bmatrix}^{(k)} = - \begin{bmatrix} \nabla_x \mathcal{L}_p \\ \nabla_\lambda \mathcal{L}_p \\ \nabla_\sigma \mathcal{L}_p \\ \nabla_s \mathcal{L}_p \end{bmatrix}, \tag{35}$$

where $\mathcal{H} = \nabla_{xx}\mathcal{L}_p$, $J_E = \nabla_x g_E(x)$ and $J_I = \nabla_x h_I(x)$. The system (35) can be further simplified—based on Gaussian elimination—by eliminating the last row. Therefore, with this elimination, it will be:

$$
\begin{bmatrix}
\hat{\mathcal{H}} & 0 & -J_E^T & J_I^T \\
0 & [\hat{\lambda}] & 0 & -I \\
J_E & 0 & 0 & 0 \\
J_I & -I & 0 & 0
\end{bmatrix}^{(k)}
\begin{bmatrix}
\Delta \mathbf{x} \\
\Delta \mathbf{s} \\
\Delta \lambda \\
\Delta \sigma
\end{bmatrix}^{(k)}
= -
\begin{bmatrix}
\nabla_x \mathcal{L}_p + [\mathbf{x}]^{-1} g_E(\mathbf{x}) \\
[\mathbf{s}]^{-1}(-\mu \mathbf{e} + \mathbf{s}\lambda) \\
-g_E(\mathbf{x}) \\
-h_I(\mathbf{x})
\end{bmatrix}
\tag{36}
$$

where $\hat{\mathcal{H}} = -\mathcal{H}^k - [\mathbf{x}^k]^{-1}[\mathbf{z}^k]$ and the diagonal matrix $[\hat{\lambda}] = -(\mathbf{s}^k)[\lambda]$. The updates for the dual variables can be obtained from Equation (37):

$$
\Delta \mathbf{z}^k = [\mathbf{x}^k]^{-1}(\mu \mathbf{e} - [\mathbf{z}^k]\Delta \mathbf{x}^k) - \mathbf{z}^k.
\tag{37}
$$

The overall resolution of the KKT system provides the search direction set for each subsequent iteration. An important condition for the Netwon step is that the Jacobian J_E is a non-singular matrix. The latter implies properties which are assigned with the constraint qualification (CQ). The QP is a critical condition that needs to be assessed along with the KKT conditions. More analytically, the LICQ necessitates that the gradients of the equality constraints and any active bound constraints (i.e., binding constraints) which are linearly independent. If this does not hold, the KKT system cannot be resolved. In [41], the authors show that, for any local optimizer, the KKT conditions with LICQ satisfied can ensure the generic existence of the Lagrangian multipliers. The inter-temporal couplings among different periods τ in some cases lead to singularities for the Jacobian of the nonlinear equalities. Section 3.4 thoroughly discusses an efficient manner to avert such issues.

3.4. Intertemporal Couplings and Singular Jacobian

In the proposed multi-period OPF, the inclusion of inter-temporal constraints (e.g., mainly due to BESS and EVs) in most cases lead to the singularity of Jacobian matrix (J_E). Whenever the gradients of the active constraints are linearly dependent, the consequence is that the Jacobian matrix for the first-order optimality conditions will be singular. This can be observed for the presented KKT systems (36) when there are some—assuming a set of j-binding constraints (from the h_I inequalities) $\mathbf{s}_j = 0$, while $\sigma_j \neq 0$. Therefore, if additionally there are gradients of the respective gradients $\nabla g_E, \nabla h_I$ are dependent when constraints are active the last three rows of the iteration matrix (36) will have dependent rows. An additional problematic condition might appear whether the binding conditions are linearly dependent; then, the Jacobi matrix is again singular according to [49].

The singularity issue to some extent occurs due to the structure of the control scheme, which essentially has no unique set of Lagrange multipliers corresponding to the dependent binding constraints. The particular case of BESS and the subsequent problems are analytically discussed in [49,50]. In these works, techniques are proposed to address the singularities; in [49], the authors simply suggest the elimination of the linearly dependent binding constraint on the Jacobian matrix is singular. Nonetheless, none can guarantee that the no constraint violation will take place and meanwhile it is tailored to mitigate certain singularities. The same authors propose further methods based on either a Moore–Penrose pseudoinverse of the Jacobian or by adding the standby losses.

Hereby, a simple technique will be presented which is model-free, which aims to correct the singularity introducing pivotal changes within the Jacobian matrix based on notions presented in [38]. To avoid the failure of singularity, a small pivot element (i.e., shadowed elements) might be added whenever the issue arises

$$
\begin{bmatrix}
\hat{\mathcal{H}} & 0 & -J_E^T & J_I^T \\
0 & [\hat{\lambda}] & 0 & -I \\
J_E & 0 & \delta_\pi \cdot I & 0 \\
J_I & -I & 0 & 0
\end{bmatrix}^{(k)}
\begin{bmatrix}
\Delta \mathbf{x} \\
\Delta \mathbf{s} \\
\Delta \lambda \\
\Delta \sigma
\end{bmatrix}^{(k)}
= -
\begin{bmatrix}
\nabla_x \mathcal{L}_p + [\mathbf{x}]^{-1} g_E(\mathbf{x}) \\
[\mathbf{s}]^{-1}(-\mu \mathbf{e} + \mathbf{s}\lambda) \\
-g_E(\mathbf{x}) - \delta_\pi \cdot \lambda \\
-h_I(\mathbf{x})
\end{bmatrix},
\tag{38}
$$

where δ_π stands for a fairly small positive number. The latter correction takes place in each iteration that the Jacobian is tracked as rank-deficient, which implies the linear dependency among constraints.

3.5. Inter-Temporal Costs

Inter-temporal costs have to be incorporated to ensure that transitions from one operating point to the next are feasible (i.e., managing flexibilities provided by the DER) and economical. These costs do apply independently to DER or assets while spanning multiple periods in particular, if multiple tariffs are defined along the horizon.

The costs for the controllable resources such as BESS and PVs follow a conditional operation regarding their mode of operation. The cost function, for instance, of charging a BESS is expressed by a linear convex function depending on the quantity of energy consumed—see the Figure 3a scenario when charging and discharging are equally priced. Any time step (e.g., charging or discharging) is dealt by imposing a Cost Constrained Variable (CCV) to represent the proper cost. The piecewise linear cost function $c_{BESS}(x)$—owned by the DSO—represented by the red line on Figure 3a—is substituted by an auxiliary variable y_{BESS} and a set of linear constraints. These linear constraints form a convex region with the $c_{BESS}(x)$, setting the y_{BESS} always to lie in the epigraph of the cost function. This auxiliary variable is onwards reflected to the objective function:

$$c_{BESS}(x) = \begin{cases} \pi_{BESS} \cdot x & \text{if } x \geq 0 \quad \text{(discharging)}, \\ -\pi_{BESS} \cdot x & \text{else } x < 0 \quad \text{(charging)}, \end{cases} \tag{39}$$

where π_{BESS} corresponds to price of utilizing the BESS either for charging or discharging. The necessary linear constraints for the auxiliary variable are the following:

$$\begin{aligned} y_{BESS} &\geq \pi_{BESS} \cdot x, \\ y_{BESS} &\geq -\pi_{BESS} \cdot x. \end{aligned} \tag{40}$$

Accordingly, the cost functions and their subsequent auxiliary variables are defined. Note that, for EV or domestic BESS, their cost function can have an arbitrage regime, in the sense that charging has a profitable impact on the objective function while discharging is penalizing it—Figure 3d. Note that all Figures 3 are indicative for a particular period τ; different functions might be regarded for different time steps implementing demand response schemes based on variable pricing schemes.

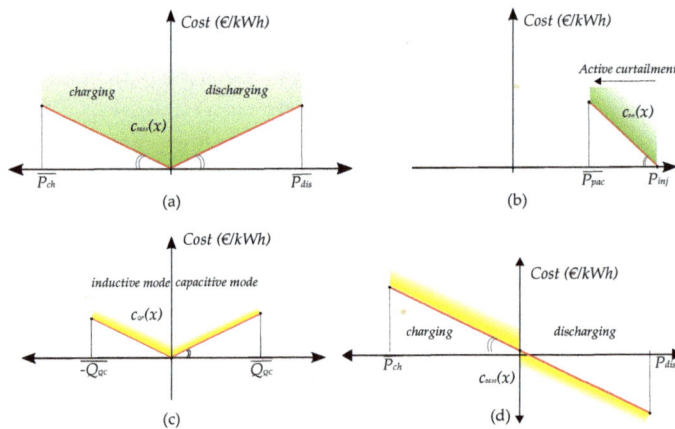

Figure 3. Cost function functions. (**a**) cost function for utilizing a BESS-owned by the DSO; (**b**) cost of active power curtailment; (**c**) cost function of reactive power control for microgeneration; (**d**) cost function of an EV with V2G operation.

4. Case Study Synopsis

The network selected as a case study to perform the validation of the proposed scheme belongs to the IEEE benchmarked LV European network [51]. The network corresponds to a real British low voltage feeder connected to the MV grid through a transformer of nominal power 800 kVA (Figure 4). The transformer is modified and considered to be 20 kV to 400 V with nominal power of 150 kVA, since only one feeder is considered in the benchmark as well. The service cables to the 55 end-consumers are also included in the grid representation.

In all scenarios, a three-phase centralized BESS is assumed to be connected at node 101 (Figure 4). The BESS capacity is 90 kWh and the maximum charging and discharging rate is 45 kW. This BESS is assumed to have an initial SoC of 0.40 with $\underline{SoC} = 0.1$ and, at the end of the optimization horizon, it has to be equal to the initial, $SoC_{H\tau} = 0.40$. The power factor of BESS$_{101}$ is considered as unitary for all the simulated cases and its charging and discharging efficiency is $\eta_{ch}, \eta_{dch} = 0.95$, [52].

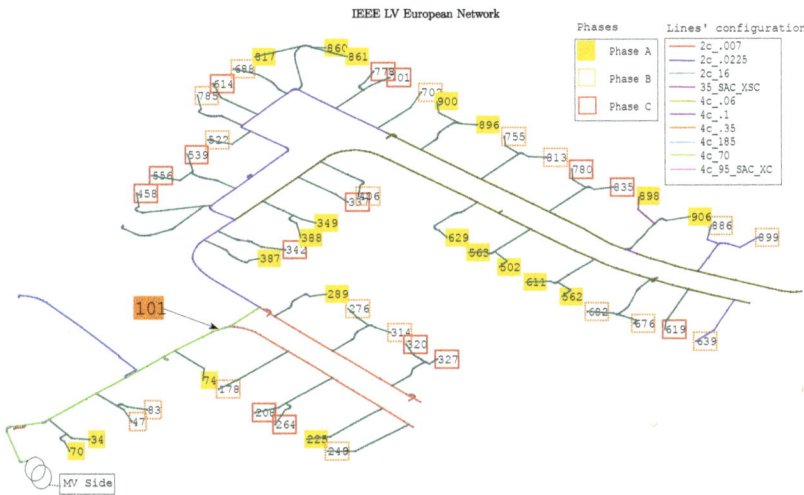

Figure 4. The IEEE European LV benchmark network. Fifty-five consumers are connected to this case network.

The load and the microgeneration profiles used correspond to daily data for a summer period, which are extrapolated from a realistic data pool provided for the benchmarked grid, which can be in [51]. All the consumers are single-phase and their phase connection is depicted in Figure 4. All the microgeneration units are considered as single-phase PV rooftop installations that are connected to the same phase as the respective residential user. The simulation day corresponds to a representative summer week day where typically peak loading conditions occur, where the load and generation profiles are illustrated in Figure 5a,b, accordingly.

Four EV models considered are based on four different EV models Nissan Leaf, Chevrolet Volt, BMW i3, Tesla S, which present different technical features regarding the Battery Capacity and charging power, in addition to their driving efficiency, which are considered from [53]. Their characteristics are listed in Table 1. All Tesla S and BMW i3 models are charged with an Efacec HomeCharger 7.4 kVA, while the rest of the EVs through an Efacec HomeCharger 3.4 kVA [54]. Therefore, the maximum charging power of each EV is limited and driven by the home charger used.

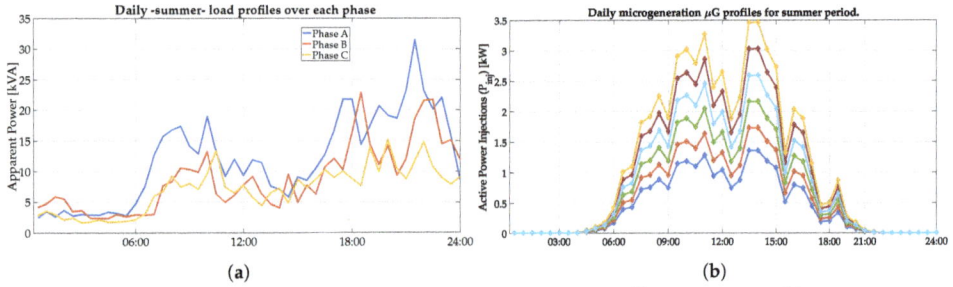

Figure 5. Data profiles: (**a**) load profiles; and (**b**) micro-generation profiles using seasonal (e.g., summer profiles) and regional data.

Table 1. Electric Vehicles' models and characteristics.

EV Model	Battery Capacity [kWh]	Charging Power [kW]	Driving Efficiency (km/kWh)	End-User Owner
Nissan Leaf	24	4	6.7	249.2/861.1/264.3/522.2
Chevrolet Volt	16	3.75	3.75	327.3/755.2/886.2/906.3/780.3/ 619.3/899.2/337.3/701.3
BMW i3	22	11	7.2	785.2/225.1/314.2/320.3/ 817.3/702.2/178.2/73.1/342.3
Tesla S	60	11	6.7	563.1/47.2/208.3/682.2/406.2/ 248.2/458.3/83.2/349.1/289.1

Concerning the EV usage, a routine is built to emulate credible scenarios, which are fed with public statistical data by [55]. This routine aims to capture patterns pertaining the EV usage upon different trip purposes as well as the trip duration (minutes) and length (km). The resulting data reflect a realistic response for the EV behavior during a day of operation, standing on the assertion that EVs charge exclusively at home. Nonetheless, the available statistical data do not explicitly provide information if the trip is from or to a destination. Therefore, the data as suggested in [56] are split into starting a trip and ending a trip, and it is assumed that a driver starts and ends every trip at home; hence, the cumulative data of trips for different days among different purposes are illustrated in Figure 6a. To assess the SoC change per each EV model used, averaged values for the purpose of each trip is used and correlated with each EV model's driving efficiency (Table 1).

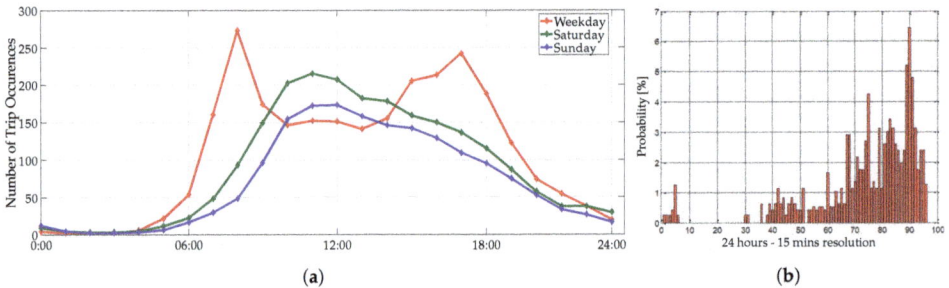

Figure 6. Data profiles: (**a**) trips in progress along a week. and (**b**) probability density function for EV charging demand, used for the dumb charging scenarios (source: [57]).

The following assumptions are also regarded for the EV:

- Initial SoC for all EV models is $SoC_0 = 0.5$, which is meant to be the same at the end of the horizon $SoC_{H\tau} = SoC_0$.

- The charging efficiency and discharging—when V2G—efficiency are considered 85% for all EV models.

Two EV charging strategies are considered in this study:

- "Dumb" charging or uncontrolled charging where the EVs are not incorporated within the proposed operational scheme. Such uncontrolled charing profiles are extracted using the data in probability density function presented in Figure 6b.
- "Smart" charging, where the EV owner communicates relevant data (i.e., flexibility as defined above) regarding their commute and accordingly its availability to be charged according to the proposed tool. The V2G mode services enable the option to utilize the EV essentially for grid services. These constraints are automatically incorporated in the multiperiod-OPF scheme as the generalized set of Equations (13d)–(13e), whenever the availability of the EV allows it. The availability of the EV to charge is considered along the day during their idle periods (i.e., parked at the owner's house).

Concerning the case where EVs follow the dumb charging, their charging occurs based on the distribution function given by [57]. The time departure and arrival as well as the daily distance traveled by each are randomly selected for a summer week day as presented in Figure 7.

According to the standard EN-50160, the 10 min mean r.m.s voltages shall not exceed the statutory limits during 95% of the week. Meanwhile, all 10 min mean r.m.s voltages shall not exceed the range of the Vn% + 10% and Vn% − 15% (which corresponds to 253 V and 195.5 V for most European grids). Given the fact that the proposed control scheme uses 30-min averaged data resolution, the voltage limits are set in [0.95, 1.05] p.u.values [58].

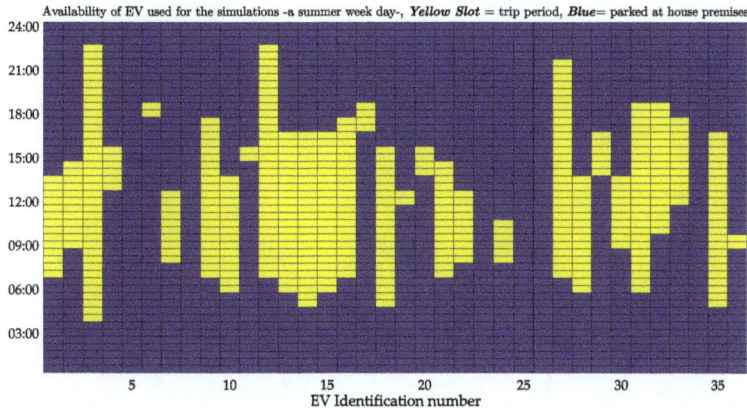

Figure 7. State of the trip for Electric Vehicles used along the simulation period. Profiles were extracted for a summer week day.

The percentage of PV and EV integration refers to the proportion of end users that own such units. For instance, 35% of EV penetration (i.e., 30 EVs), where the charging point of the EV are indicated in the last column of Table 1.

The use of DER is prioritized through the weighted terms $c_k(\tau)$, in the sense that the operational tool attempts to manage the flexibilities by addressing any voltage issues and respects the secondary transformer's rated power with the controllable DER assigned with the more economical combination of $c_k(\tau)$. These $c_k(\tau)$ can be attributed with real operational cost values to reflect monetary values for the use of flexibility. For this study, the weighted terms for the use of flexibility for each type of DER derive a merit order scheme settled as $c_{BESS} < c_{EV} < c_{V2G} < c_{QR} < c_{APC}$, which present respectively the price of using the BESS, incorporating an EV in the coordinated charging, the use of V2G mode and finally the use of reactive and active power by the microgeneration. In this way, the tool prioritizes the use of the centralized BESS that is owned by the DSO, avoids excessive active power curtailments, and the presence of the EVs restrains the dispatch of reactive power by the PVs, which is rather not effective for addressing voltage issues in LV grids (i.e., $R > X$). Note that, in this study, the V2G operation is set slightly cheaper than any control of the microgeneration (i.e., APC or QR).

Prior to the presentation of the results within the proposed control approach, an exploration of discrete incremental integrations of PV and EV are given. For both cases, no control was deployed. In Figure 8a, the impact in voltage magnitudes is depicted; notable voltage drops can be observed for an EV integration about 35% (i.e., 20 EVs). Higher EV integration of 55–65% present severe voltage drops as well as the overloaded condition for the secondary substations (see Figure 8b). In Figure 9a, the analogous scenario for PV integration is presented. Overvoltages appear in scenarios with more than 35% of PV integration, where the reverse power flow towards the secondary substation can be also viewed through Figure 9b. One can notice that, in a mixed scenario with PVs and EVs, overvoltages will typically appear during morning and evening hours, whereas voltage sags will arise at late hours when most regular charging of EVs occur (see Figure 6a).

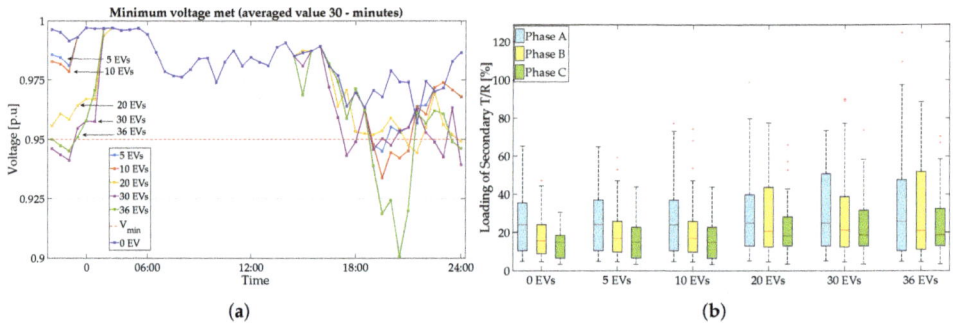

Figure 8. Incremental integration of EV scenarios: (**a**) minimum voltage range over all phases and buses and (**b**) secondary transformer loading for each case of EV integration (the increase in loading is observed up to 120%).

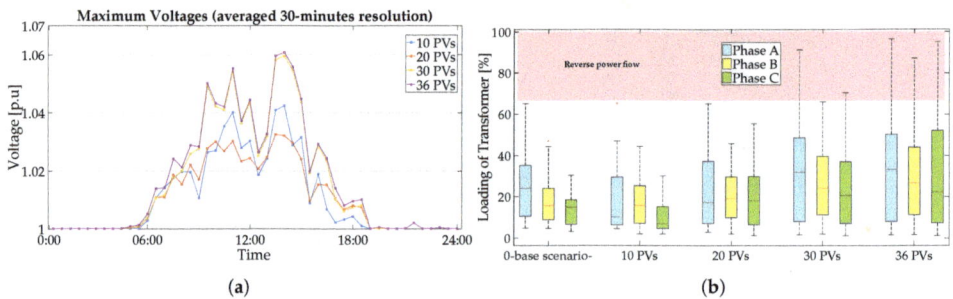

Figure 9. Incremental integration of EV scenarios: (**a**) minimum voltage range over all phases and buses and (**b**) secondary transformer loading for each case of EV integration (the increase in loading is observable up to 120%).

The explored scenarios in which the proposed control scheme is validated are defined in Table 2. Analytical information regarding the points of connection for the micro-generation and EVs are given in Appendix A. The scenarios are selected in order to validate that the MACOPF could allow high integration of microgeneration avoiding overvoltages and overloading of the transformer; secondly, mixed scenarios of DER integration including EVs are assessed by the coordinated control amongst them.

Table 2. Scenarios' description.

Scenario	Case 01 (C1)	Case 02 (C2)	Case 03 (C3)	Case 04 (C4)	Case 05 (C5)	Case 06 (C6)
EV [%]	0	0	0	35	55	65
PV [%]	55	73	85	0	0	55
BESS 101	✓	✓	✓	✓	✓	✓

5. Results

The assessment of the proposed control framework takes places in all the scenarios for a day-ahead deterministic planning of operation with a time-step of $\Delta\tau = 30$ min, (i.e., 48 time steps). The available active measures for operational purposes are active power control of the centralized three-phase BESS, the coordinated charging of the EVs in addition to V2G mode of operation where both are considered once the EV is available (i.e., parked at house premises), and the control active and reactive power of the microgeneration. The setup for the controller considers minimum power factor for microgeneration 0.9 and maximum allowed curtailment 15% of the injected power by each unit.

5.1. Cases C01–C03

In scenario C01, with 65% (36 PV units) overvoltages (up to 1.062 p.u) do arise along the grid due to notable active power injections. Additionally, reversed power flow is also present, increasing the loading conditions of the secondary transformer up to about 90%. In Figure 10a, the actions taken along the horizon period are illustrated. Among the decisions, BESS$_{101}$ is essentially charged during sunny periods, reducing the reversed flows to the substation and restraining the voltages. Ultimately, the voltages issues are addressed in coordination with reactive power dispatch by the microgeneration (Figure 10a). Note that, for the same scenario, the use of the control approach solely with APC derives 5.3 kWh enabled, which corresponds to minimum curtailments since the controller is centralized (Figure 10b); hence, dealing with local P–V droop control would lead to higher curtailments since there is no topological confluence among PV and the decisions are merely based on the voltage at the point of connection of the inverter.

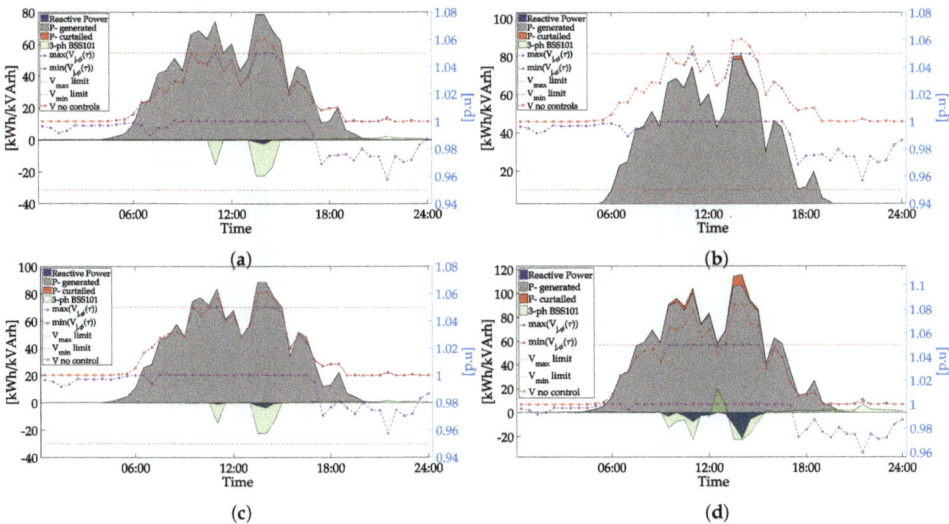

Figure 10. Incremental integration of PV, Cases 01–03: (**a**) Case 01; (**b**) Case 01, solely APC was selected within the controller; (**c**) Case 02; (**d**) Case 03.

Scenario 02 refers to the point of PV integration where overvoltages and secondary substation are within admissible bounds without curtailing any active power by the PV units. The subsequent controls derived in this case refer to the coordinated operation of the centralized BESS with reactive power dispatch of a set of microgeneration units. The control setpoint of the BESS and its SoC for the day of operation are illustrated in Figure 11a–c.

Given the selected options for the controller (i.e., minimum PF = 0.9 and maximum APC 15%), the proposed control is capable of mitigating overvoltages and ensuring rated power for the secondary substation up to 85% of PV integration. Analytically, the actions taken for this scenario are illustrated in Figure 10d. Table 3 presents the anticipated curtailed power and dispatch reactive power within the day of operation for C01–C03.

Figure 12 presents all cases C01–C03 and the corresponding transformer loading conditions if the decision derived by the MACOPF are followed. It is noted that in all cases the MACOPF delivers actions which restrain the reversed power to the secondary substation. Only in scenario C03 is there a slight relaxation of the respective constraint since all of the active measures have been optimally utilized.

Excessive active power curtailment may indirectly mean high compensation costs for the DSO. The installation of a BESS can reduce the need of APC as shown in scenarios 01–03. In particular, scenario 01 presented that, solely based on APC, 5.3 kWh at least would be curtailed with local P–V droop based control. This fact may justify the investment from the DSO side when a high integration of PV prevails.

As a remark stands, the current analysis leans on averaged values of 30-min resolution; therefore, notable overvoltages may appear in higher resolution analysis for less PV integration. The latter justifies the tight limits posed for the voltage magnitudes for the purpose of this study. Concurrently, the importance of coordinated actions will be observable even in scenarios with less PV integration.

Table 3. Resulting curtailed active power and dispatched reactive power for the microgeneration units.

Scenario	Curtailed Energy [kWh]	Reactive Energy [kVArh]
Case 01	0	3.94
Case 02	0	7.43
Case 03	34.5	59.3

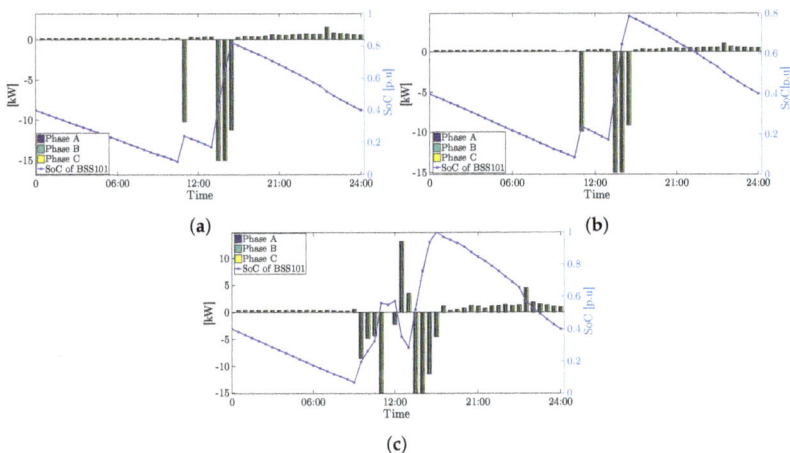

(a)   (b)

(c)

Figure 11. Control set-points and SoC derived by the MACOPF for centralized three-phase BESS$_{101}$ for: (**a**) Case 01; (**b**) Case 02; (**c**) Case 03.

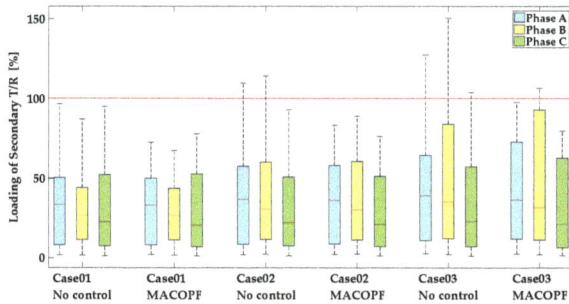

Figure 12. Secondary transformer loading conditions for Cases 01–03, overloaded conditions are noticed due to reversed power injected by microgeneration units; admissible conditions are obtained applying the proposed coordinated operation.

5.2. Cases C04–C06

In this part of the study, the EV integration is introduced within the LV grid. Initially, cases 04–05 examine the integration of 20 and 30 EV accordingly. The EV are randomly distributed along the end-consumers, only following the assumption that no more than one EV can be connected at each consumer. The point of connections for each scenario are given in Appendix A.

Case 04 presents voltage drops along the early hours (12:30 a.m. to 4:00 a.m.) but most significantly and below the posed lower limits in the late hours where the peak load is met (9:00 p.m.–10:00 p.m.). In this scenario, the MACOPF derives an optimal scheduling (Figure 13b) of the DSO's BESS in order to address the undervoltages in the late hours. The coordinated charging for the current case is not deemed to be needed according to the proposed control. Nonetheless, it can be seen from Figure 13a that a slight increase of the peak load (i.e., hereby 6:30 p.m. to 11:00 p.m.) might trigger critical voltage sags and possible transformer loading conditions. It should be stressed that the 35% integration of EV, following dumb charging, leads to an increase of 22% of the peak load compared to the base case.

The following scenario C4 emulates a case with 55% of EV integration. In this case, the importance of multiple DER coordination is stressed since critical voltage sags can be noticed if EVs follow the dumb charging operation. The MACOPF yields a scheduling for the BESS$_{101}$, which essentially supports during the late hours (i.e., 8:00 p.m.–9:30 p.m.) in addition to some EVs that are decided to operate in V2G mode. The peak load conditions along this interval are deteriorated due to the EVs that are added to consumption while returning from their trips. The coordinated charging during the early hours avoids the voltage drop along the distribution feeder, which would be noticed in the uncoordinated charging (Figure 14a).

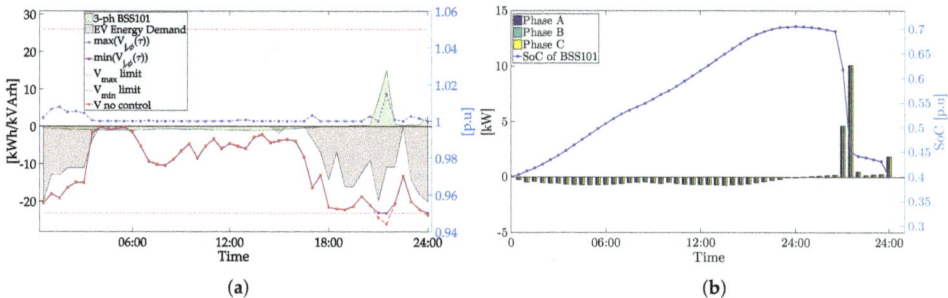

(a)

(b)

Figure 13. Case 04: (**a**) resulting voltage ranges and actions yielded for BESS$_{101}$; (**b**) control set-points and SoC for BESS$_{101}$.

The centralized BESS$_{101}$ is not adequate to entirely mitigate voltage problems due to the fact that some of the EVs are connected to the ending point of the feeder. Therefore, V2G mode of operation (i.e., for EVs that are willing to provide it) and coordinated charging efficiently added to the coordinated operation to maintain the voltages within the bounds. The SoC for all EVs are illustrated in Figure 14b. The V2G mode of operation comprises generally an effective measure to address undervoltages in LV grids due to the fact that active power is purely injected ($R > X$). Nevertheless, proper spatial and temporal distribution of the available EVs to provide such service is necessary. The spatial distribution refers to the bus and the phase of connection of the EV, whereas the temporal distribution to the availability along the horizon.

The following scenario C06 presents a case with increased PV and EV integration (i.e., 55% and 65%, respectively). The presence of PV installation injects notable active power during the sunny period (10:00 a.m. to 3:00 p.m.), whereas, in the rest of the time slots, voltage drops are observed due to the dumb charging. Notice that, in the current study, the PVs and EVs are randomly distributed, in the sense that a PV installation is not necessarily connected at a point where an EV is placed (see Tables A1 and A2). In addition, the fact that the majority of EVs progress in the sunny hours essentially implies that only a few could be available to be shifted for charging in these time slots. The subsequent results for this case are illustrated in Figure 15a. In this scenario, as it can be seen from the SoC of EVs in Figure 15b, the EVs are charged within a coordinated way from 12:00 a.m. to 5:00 a.m. where light loading conditions prevail in the grid avoiding the undervoltages that are present within the dumb charging. Nonetheless, due to increased charging demand, idle EVs provide support at the slot 5:30 a.m. as well as at 5:00 p.m. to 6:00 p.m., maintaining the voltages within the bounds. Accordingly, during sunny periods, shifted EV charging can be noticed as well.

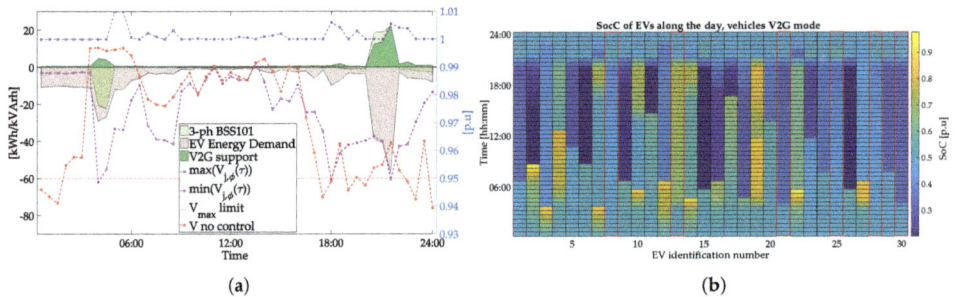

Figure 14. Case 05: (**a**) resulting voltage ranges, coordinated charging in comparison with dumb charging; BESS$_{101}$ scheduling of operation and V2G actions; (**b**) SoC for all EVs; circled by red line correspond to V2G mode of operation.

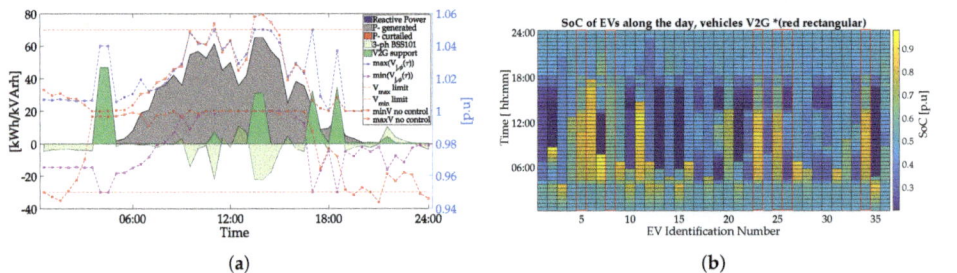

Figure 15. Case 06: (**a**) resulting voltage ranges, coordinated charging in comparison with dumb charging; BESS$_{101}$ scheduling of operation and V2G actions; (**b**) SoC for all EVs, circled with a red line, correspond to V2G mode of operation.

6. Conclusions

This work presents a tool that is capable of providing support to the DSO decision for the operation of LV three-phase distribution grids with increased integration of DER. The computational techniques proposed based on the explicit calculation of the first and second order derivatives (i.e., Jacobian and Hessian of the Lagrangian and the objective function), in addition to pivotal adjustments in the Jacobian, ensure a tractable optimal control based on the exact AC power flows.

The proposed centralized scheme ensures admissible voltage profiles by minimizing the active power curtailments of microgeneration through the efficient coordination of DER—maximizing, in this sense, the integration of microgeneration. The coordinated operation among the DER units reassures in the presented study up to 73% integration of microgeneration avoiding any curtailment brought by PV units. The EV integration can be also maximized if coordinated charging is adopted within the MACOPF, which essentially can ensure admissible voltages and normal loading condition for the transformer. The V2G mode of operation can be regarded as important when a high integration of EV takes place with increased peak load conditions in the grid.

For all simulated cases, one can conclude that the radial configuration of LV networks present overvoltage issues particularly in buses—electrically—furthest from the substation since the high resistance of the lines leads to the aggravation of them. In addition, distant nodes at the ending point might face significant voltage drops (i.e., if EVs are present) or even overvoltages (i.e., if PVs are present). Therefore, the placement of a BESS in proper location (i.e., adjacent to nodes with voltage issues) along the grid is rather crucial since the installation adjacent to the secondary substation provides mainly reduced loading for the transformer rather than voltage support to the furthest points.

The proposed control scheme is based on deterministic analysis for the planning in a day-ahead scale for the operation of the grid. Meanwhile, the scheme can be deployed only by the subsequent communication technologies (for online implementation), together with forecasted data and power flow-state estimation tools (i.e., if topology is not known).

Author Contributions: Formal analysis, K.K.; Methodology, K.K.; Software, K.K.; Supervision, I.M., N.S. and H.L.; Validation, K.K.; Writing—original draft, K.K.; Writing—review & editing, I.M., N.S. and H.L.

Funding: This project has received funding from the European Unions Horizon 2020 research and innovation programme under the Marie Skodowska-Curie Grant No. 675318 (INCITE).

Conflicts of Interest: The authors declare no conflict of interest.

Abbreviations

The following abbreviations are used in this manuscript:

DER	Distributed Energy Resources
LV	Low Voltage
OPF	Optimal Power Flow
PV	Photovoltaic
BESS	Battery Storage System
EV	Electric Vehicles
DSO	Distribution System Operator
A-DMS	Advanced Distribution Management Systems
CL	Controllable Loads
OLTC	On Load Tap Changer
MACOPF	Multiperiod AC-OPF
IP	Interior-Point
SoC	State-of-Charge
V2G	Vehicle-to-Grid
APC	Active Power Curtailment
QR	Reactive Power Control
BFS	Backward-Forward Sweep
KKT	Karush–Kuhn–Tucker
LICQ	Linear Constraint Qualification
CQ	Constraint Qualification
CCV	Cost Constrained Variable

Appendix A. Complementary Data for Case Study

Table A1. SPoints of PV Connections for the Presented Scenarios.

Points of PV connections	522.2/388.1/178.2/ 676.2/639.2/337.3/ 701.3/614.3/562.1/ 682.2/70.1/556.3/ 629.1/47.2/349.1/ 563.1/264.3/458.3/ 249.2/289.1	611.1/74.1/320.3/ 73.1/276.2/225.1/ 327.3/387.1/619.3/ 702.2	702.2/502.1/342.3/ 208.3/539.3/ 688.2/406.2	248.2/83.2/314.2/ 896.1/785.2/900.1/ 899.2/755.2/780.3/ 898.1/813.2

Scenarios [% of PV]	30%			
		55%		
			65%	
				85%

Table A2. Points of EV Connections for the Presented Scenarios.

Points of EV connections	327.3/835.3/785.2/563.1/ 755.2/249.2/225.1/47.2/ 886.2/898.1/314.2/208.3/ 906.1/861.1/320.3/682.2/ 780.3/406.2/817.1/248.2	619.3/860.1/702.2/ 458.3/899.2/264.3/ 178.2/83.2/337.3/ 556.3	73.1/349.1/701.3/ 522.2/342.3/289.1

Scenarios [% of EV]	30%		
		55%	
			65%

Appendix B. First and Second Order Derivatives for Multi-Variable Function

In this section, a brief description of the calculations of the first and second order gradients for the nonlinear constraints and the objective function. In general, for a scalar function $f : \mathbb{R}^k \to \mathbb{R}$ of a real vector $X = [x_1, x_2, \ldots, x_k]^T$, the Jacobian matrix that corresponds to the first-order derivatives is given as:

$$\frac{\partial f}{\partial X} = \begin{bmatrix} \frac{\partial f}{\partial x_1} & \cdots & \frac{\partial f}{\partial x_k} \end{bmatrix}. \tag{A1}$$

The Hessian matrix of f that corresponds to the second-order derivatives is:

$$\frac{\partial^2 f}{\partial X^2} = \frac{\partial}{\partial X}\left(\frac{\partial f}{\partial X}\right)^T = \begin{bmatrix} \frac{\partial^2 f}{\partial x_1^2} & \cdots & \frac{\partial f}{\partial x_1 \partial^2 x_k} \\ \vdots & \ddots & \vdots \\ \frac{\partial^2 f}{\partial x_k \partial x_1} & \cdots & \frac{\partial^2 f}{\partial x_{x_k}^2} \end{bmatrix}. \tag{A2}$$

For a vector function $G : \mathbb{R}^k \to \mathbb{R}^m$ of a vector X, the first derivatives that form the Jacobian matrix is:

$$G_X = \frac{\partial G}{\partial X} = \begin{bmatrix} \frac{\partial g_1}{\partial x_1} & \cdots & \frac{\partial g_1}{\partial x_k} \\ \vdots & \ddots & \vdots \\ \frac{\partial g_m}{\partial x_1} & \cdots & \frac{\partial g_m}{\partial x_k} \end{bmatrix}. \tag{A3}$$

References

1. Eid, C.; Codani, P.; Perez, Y.; Reneses, J.; Hakvoort, R. Managing electric flexibility from Distributed Energy Resources: A review of incentives for market design. *Renew. Sustain. Energy Rev.* **2016**, *64*, 237–247, doi:10.1016/j.rser.2016.06.008. [CrossRef]

2. Lotfi, M.; Monteiro, C.; Shafie-khah, M.; Catalão, J.P.S. Evolution of Demand Response: A Historical Analysis of Legislation and Research Trends. In Proceedings of the 2018 Twentieth International Middle East Power Systems Conference (MEPCON), Cairo, Egypt, 18–20 December2018; pp. 968–973, doi:10.1109/MEPCON.2018.8635264. [CrossRef]
3. Ochoa, L.F.; Mancarella, P. Low-carbon LV networks: Challenges for planning and operation. In Proceedings of the 2012 IEEE Power and Energy Society General Meeting, San Diego, CA, USA, 22–26 July 2012; pp. 1–2.
4. Bruno, S.; La Scala, M. Unbalanced Three-Phase Optimal Power Flow for the Optimization of MV and LV Distribution Grids. In *From Smart Grids to Smart Cities: New Challenges in Optimizing Energy Grids*; Wiley: Hoboken, NJ, USA, 2016; pp. 1–42. doi:10.1002/9781119116080.ch1.
5. Control and Management Architectures. *Smart Grid Handbook*; Wiley: Hoboken, NJ, USA, 2016. doi:10.1002/9781118755471.sgd058. [CrossRef]
6. Karagiannopoulos, S.; Roald, L.; Aristidou, P.; Hug, G. Operational Planning of Active Distribution Grids under Uncertainty. Available online: http://eprints.whiterose.ac.uk/120043/ (accessed on 31 January 2018).
7. Bruno, S.; Lamonaca, S.; Rotondo, G.; Stecchi, U.; La Scala, M. Unbalanced three-phase optimal power flow for smart grids. *IEEE Trans. Ind. Electron.* **2011**, *58*, 4504–4513. [CrossRef]
8. Faria, P.; Spínola, J.; Vale, Z. Distributed Energy Resources Scheduling and Aggregation in the Context of Demand Response Programs. *Energies* **2018**, *11*, 1987. [CrossRef]
9. Soares, T.; Silva, M.; Sousa, T.; Morais, H.; Vale, Z. Energy and Reserve under Distributed Energy Resources Management—Day-Ahead, Hour-Ahead and Real-Time. *Energies* **2017**, *10*, 1778. [CrossRef]
10. Tonkoski, R.; Lopes, L.A.C.; El-Fouly, T.H.M. Coordinated Active Power Curtailment of Grid Connected PV Inverters for Overvoltage Prevention. *IEEE Trans. Sustain. Energy* **2011**, *2*, 139–147, doi:10.1109/TSTE.2010.2098483. [CrossRef]
11. Weckx, S.; Gonzalez, C.; Driesen, J. Combined Central and Local Active and Reactive Power Control of PV Inverters. *IEEE Trans. Sustain. Energy* **2014**, *5*, 776–784, doi:10.1109/TSTE.2014.2300934. [CrossRef]
12. Demirok, E.; González, P.C.; Frederiksen, K.H.B.; Sera, D.; Rodriguez, P.; Teodorescu, R. Local Reactive Power Control Methods for Overvoltage Prevention of Distributed Solar Inverters in Low-Voltage Grids. *IEEE J. Photovolt.* **2011**, *1*, 174–182, doi:10.1109/JPHOTOV.2011.2174821. [CrossRef]
13. Heleno, M.; Rua, D.; Gouveia, C.; Madureira, A.; Matos, M.A.; Lopes, J.P.; Silva, N.; Salustio, S. Optimizing PV self-consumption through electric water heater modeling and scheduling. In Proceedings of the 2015 IEEE Eindhoven PowerTech, Eindhoven, The Netherlands, 29 June–2 July 2015; pp. 1–6, doi:10.1109/PTC.2015.7232636. [CrossRef]
14. Olivier, F.; Aristidou, P.; Ernst, D.; Van Cutsem, T. Active Management of Low-Voltage Networks for Mitigating Overvoltages Due to Photovoltaic Units. *IEEE Trans. Smart Grid* **2016**, *7*, 926–936, doi:10.1109/tsg.2015.2410171. [CrossRef]
15. Degroote, L.; Renders, B.; Meersman, B.; Vandevelde, L. Neutral-point shifting and voltage unbalance due to single-phase DG units in low voltage distribution networks. In Proceedings of the 2009 IEEE Bucharest PowerTech, Bucharest, Romania, 28 June–2 July 2009; pp. 1–8, doi:10.1109/PTC.2009.5281998. [CrossRef]
16. Miranda, I.; Leite, H.; Silva, N. Coordination of multifunctional distributed energy storage systems in distribution networks. *IET Gener. Transm. Distrib.* **2016**, *10*, 726–735, doi:10.1049/iet-gtd.2015.0398. [CrossRef]
17. Fortenbacher, P.; Zellner, M.; Andersson, G. Optimal sizing and placement of distributed storage in low voltage networks. In Proceedings of the 2016 Power Systems Computation Conference (PSCC), Genoa, Italy, 20–24 June 2016; pp. 1–7, doi:10.1109/PSCC.2016.7540850. [CrossRef]
18. Hoppmann, J.; Volland, J.; Schmidt, T.S.; Hoffmann, V.H. The economic viability of battery storage for residential solar photovoltaic systems—A review and a simulation model. *Renew. Sustain. Energy Rev.* **2014**, *39*, 1101–1118. [CrossRef]
19. European Union. Directive 2009/72/EC of the European Parliament and of the Council of 13 July 2009 Concerning Common Rules for the Internal Market in Electricity and Repealing Directive 2003/54/EC. *Off. J. Eur. Union* **2009**, *211*, 55–93.
20. Efkarpidis, N.; De Rybel, T.; Driesen, J. Optimization control scheme utilizing small-scale distributed generators and OLTC distribution transformers. *Sustain. Energy Grids Netw.* **2016**, *8*, 74–84, doi:10.1016/j.segan.2016.09.002. [CrossRef]

21. Costa, H.M.; Sumaili, J.; Madureira, A.G.; Gouveia, C. A multi-temporal optimal power flow for managing storage and demand flexibility in LV networks. In Proceedings of the 2017 IEEE Manchester PowerTech, Manchester, UK, 18–22 June 2017; pp. 1–6, doi:10.1109/PTC.2017.7981166. [CrossRef]

22. Olival, P.C.; Madureira, A.G.; Matos, M. Advanced voltage control for smart microgrids using distributed energy resources. *Electr. Power Syst. Res.* **2017**, *146*, 132–140, doi:10.1016/j.epsr.2017.01.027. [CrossRef]

23. Madureira, A.; Gouveia, C.; Moreira, C.; Seca, L.; Lopes, J.P. Coordinated management of distributed energy resources in electrical distribution systems. In Proceedings of the 2013 IEEE PES Conference on Innovative Smart Grid Technologies (ISGT Latin America), Sao Paulo, Brazil, 5–17 April 2013; pp. 1–8, doi:10.1109/ISGT-LA.2013.6554446. [CrossRef]

24. Kotsalos, K.; Silva, N.; Miranda, I.; Leite, H. Scheduling of operation in Low Voltage distribution networks with multiple Distributed Energy Resources. In Proceedings of the CIRED Workshop, Ljubljana, Slovenia, 7–8 June 2018.

25. Connell, A.O.; Flynn, D.; Keane, A. Rolling Multi-Period Optimization to Control Electric Vehicle Charging in Distribution Networks. *IEEE Trans. Power Syst.* **2014**, *29*, 340–348, doi:10.1109/TPWRS.2013.2279276. [CrossRef]

26. Campos, F.; Marques, L.; Kotsalos, K. Electric Vehicle CPMS and Secondary Substation Management. In Proceedings of the 8th Solar & 17th Wind Integration Workshop, Stockholm, Sweden, 16–17 October 2018.

27. Efkarpidis, N.; De Rybel, T.; Driesen, J. Technical assessment of centralized and localized voltage control strategies in low voltage networks. *Sustain. Energy Grids Netw.* **2016**, *8*, 85–97, doi:10.1016/j.segan.2016.09.003. [CrossRef]

28. Sperstad, I.B.; Marthinsen, H. *Optimal Power Flow Methods and Their Application to Distribution Systems with Energy Storage: A Survey of Available Tools and Methods*; SINTEF Energi. Rapport; SINTEF: Trondheim, Norway, 2016.

29. Sereeter, B.; Vuik, K.; Witteveen, C. Newton Power Flow Methods for Unbalanced Three-Phase Distribution Networks. *Energies* **2017**, *10*, 1658. [CrossRef]

30. Karagiannopoulos, S.; Aristidou, P.; Hug, G. Co-optimisation of Planning and Operation forActive Distribution Grids. In Proceedings of the 2017 IEEE Manchester PowerTech, Manchester, UK, 18–22 June 2017.

31. Christakou, K.; Tomozei, D.C.; Le Boudec, J.Y.; Paolone, M. AC OPF in radial distribution networks–Part I: On the limits of the branch flow convexification and the alternating direction method of multipliers. *Electr. Power Syst. Res.* **2017**, *143*, 438–450. [CrossRef]

32. Ciric, R.M.; Feltrin, A.P.; Ochoa, L.F. Power flow in four-wire distribution networks-general approach. *IEEE Trans. Power Syst.* **2003**, *18*, 1283–1290, doi:10.1109/tpwrs.2003.818597. [CrossRef]

33. Cheng, C.S.; Shirmohammadi, D. A three-phase power flow method for real-time distribution system analysis. *IEEE Trans. Power Syst.* **1995**, *10*, 671–679. [CrossRef]

34. Bazrafshan, M.; Gatsis, N. Comprehensive Modeling of Three-Phase Distribution Systems via the Bus Admittance Matrix. *IEEE Trans. Power Syst.* **2018**, *33*, 2015–2029, doi:10.1109/TPWRS.2017.2728618. [CrossRef]

35. Gorman, M.; Grainger, J. Transformer modelling for distribution system studies. II. Addition of models to Y/sub BUS/and Z/sub BUS. *IEEE Trans. Power Deliv.* **1992**, *7*, 575–580. [CrossRef]

36. Shirmohammadi, D.; Hong, H.W.; Semlyen, A.; Luo, G.X. A compensation-based power flow method for weakly meshed distribution and transmission networks. *IEEE Trans. Power Syst.* **1988**, *3*, 753–762, doi:10.1109/59.192932. [CrossRef]

37. Kourounis, D.; Fuchs, A.; Schenk, O. Towards the Next Generation of Multiperiod Optimal Power Flow Solvers. *IEEE Trans. Power Syst.* **2018**, doi:10.1109/TPWRS.2017.2789187. [CrossRef]

38. Nocedal, J.; Wright, S.J. *Numerical Optimization*, 2nd ed.; Springer: New York, NY, USA ,2006; pp. 497–528.

39. Zhu, J. *Optimization of Power System Operation*; John Wiley & Sons, Inc.: Hoboken, NJ, USA, 2015; pp. 1–12, doi:10.1002/9781118887004.ch1.

40. Wachter, A. An Interior Point Algorithm for Large-Scale Nonlinear Optimization with Applications in Process Engineering. Ph.D. Thesis, Carnegie Mellon University, Pittsburgh, PA, USA, 2003.

41. Hauswirth, A.; Bolognani, S.; Hug, G.; Dörfler, F. Generic Existence of Unique Lagrange Multipliers in AC Optimal Power Flow. *arXiv* **2018**, arXiv:1806.06615.

42. Torres, G.L.; Quintana, V.H. An interior-point method for nonlinear optimal power flow using voltage rectangular coordinates. *IEEE Trans. Power Syst.* **1998**, *13*, 1211–1218, doi:10.1109/59.736231. [CrossRef]

43. Wächter, A.; Biegler, L.T. On the implementation of an interior-point filter line-search algorithm for large-scale nonlinear programming. *Math. Program.* **2006**, *106*, 25–57, doi:10.1007/s10107-004-0559-y. [CrossRef]

44. Coleman, T.; Branch, M.A.; Grace, A. Optimization toolbox. In *For Use with MATLAB. User's Guide for MATLAB 5, Version 2, Relaese II*; MATLAB, MathWorks: Natick, MA, USA, 1999.

45. Zimmerman, R.D. *AC Power Flows, Generalized OPF Costs and Their Derivatives Using Complex Matrix Notation*; Report; PSERC: Tempe, AZ, USA, 2010.

46. Frank, S.; Rebennack, S. An introduction to optimal power flow: Theory, formulation, and examples. *IIE Trans.* **2016**, *48*, 1172–1197, doi:10.1080/0740817X.2016.1189626. [CrossRef]

47. Wood, A.J.; Wollenberg, B.F.; Sheblé, G.B. *Power Generation, Operation, and Control*, 3rd ed.; John Wiley & Sons: Hoboken, NJ, USA, 2013; book section 8.

48. Gilbert, J.C.; Josz, C. Plea for a Semidefinite Optimization Solver in Complex Numbers. Available online: https://hal.inria.fr/hal-01497173/document (accessed on 31 January 2018)

49. Baker, K.; Hug, G.; Xin, L. Inclusion of inter-temporal constraints into a distributed Newton-Raphson method. In Proceedings of the 2012 North American Power Symposium (NAPS), Boston, MA, USA, 4–6 August 2011; pp. 1–6, doi:10.1109/NAPS.2012.6336344. [CrossRef]

50. Baker, K.; Zhu, D.; Hug, G.; Li, X. Jacobian singularities in optimal power flow problems caused by intertemporal constraints. In Proceedings of the 2013 North American Power Symposium (NAPS), Manhattan, KS, USA, 22–24 September 2013; pp. 1–6, doi:10.1109/NAPS.2013.6666876. [CrossRef]

51. Espinosa, A.N. *Dissemination Document "Low Voltage Networks Models and Low Carbon Technology Profiles"*; The University of Manchester: Manchester, UK, 2015.

52. Palizban, O.; Kauhaniemi, K. Energy storage systems in modern grids—Matrix of technologies and applications. *J. Energy Storage* **2016**, *6*, 248–259, doi:10.1016/j.est.2016.02.001. [CrossRef]

53. Osório, G.; Shafie-khah, M.; Coimbra, P.; Lotfi, M.; Catalão, J. Distribution system operation with electric vehicle charging schedules and renewable energy resources. *Energies* **2018**, *11*, 3117. [CrossRef]

54. Efacec. EV Homecharger. Available online: http://electricmobility.efacec.com/wp-content/uploads/2016/10/CS195I1404C1_HC.pdf (accessed on 30 November 2018).

55. Survey, N.T. *National Travel Survey: England 2016*; Report; National Travel Survey: Great Britain, UK, 2016.

56. Pedersen, R.; Sloth, C.; Andresen, G.B.; Wisniewski, R. DiSC: A simulation framework for distribution system voltage control. In Proceedings of the 2015 European Control Conference (ECC), Linz, Austria, 15–17 July 2015; pp. 1056–1063.

57. Richardson, P.; Moran, M.; Taylor, J.; Maitra, A.; Keane, A. Impact of electric vehicle charging on residential distribution networks: An Irish demonstration initiative. In Proceedings of the 22nd International Conference and Exhibition on Electricity Distribution (CIRED 2013), Stockholm, Sweden, 10–13 June 2013; Volume 2013, pp. 1–4, doi:10.1049/cp.2013.0873. [CrossRef]

58. Masetti, C. Revision of European Standard EN 50160 on power quality: Reasons and solutions. In Proceedings of the 2010 14th International Conference on Harmonics and Quality of Power (ICHQP), Bergamo, Italy, 26–29 September 2010; pp. 1–7.

![energies logo] *energies*

MDPI

Review

A Healthy, Energy-Efficient and Comfortable Indoor Environment, a Review

Paulína Šujanová [1,2,*], Monika Rychtáriková [1,3], Tiago Sotto Mayor [4] and Affan Hyder [5]

1   Faculty of Civil Engineering, Slovak University of Technology in Bratislava, Radlinského 11,
    810 05 Bratislava, Slovakia; monika.rychtarikova@kuleuven.be
2   A&Z Acoustics s.r.o., Letisko M.R. Štefánika 63, 820 01 Bratislava, Slovakia
3   Faculty of Architecture, KU Leuven, Hoogstraat 51, 9000 Gent/ Paleizenstraat 65, 1030 Brussels, Belgium
4   Transport Phenomena Research Centre, Engineering Faculty of Porto University, Rua Dr. Roberto Frias,
    4200-465 Porto, Portugal; tiago.sottomayor@fe.up
5   Vesalius College Brussels, Boulevard de la Plaine 5, 1050 Brussels, Belgium; affanhyder@hotmail.com
*   Correspondence: paulina.sujanova@stuba.sk

Received: 28 February 2019; Accepted: 3 April 2019; Published: 12 April 2019

Abstract: Design strategies for sustainable buildings, that improve building performance and avoid extensive resource utilization, should also promote healthy indoor environments. The following paper contains a review of the couplings between (1) building design, (2) indoor environmental quality and (3) occupant behavior. The paper focuses on defining the limits of adaptation on the three aforementioned levels to ensure the energy efficiency of the whole system and healthy environments. The adaptation limits are described for measurable physical parameters and the relevant responsible human sensory systems, evaluating thermal comfort, visual comfort, indoor air quality and acoustical quality. The goal is to describe the interactions between the three levels where none is a passive participant, but rather an active agent of a wider human-built environment system. The conclusions are drawn in regard to the comfort of the occupant. The study reviews more than 300 sources, ranging from journals, books, conference proceedings, and reports complemented by a review of standards and directives.

Keywords: indoor environment quality; occupant comfort; building climate control; healthy building; energy efficiency; adaptability

---

1. Introduction

On average, people spend around 80–90% of their lives inside buildings [1]. Therefore, buildings have to provide a healthy and comfortable environment for humans. Buildings account for approximately 19% of all global greenhouse gas (GHG) emissions in the world [2], and about 31% of global final energy demand [3]. In recent years, much effort has been put into the development of efficient and cost-effective technologies, to ensure sustainability of the built environment [4,5]. Reduction of energy demand and the increase of energy efficiency is considered to provide a dominant contribution to tackle global climate change [6]. However, the amount of energy consumed depends not only on the criteria set for the indoor environment and applied technology but also on the behavior of occupants [7]. This may create a conflict between strategies that focus on the reduction of energy consumption and those to maintain a healthy and comfortable indoor environment. To achieve a balance between comfort and efficiency, synergies between building design, building climate control and occupant needs have to be established. Future buildings have to be able to quickly react and adapt themselves to immediate requirements (weather, occupancy, function) [8].

In this paper, we will describe the process of adaptation, that can take place in the human-built environment system, at three different levels (Figure 1):

- at the building design and construction level (passive and active designs),
- at the indoor environmental quality control level (management of indoor environment controls),
- and at the occupant level.

The goal is to raise awareness about the importance of the interdependencies between these levels, which is done by defining limits for each identified adaptation level. The paper is organized in six sections (Figure 2). In Section 2 a short description of the used research methods is presented along with a holistic analysis of selected review papers. In Section 3 we focus on the indoor environmental factors (IEQ). We start with a description of healthy environments and continue with standard requirements for IEQ performance. This is followed by a description of requirements for occupants' well-being. The section ends with a description of the negative impact of poor IEQ on occupants' health and performance. In Section 4 we focus on human physical, behavioral and physiological limits in relation to the ability to react to environmental conditions. In Section 5 a short introduction in building energy efficiency policy is given before a research review summary is presented focused on operational and technological limits of building design and building climate control system that lead to healthy, energy efficient and comfortable indoor environments. The section concludes with the description of factors influencing the quality of interactions between the occupant and the building environment.

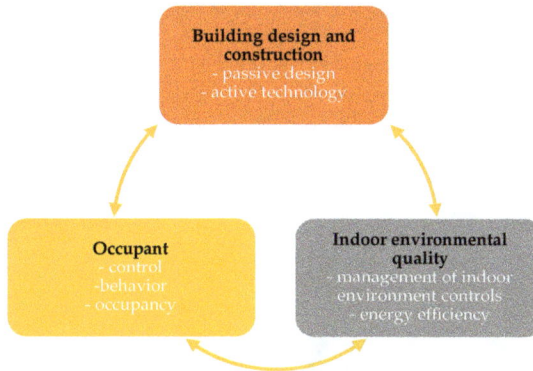

Figure 1. Connections between three levels (Building-Indoor environmental factors (IEQ)-Occupant) of a healthy and energy efficient building concept.

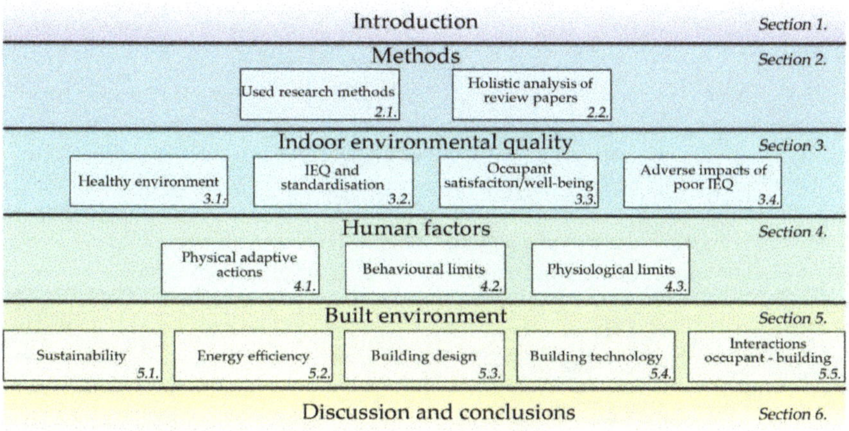

Figure 2. Organization and sections of the review paper.

2. Methods

2.1. Research Methods Used

The text of this paper is drafted based on a literature review, presenting results of studies on healthy indoor environment and energy efficiency. This was achieved using an internet-based search on relevant scholarly articles in the databases of Web of Science, Scopus and Google Scholar. In these databases a search for identifiable keywords was performed, to identify articles within the scope of our literature review, whereas the goal was to gain more knowledge about the relationship between occupant, indoor environment and energy efficiency. The search terms used were, "indoor environmental quality, thermal comfort, visual comfort, air quality, acoustic quality, healthy environment, energy efficiency, nearly zero energy buildings, occupant comfort, human sensory system, adaptive behavior, building climate control, building management systems and predictive technologies". The articles found were then briefly scanned for relevancy, and articles considered unfit for the purpose of this study were removed from the created collection. This resulted in a database of journal articles and conference proceedings, that were further uploaded to a reference management system and checked for duplicates. Further, the search results were briefly scanned and divided into groups to identify potential patterns. When necessary the review was complemented by adding references to directives, standards, and databases. As a result, we present in this paper a review of more than 212 journal articles, 17 conference proceedings, 32 reports, 17 books, six directives, 15 standards, four declarations, six web-pages, four databases and one constitution published between 1960 and 2019 (Figure 3).

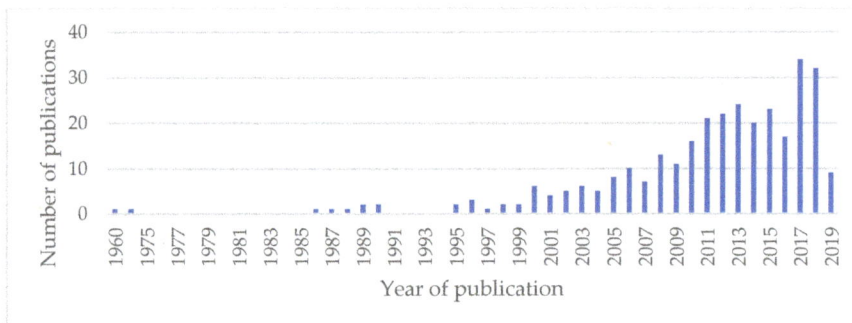

Figure 3. Number of publications analyzed according to publication year (based on publications used in the presented paper).

2.2. Holistic Analysis of Review Papers

To check whether or not there was already a review paper on the topic of the present work, and to find out trends in the current publications, we made a holistic analysis of the identified review papers focusing on defining the relationship between the occupant and the indoor environment, and further about the technological approaches to ensure the IEQ. We have identified 93 such papers published between 1998 and 2019. The identified review papers focused on the human-built environment relationship, describing IEQ and its effect on health, well-being and comfort, occupant productivity, and user satisfaction [9–14]. In addition to that, the papers described factors influencing human comfort and health [15–20], driving factors for occupant behaviors in buildings [21–24], occupant productivity [25], effect of behavior on energy consumption [7,26,27], and links between user satisfaction and adaptive behavior [28,29]. Literature reviews describing technological aspects of indoor environment control focus on advances in energy and environmental performance of buildings [6,30–39], principles of smart buildings [40–42], passive and green buildings design [43–45], intelligent control systems [46–53], building and façade design [8,54–59], and effects of climate change on indoor environmental quality [60]. Further reviews have addressed occupant behavior modeling [61,62] and occupancy based model predictive control for building indoor climate control [63,64]. Lastly, papers focused on indoor environmental quality factors in general [65–67], and individually on thermal comfort [68–78], visual comfort [79–84], acoustic comfort [85–88], indoor air quality [89–98].

These previous reviews show a trend in increased technologization of indoor environment control, whereas a predominance in studies focused on thermal comfort and indoor air quality was observed. It has to be noted, that in the reviews it was repeatedly stated that occupant satisfaction with one physical parameter of IEQ is strongly dependent on satisfaction with all other IEQ indicators and that it is necessary to take into account all IEQ factors when evaluating indoor comfort otherwise the evaluation could lead to false assumptions. While reviewing literature surveys on building climate control, these focus on thermal comfort, visual comfort, indoor air quality, and energy efficiency, while neglecting acoustic comfort, and the overall impact on the health of the occupants. Furthermore, there is a gap in describing the connections between user, indoor environment quality and building design. Studies generally focus on each aspect individually, but a comprehensive definition of the couplings is not provided. Therefore, we have decided to focus on these couplings, with the aim of providing a framework within which the factors that limit the interactions between occupants, IEQ and building design are identified.

On these identified review papers, a data-driven analysis on bibliographic data, such as keyword analysis was performed. This enables the visualization of trends in research publications via analysis of co-occurrence of field-specific terms. This was done using a freely available software tool for analyzing bibliographic data, VOSviewer [99]. The mapping approach employed was keyword network visualization on co-occurrence and link strength of terms extracted from abstracts of the identified review papers (Figure 3). Out of a total of 2331 terms identified occurring in the abstracts, the terms with a minimum of five occurrences and the top 52 according to relevance were extracted, whereas unrelated words were eliminated (e.g., generic terms such as review, field, literature) and abbreviations were replaced by full terms with the use of a self-defined thesaurus file. This resulted in a network visualization of terms, where the size of the label and circle of a term determines its importance (the higher the importance the larger the label and circle), the link represents a connection between two terms and its strength represents the strength of co-occurrence, and the distance between two terms is representative of their relative co-occurrence. Lastly, the network was color-overlaid (see a legend in Figure 4) where the color assigned represents the average publication year of a term, enabling the analysis of trends in the publications.

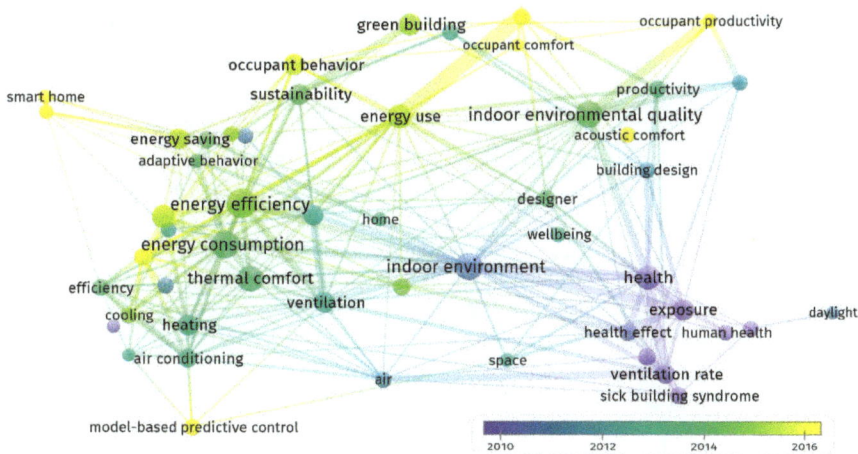

Figure 4. Keyword network visualization on co-occurrence and link strength of terms extracted from abstracts of review papers, color scale based on the average year of publication.

We can observe a shift in trends in the topics on which the publications focus. Firstly, before 2010 the publications focused mainly on the health effects of the indoor environment. More recent publications, published in 2014 or later, focus on control strategies and connections between energy use and occupant comfort. After a detailed analysis, a strong link between the terms energy use and human dimension can be observed with increasing occurrence of these terms in more recent publications, indicating a rise in awareness of the importance of this coupling. The term with the highest number of links can be found for the term indoor environment (42).

3. Definition of Indoor Environmental Quality

3.1. Healthy Environment

The World Health Organization (WHO) defines the environment, as it relates to health, as *"all the physical, chemical, and biological factors external to a person, and all the related behaviours"* [100]. The WHO report also confirms that *"approximately one-quarter of the global disease burden is due to modifiable environmental factors."* In our research, we will focus only on the information about building environmental performance.

The research of the last 20 years has identified a direct link between building design, human health, and well-being. It has been established, that the IEQ has a direct impact on the human health and work and/or study performance [10,14,101] and might even influence the development of learning disorders such as dyslexia, or voice problems, due to insufficient acoustic conditions [102]. Recent research on improved building environments is focused on the development of appropriate actions to reduce or eliminate harmful health effects [103]. The concept of a "healthy building" was identified by the WHO which established first guidelines to ensure well-being [104]. To assess the performance of the building environment, various approaches have been developed [43,105] but interactions occurring at different levels (at the building design level, IEQ control level, and occupant level) are still largely ignored [16,106].

According to HOPE [107], Health Optimization Protocol for Energy-efficient buildings, a building is defined as *"healthy and energy efficient"* if:

"it does not cause or aggravate illnesses in the building occupants; it assures a high level of comfort for the building occupants; it minimises the use of energy used to achieve desired internal conditions, taking into account available state-of-the art technology and non-technical measures."

The criteria include a set of measurable physical parameters, called indoor environmental factors, evaluating thermal comfort, visual comfort, indoor air quality, and acoustical quality.

3.2. General Definition of IEQ and Approaches of Standardization

IEQ refers to the quality of a building's indoor environment whereas the health and well-being of building occupants should be promoted. Subjective perception of comfort of occupants in a room depends on many factors. According to Sarbu and Sebarchievici these are [108]: *"temperature, humidity and air circulation, smell and respiration, touch and touching, acoustic factors, sight and colours effect, building vibrations, special factors (solar gain, ionization), safety factors, economic factors, unpredictable risks and design."* This review paper will focus only on measurable physical parameters, evaluating thermal comfort, visual comfort, indoor air quality and acoustical quality, Table 1. Further, the connection between these physical parameters and the human physiology will be described. A detailed overview of performance criteria of buildings for health and comfort can be found in [15,109,110].

Table 1. Indoor environment quality factors and their parameters.

	Thermal Comfort	Visual Comfort	Acoustic Comfort	Indoor Air Quality
Note		Requirements for lighting and daylighting		Close connection to thermal environment
Parameters	PMV/PPD or operative temperature; Humidity; Air velocity; Draught; Vertical air temperature differences; Radiant temperature asymmetry; Surface temperature of the floor.	Daylight provision; View out; Exposure to sunlight, Illuminance; Luminance; Light uniformity; Glare; Color (color rendering, light source color);	Sound level difference; Sound insulation; Sound absorption; Noise level(s) Frequency; Reverberation time.	Indoor sources of pollution; Outdoor sources of pollution; Ventilation parameters; Airflow patterns and pressure relationships; Air filtration system.
Influence factors	Clothing; Activity	Light source, Visual task; Use of room		Sources of heat gains; Outdoor conditions and outdoor air ventilation rate

3.2.1. Thermal Comfort

According to ASHRAE 55 [111] thermal comfort is *"that state of mind which expresses satisfaction with the thermal environment"*. Thermal comfort comprises parameters, such as temperature, humidity, and air velocity (see Table 1). The recommended criteria are specified by international standards such as ASHRAE 55 [111], ISO 17772-1 [112], EN ISO 7730 [113]. These documents specify methods for the measurement and evaluation of thermal environments and provide calculation tools that enable prediction of the overall satisfaction of occupants with the indoor thermal conditions (i.e., predicted mean vote, predicted percentage of dissatisfied and adaptive predicted mean vote).

The thermal comfort parameters can be controlled via heating, cooling and air conditioning systems. It has to be mentioned that all of the technologies controlling thermal comfort are slow-reacting technologies, meaning their adaptation to requirements (e.g., new air temperature) takes some time.

3.2.2. Visual Comfort

Visual comfort is defined according to EN 12665 [114] as *"the subjective condition of visual well-being induced by the visual environment"*. A well-designed lighting system must provide adequate illumination to ensure safety and enable movement, contribute to visual comfort and facilitate visual performance and color perception. Monitored parameters include quantitative physical measures of the luminous environment (illuminance, luminance, daylight provision and glare) and qualitative aspects of vision (distribution, uniformity, color rendering, the spectral composition of radiation). A significant part of the illumination of spaces should be provided by daylight with daylight openings, which provide a view to the outside, contributing to the psychological well-being of the occupants [115].

The lighting properties are based on photopic requirements, psychological and photobiological stimulation influencing the health of the occupant. The research of the last 30 years in the field of daylighting has proven a direct relationship between natural light and health. It has been proven that the luminous environment has to ensure visual as well as biophysical human well-being [116–118]. Insufficient or inappropriate light can lead to distortions of internal biological rhythms. This may have an impact on performance, safety, and health. Exposure to adequate light promotes the synchronization of the internal human circadian rhythms linked to hormone secretion.

It should be noted that current international standards are based only on photopic (diurnal) sensitivity of the human eye, ignoring scotopic (nocturnal) vision and non-visual effects of light [119]. A review of daylighting and lighting control strategies can be found in [81,119–121].

Lighting systems and solar shading systems belong to the fast-reacting systems. Consequentially the change in conditions can take place nearly immediately after a change of requirements. They also belong to the low-energy consumption systems. Nevertheless, it can account for up to 25% of the total building energy consumption [122]. Here, the use of daylight sensors and photocells leads to a major reduction of the energy consumption for lighting [123].

3.2.3. Indoor Air Quality

Standards related to IEQ define ventilation rates, humidity and exposure limits for air pollutants. Measurements of indoor air quality are based on an indirect approach of measuring the intensity of the ventilation, whereas the recommended ventilation rates can be found in ISO 1772-1 [112]. Only when the requirements for ventilation rates are reached, the measurements for specific pollutants can be made. Some recommended values according to the type of building can be found in Guidelines for indoor air quality for selected air pollutants [124]. EN 16798-3 [125] provides performance requirements for non-residential buildings. ASHRAE 62.1 and 62.2 and CR 1752 [126–128] describe general ventilation requirements for acceptable indoor air quality, further complemented by a guide for design, construction and commissioning [129]. In all current ventilation standards, only two methods of evaluation are described, a *"prescriptive method"* and an *"analytical procedure"* [110]. Indoor air quality is closely related to thermal comfort. Thus, some recommendations such as humidity can be found in standards defining thermal comfort.

Indoor air quality can be controlled in natural ways (windows, wall openings), and artificial ways (air conditioning system). Meaning that the change of air can be either supplied naturally through windows, or it can be introduced by an air conditioning system. Ventilation belongs to the fast reacting systems, so the adaptation to changed requirements can take place immediately.

3.2.4. Acoustic Quality

Acoustic comfort is often understood to be a situation where there is an acceptable level of noise. However, the perception of sound is a much more complex issue, that depends not only on sound intensity and its temporal and spectral features, but also on the activity of a person, state of mind, and expectations among other factors. Special attention to the perception of complex sounds in this context is paid to by the so-called Soundscape research [130].

Noise is clearly defined in DIN 1320 [131] as *"the sound occurring within the frequency range of the human hearing which disturbs silence or an intended sound perception and results in annoyance or endangers the health"*. Sound can be defined as a wave motion from a sound-producing object. It varies according to frequency and pressure.

According to Genuit [132], acoustic quality *"is the degree to which the totality of the individual requirements made on an auditory event are met"*. It comprises three different kinds of influencing variables: physical (sound field), psychological (auditory evaluation) and psychoacoustic (auditory perception), so it is multidimensional. Navai and Veitch [86] defined acoustic satisfaction *"as a state of contentment with acoustic conditions; it is inclusive of annoyance, loudness, and distraction"*. However, there is no standard definition neither for acoustic quality nor acoustic comfort/satisfaction. A good acoustic environment is typically associated with the isolation of unwanted sounds and presence of pleasant sound.

In general, acoustic quality in buildings can be influenced and reached by (1) sound insulation and (2) sound absorption [133]. (1) The sound insulation of a building façade, roof and windows is important in terms of the protection of the building interior from unwanted outdoor noise (such as traffic noise). Sound insulation of floors, wall partitions and doors needs to be considered once indoor sources (neighbours' noise, air-conditioning units) are present. (2) Once a room is well insulated, room acoustic aspects come into attention. Sound produced inside a room (such as restaurant ambiance, classes at school) can be enhanced due to multiple reflections, causing deterioration of speech intelligibility and the presence of unacceptable noise levels from sources in the interior [85]. In such cases, volume, shape and the total amount of sound absorption play a role. There are many standards that describe sound insulation and sound absorption measurement and simulation procedures. To mention just a few, guidelines can be found in the ISO 10140 series [134], ISO 717 series [135], ISO 3382 series [136], and ISO 12354 series [137].

3.3. Occupant Satisfaction/ Well-Being

Standards set the minimum requirements for the IEQ. However, well-being and occupant satisfaction cannot be achieved just by meeting these minimum requirements. The overall well-being does not depend only on physical factors but also on psychological factors in which social and cultural context plays a big role. While there are several definitions for well-being [17], there is not a generally acknowledged definition. The authors have therefore decided to use and adopt the WHO definition of health [138]:

"Health is a state of complete physical, mental and social well-being and not merely the absence of disease or infirmity."

This means that well-being, consists of physical, mental and social factors, and their overall satisfaction promotes health. It also means that solely the absence of detrimental influences does not lead to well-being, as the later depends on the overall perception of the surrounding environment and other conditions. However, factors unrelated to the IEQ influence the overall comfort perception.

These include climate and weather conditions, building related factors as design, disposition and control, and individual characteristics of the occupants (gender, age, type of work).

Several studies highlighted the differences between the IEQ factors and the overall comfort perception and emphasize the multidimensional nature of comfort perception [11,16]. The results of different studies on the relative impact of different IEQ conditions on the overall satisfaction have brought mixed results [139,140]. A number of studies have tried to create a ranking of importance of IEQ factors for overall satisfaction with the IEQ. An overview of the results of 12 studies and their assigned ranking is given in Table 2 and Figure 5. The studies have resulted in different ranking orders of IEQ factors. This might be due to the fact that factors such as age [141–144], socio-cultural factors [145] and space-type [67,143] might have an impact on the evaluation of the perceived importance concerning satisfaction with a certain IEQ factor. For example in studies done by Chiang et al. [141], Zalejska-Jonsson and Wilhelmsson [142], and Frontczak et al. [143] the importance of satisfaction with the noise level was reported to be higher for elderly people. Further, out of the 12 studies, a maximum of two have resulted in the same ranking order, and studies focusing on the same building type came to different conclusions. Eleven of the studies based their research on questionnaire survey answers (QS) among occupants and one compares measured data to the requirements of EN 15 251. The use of weighting schemes for IEQ evaluation models was criticized multiple times. Moreover, the relevance of the use of QS to measure effects of IEQ was concluded to be unreliable [25,65,67]. Current IEQ weighting schemes should be, therefore, used with caution.

Table 2. List of studies that have created a ranking of importance of IEQ factors.

Study	Building Type	Type of Survey
Chiang et al., 2001 [141]	Senior house	QS among occupants of 12 senior houses
Chiang and Lai, 2002 [146]	General dwelling and office buildings	12 QS on experts
Humphreys, 2005 [144]	Offices	4655 QS among occupants
Wong et al., 2008 [147]	Offices	293 QS among occupants
Astolfi and Pellerey, 2008 [148]	Classrooms	1006 QS among occupants
Lai et al., 2009 [149]	Residential buildings	125 QS among occupants
Cao et al., 2012 [150]	Public buildings	500 QS among occupants
Marino et al., 2012 [151]	Offices	Measured values compared to EN 15251
Ncube and Riffat, 2012 [152]	Offices	68 QS among occupants
Frontczak et al., 2012 [142]	Offices	50000 QS among occupants
Heinzerling et al., 2013 [67]	Offices	52980 QS among occupants
Zalejska-Jonsson and Wilhelmsson, 2013 [141]	Residential buildings	5756 QS among occupants

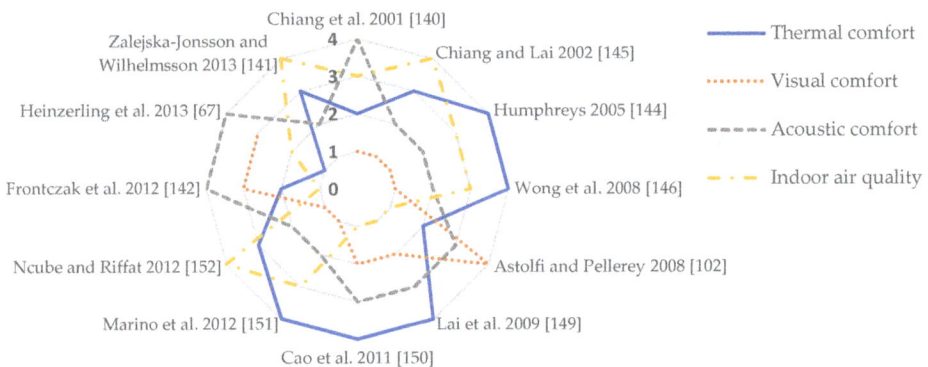

Figure 5. Ranking according to the importance of individual IEQ factors and their contribution towards overall satisfaction with the IEQ. The higher the number, the higher the importance.

More research is required to apply IEQ weighting schemes for the evaluation of the quality of the indoor environment and for benchmarking. Multiple studies have revealed that other parameters, such as the amount of space, visual privacy, working environment, and addition of *"personal control"* leads to increased satisfaction [65,153–156], whereas the absence of adjustment mechanisms or their ineffectiveness can lead to discomfort and frustration [157,158]. Similar results can be found in the aforementioned HOPE project [107,159], where a significant relationship between combinations of view type, view quality, and social density and perceived discomfort was found. Further, studies focused on the relationship between occupant productivity and IEQ identified temperature, noise level and air quality as the most significant for job performance [11,143].

3.4. Adverse Impacts of Poor IEQ

Poor IEQ can have an adverse impact on human health and performance. Poor building design and management, as well as the presence of harmful substances (in furniture, carpets, and air), can cause the occurrence of related comfort complaints, which can result in decreased performance and increased amounts of sick leave [160–162]. Symptoms characterizing poor IEQ are throat, nose and eye irritations, eczema, headaches, fatigue and lack of concentration. These are commonly referred to as sick building syndrome (SBS) [18,101]. The first reports of SBS date back to the 1970s as a result of mechanically ventilated buildings where ventilation and stress-related factors were identified as the main causes of sickness absence [163]. Among additional risk factors are moisture in buildings, higher air temperature, and a high concentration of specific volatile compounds [14]. Some of the causes can be quickly eliminated (old carpets and furniture) while others connected to the building construction involve major investments. There was not yet defined an exact cause for SBS, but in general, the SBS symptoms only occur when there are more than two causing factors [12].

3.4.1. Thermal Comfort

A study on the relationship between performance and interior temperature [164], showed a decrease in performance by 2% per °C increase for temperature ranging between 25 and 35 °C, and no effect on the performance for temperature ranging between 21 and 25 °C. In a different study, Seppänen and Fisk [14] observed an increase in performance for temperature reaching up to 20–23 °C, and a decrease in performance with temperature above 23–24 °C. Nicol and Humphreys [165] identified three contextual variables which influence the thermal comfort perception. The first is the climate (with the corresponding culture and thermal attitudes), the second is the nature of a building (and its services) and the third is time (defining the rate at which the changes occur to which the occupants have to adapt). A relationship between the overall indoor temperature and the SBS was shown in a study published by Wyon [166]. In this study, it was observed that by systematically increasing the temperature from 20 °C to 24.5 °C complains related to the SBS increased from 10% to 60%. The increased temperature led to a higher concentration of harmful substances in the air originating from fixtures. This indicates that there is a high dependency between temperature and air quality.

The link between temperature and performance is particularly relevant in occupational settings because the overall output of workers and their productivity may be strongly affected by increasing workplace temperature and extreme weather events (e.g., heat-waves). This problematic paradigm is receiving increasing attention from different members of society, as the detrimental impact of climate change becomes more evident for the research community, industrial actors, policymakers, and health providers. It is thus not surprising to see a growing number of international initiatives to address these problems and minimize their effect on the population [167].

3.4.2. Visual Comfort

Studies have shown a positive effect of daylighting on health [83]. Discomfort glare belongs to one of the most reported complaints, while blinds operation, its limited functionality, or disturbance by automated operation lead to frustration [168]. A statistically significant relationship between view type, view quality, social density, and perceived discomfort has been reported in Aries, Veitch and Newsham [169]. This is also supported by findings of the HOPE project of positive statistical correlation between view and comfort [159]. These have been identified, to have a positive effect on human psychology, sleep quality and mood. Some other relationships between luminous conditions and occupant satisfaction can be found in [81,84,119].

3.4.3. Indoor Air Quality

To the well-known building-related illnesses attributable to the indoor air quality belong among others CO intoxication, allergic diseases such as rhinitis and asthma, and sensitivity to chemicals in indoor air [170,171]. The effect of poor indoor air quality further depends on intensity, length, and source of noxious exposure. However, risk characterization for air pollutants is mostly done on a single chemical basis, whereas exposures are always a complex mixture of substances [172]. The link between poor indoor air quality and productivity losses has been summarized in [14,108]. In a recent report on noncommunicable diseases WHO explicitly singled out indoor air quality as a cause for 3.7 million deaths in 2012 [173].

3.4.4. Acoustic Comfort

Noise might cause distraction, stress, annoyance, leading to fatigue or it may even damage our hearing. It is well known that a single sound event with a high level, such as 130 dB will render one deaf and that continuous exposure to sound levels higher than 85 dB may lead to a gradual hearing loss. Moreover, much lower sound levels might affect human health once a person is exposed to inappropriate sound levels for a long time. Teachers, for instance, might develop cardiovascular problems or voice disorders if they have to teach in environments with increased sound levels, and even very low levels of 35 dB can paradoxically affect our health in terms of insomnia if they appear in periods of silence and disturb sleep patterns. Studies have shown that not only the difference in sound pressure level but also frequency characteristics affect the resulting degree of annoyance and thus discomfort [174]. For the indoor sound environment in general, sound level and speech intelligibility, and in particular speech privacy belong to be the most important quantifiers [85,175,176]. Finally, in terms of noise and health issues, WHO LARES (Large Analysis and Review of Housing and Health) analysis based on a survey about European housing in 2002–2003 concluded that neighbor noise might also cause health problems [177]. The noise effects have been associated in the past only with damages to hearing. Nowadays it is also established that chronic exposure to noise can cause a variety of health problems, such as hearing impairments, hypertension, cardiovascular problems, sleep disturbances and annoyance. Noise exposure can also have psychological effects such as stress, but there are also detrimental effects on cognitive performance and attention [178].

4. Human Factors in Adaptation

"If a change occurs such as to produce discomfort, people react in ways which tend to restore their comfort." This definition of adaptive principle, originally formulated for occupant behavior in regard to thermal discomfort [165], is also applicable for all other measurable physical parameters defining IEQ. On the user level, this adaptation can be represented by performing *"adaptive actions"*. Enabled through technology *"adaptive actions"* can take place also on building indoor climate control and on the building design level.

Other than the natural/technological limits of adaptation available on each of the three levels of adaptation (the building design and construction level, the IEQ control, the occupant level), time can be identified as one of the decisive factors determining the level on which the adaptation should take place. The time factor could be defined as how fast the environment is able to change, or the period necessary for occupant adaptation/acclimatization, or the time between last two adaptations (in certain cases the frequency of change). This should theoretically follow the natural principle of least resistance, where actions with smaller overall impact and lower energy input should take precedence. Whereas this can be guaranteed for the adaptations at the building indoor climate control level and also on the building design level, the occupant behavior introduces significant uncertainty, because *"occupants adapt their environment and personal characteristics to achieve their comfort in ways that are convenient to them rather than being necessarily energy-conserving"* [61]. Occupants can adapt to their environment through [19,179]:

- physical adaptive actions (e.g., increase their physical activity, change their clothes, altering their environment);
- behavioral adjustment (e.g., change of habits or tolerance limits);
- physiological adaptation (e.g., accommodation of the eye); where limits of the adaptation are imposed by the human physiology in it of itself.

These adaptive actions can take place consciously (most of the physical adaptive actions), subconsciously (some of the behavioral changes) or be automatically steered by the peripheral nervous system (physiological adaptation). These are defined by their natural limits ranging from ability limits to perception thresholds.

4.1. Limits of Human Physical Adaptive Actions

When talking about physical adaptative actions in the framework of human adaptation to the indoor environmental conditions we refer to all the adaptive actions that require physical activity. The goal of the actions performed by the occupants is to alter their immediate environment to achieve comfort. According to Nikolopulou et al. [179] we distinguish reactive (referring to personal changes such as clothes) and interactive adaptive actions (referring to interactive actions with the environment, such as the closing of curtains). Physical adaptive actions are therefore limited by occupants' abilities [180] and the accessibility to environmental altering control tools but can take place only while respecting the human behavioral limits and physiological limits.

4.2. Human Behavioral Limits

The necessity of behavioral change has been identified as a significant barrier when taking actions related to the mitigation of climate change, such as a change in energy consumption habits. The required change is perceived to be only reachable if discomfort, the sacrifice of living standards and social image is involved [181]. The indoor environment should provide conditions in which people can function effectively, efficiently and comfortably. However, human global experience of comfort does not solely depend on physiological limits but also psychological factors influencing behavior patterns. These patterns depend on a person's attitude, initiative, motivation and so on, and are based on the needs and expectations of the individual [182,183]. Environmental stressors impact human psychology and human physiology. A psychological stressor can exceed the limits of tolerance and disturb the psychological equilibrium of an individual, whereas the stressor may occur in the form of over- and under-stimulation [184]. Each act of perception (sensory trigger) followed by a reaction is a hypothesis based on prior experience and any new sensory event is perceived in relation to a former experience. Based on the demands of the environment, a shift in the distribution of the responses can take place. This process is called adaptation (Figure 6). Behavioral adjustment enables people to actively maintain their own comfort, and it plays a crucial role in energy conservation. This, for example, can be observed bet the change of clothing during different seasons. Behavior plays a significant role in long-term changes in consumption patterns and attitudes towards the environment, but psychological barriers hinder this process of behavioral adaptation. These barriers, according to Gifford include [185]: limited cognition of the problem, ideological worldviews, comparisons with other people (i.e., why should I if they do not?), sunk cost and behavioral momentum (e.g., it is cheaper to ignore than to change, and its simpler to repeat than to change), disregard towards experts and authorities (i.e., mistrust to other people leads to ignorance of advice), perceived risk of change (i.e., uncertainty of investment into energy efficient technologies), and positive but inadequate behavior change (i.e., installing solar panels, but increasing overall energy consumption). There are three critical factors determining successful behavioral adaptation when it comes to energy saving and indoor environment [35]:

- means in the form of knowledge about how to or which technology to use,
- motive for change in form of reason or incentive,
- and opportunity in form of available resources such as time and money, further strengthened by the ease of use, general acceptance of changed behavior by other people and maintenance of comfort and health.

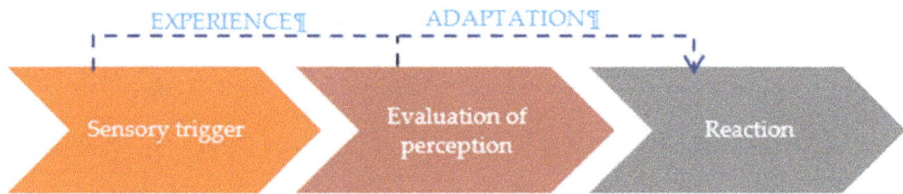

Figure 6. The processing of sensory triggers and subsequent human response.

Further, the process of adaptation can only take place in between physiological limits of the human sensory system, particularly the limits of human capacity to experience temperature and touch, light and color, sound and smell.

4.3. Human Physiological Limits

4.3.1. Physiological Limits of Human Somatosensory System

The somatosensory system is responsible for the sensation experience labeled as temperature. It governs other than the sense of touch, also the sensation of pain, movement, and posture, pressure, vibration and temperature. Through mechanoreceptors in the skin, low vibrations with 5–250 Hz frequency can be perceived [186]. For the perception of the temperature, the responsibility lies with cold- and warm- sensitive sensory fibres in the skin, responding to changes in skin temperature, whereas the first only respond to cooling, and the latter only to warming. The cool and cold sensitive fibers respond to temperatures lower than 17 °C to a maximum of 34 °C, while warm receptors respond to a range of 33 °C to 46 °C. The temperature of 32 °C (approximate temperature of human skin in thermal-neutral condition) produces no sensation at all (known as physiological zero). Temperatures above and around 43 °C will begin to cause skin damage if the skin contact is of prolonged duration, and temperatures under 17 °C are reported as painfully cold by humans [187]. The upper limit for thermal radiation exposure is defined at approximately 3500 W/m^2 for exposure shorter than 10 seconds, before the onset of skin burn [188], Figure 7. However, the perception of thermal comfort does not depend only on temperature (air, radiant, surface) but also on air humidity, air velocity and individual human parameters like clothing, and metabolic rate which depends on age and activity level [189]. It has to be mentioned that continued exposure to warm or cold stimuli, leads to a gradual desensitization and an increase in the threshold values. The theoretical limit is in this case presented by the core body temperature. If the body temperature exceeds 40 °C, an acute risk of heat stroke arises and if the core temperature falls under 35 °C a person is at risk of hypothermia [190].

Several studies suggested using the thermal sensation instead of the temperature *per se* to ensure the well-being of occupants [191]. A contemporary state-of-the-art on thermal comfort approaches, physiological basis of comfort and mathematical modeling of heat exchanged between the human body and its environment can be found in [69].

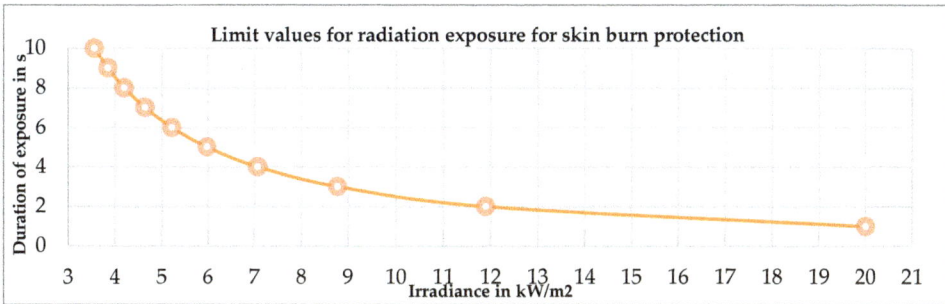

Figure 7. Limit values for radiation exposure for skin burn protection based on [188].

4.3.2. Physiological Limits of Human Vision

Visual perception is the ability to interpret information from light in the visible spectrum reflected by the objects in the environment, that reach the human eye. The optical performance of the human eye decreases with age. The human visual system (of a young and healthy individual) perceives visible radiation (light) in the range of wavelengths between 370 and 730 nm, with the highest sensitivity for photopic sight at 555 nm [192]. The normal visual acuity, based on the smallest separation at which two parallel lines can be discriminated, for human vision, corresponds to ~1 arc-minute of visual angle (a 2 Euro coin subtends an angle of 1 arc-minute at a distance of ~10 m) [193]. The range of light levels perceptible spans over 11 orders of magnitude from an absolute threshold of $3{,}10 \times 10^{-6}$ cd/m^2 for "scotopic" vision, to 3×10^5 cd/m^2 for "photopic" vision [194]. The flicker fusion threshold at which an intermittent light stimulus appears to be completely steady lies between 50 and 90 Hz, but visual flicker artifacts can be in certain cases observed at rates over 500 Hz [195]. The theoretical maximal resolution is estimated to be approximately 15 million variable resolution pixels per eye [196]. Depending on the bone structure, the maximum human stereoscopic field of view can be up to 180–200° horizontally and up to 135° vertically, although objects can be seen in detail only within the central area of the visual field (near periphery) [197].

The eye is able to refocus and sharpen the image within a certain range, through an effect known as accommodation of the eye. Further to maintain visual acuity, the human visual system is capable to adapt to degraded visual conditions, such as low illumination and glare. In order to adapt (or avoid damage) to the radiant flux that reaches the retina, regulating mechanisms take place in the eye, ranging from a variation of the pupil diameter, to neuronal and chemical processes in the pigments within the retina. A full sensitivity adaption can take up to 20 minutes (a dark adaptation of the rods) [192]. Adaptation can also take place through eye movement and neck rotation (change of focus point, gaze diversion) and a change of position. Human eyes can pitch 20° upward and 25° downward, while eye yaw is symmetric by allowing 35° to the left or right. Beginning in a neutral position the normal neck range of motion equals 60° for neck flexion up and down, 45° for lateral neck flexion to the right and left and 80° for neck rotation, turning head to the right and to the left [198].

4.3.3. Physiological Limits of Human Auditory System

Hearing is the first sense to be developed in a human fetus, this takes place in the 12th week after conception. The ears have not been formed yet at that time, but the fetus is already able to perceive different vibrations and resonances. Sound can be perceived through air and bone conduction. The frequency spectrum of hearing lies between 20 Hz to 20 kHz, whereas the upper hearing limit decreases with age (most adults do not hear over 16 kHz). The human hearing system can sense a sound pressure ranging from 10^{-5} Pa to 100 Pa, defined by 0 dB as the threshold of hearing and 130 dB as the threshold of pain. Further, we are able to distinguish around 1300 tones, with the ability to distinguish intervals, defined as the difference between two audible frequencies, such as an octave (factor of 2). The auditory system is capable, aside from the perception of loudness, pitch, and timbre, to evaluate the spatial position of a sound source in terms of direction and distance [199]. Based on the physiology of the ear and head, the ability to localize a source of sound starts at 500 Hz, corresponding to 69 cm of wavelength, which is four times the diameter of an average human head. At 4 kHz, with a wavelength below 10 cm, the head and especially its details (outer ear, especially pinna) start to cast a shadow and will scatter sound waves [200].

4.3.4. Physiological Limits of Human Sensitivity to Inhalation of Pollutants

The human perception of the air quality involves two human senses: the general chemical sense (somesthesia), located all over the mucous membranes, in the eyes and nose sensitive to irritants, whereas the olfactory sense, located in the nose, is responsible for odor detection and pheromone detection [201]. These senses interact, therefore substances can evoke odor and an irritating sensation, while odor sensitivity decreases with time, discomfort from irritation will normally increase with time [110,202]. Further, all the chemical substances that enter the nose (or mouth) while breathing then continue further down the respiratory tract to the lungs. Olfactory receptor cells contain more than 1000 receptor proteins, making humans capable of detecting around one trillion odors [203]. Inhaled toxins represent a health risk and can result in i.e., in chemically induced nasal lesions, sensory irritation, olfactory and trigeminal nerve toxicity [204]. The exposure limits for substances are based on their overall effects on health, defined as the concentration of the pollutants over time [205,206].

5. Built Environment

In Section 3 we have described the minimum performance requirements for the IEQ defined in standards. These represent a basic framework for occupants' well-being. In Section 4 a description of human physical, behavioral and physiological limits for adaptation to environmental conditions were discussed. These all have to be taken into consideration while designing an energy efficient and healthy building environment, as the overall goal is not just the deployment of such environments but also their overall acceptance by their occupants [207]. Design principles, initiatives, and tools that encourage the culture of energy efficiency [208] and active engagement of occupants should be employed [209], otherwise the potential of an innovative build environment remains largely untapped [24]. The main purpose of more energy efficient buildings is to increase the sustainability of the built environment while ensuring the long-term well-being of occupants. Ignorance of the before described *"human factors"* and requirements for well-being can reduce the adoption and use of energy efficient measures and can lead to disinterest and disenchantment of building occupants [40,41,210].

5.1. Sustainability of the Built Environment

Whereas the quality of the indoor environment is noticeable within a short period of time (immediately or after a few days), its impact is not limited solely on the human sensory system. The means with which the level of comfort has been achieved plays a significant role in the consumption and distribution of resources. These have an impact on the overall sustainability of the built environment as well as the resulting energy efficiency. The sustainability concept as we know it first appeared in 1987 in the Burtland Report, which was a result of a paramount need to study and identify what impact human activity has on the environment, and was defined as [211]:

"Sustainable development is development that meets the needs of the present without compromising the ability of future generations to meet their own needs."

The report further continues with:

" ... sustainable development is not a fixed state of harmony, but rather a process of change in which the exploitation of resources, the direction of investments, the orientation of technological development, and institutional change are made consistent with future as well as present needs."

The built environment dominates humanity's impact on climate change, results in resource depletion, loss of biodiversity, contributes directly to health and well-being, and has socio-economical underlining consequences. The issues of human environmental impact and the natural footprint are approached by two major international agreements that address climate change: United Nations Framework Convention on Climate Change (UNFCCC), and the Kyoto Protocol, with its follow-up agreement at COP21 in Paris 2015 which entered into force on 4 November, 2016 [212]. Based on these agreements the signatory countries have developed their own climate and energy policies, with the aim of holding warming well below 2 °C, and striving to limit warming to 1.5 °C [213].

5.2. Energy Efficiency of the Built Environment

The building sector accounts for approximately 19% of all global greenhouse gas (GHG) emissions in the world, whereas most of GHG emissions belong to indirect CO_2 emissions from electricity use in buildings. These also show a dynamic growth in the last four decades (quintupled for residential, and quadrupled for commercial), and it is estimated that the global building energy use, may double or triple by the mid-century. Buildings embody the biggest unmet need for basic energy services and represent a critical piece for low-carbon future [2].

The reduction of energy demand and the increase of energy efficiency is considered to be a dominant contribution to tackle global climate change. However, caution is advised against the oversimplification of the problematics of energy demand reduction [36]. The decoupling of energy consumption from global economic growth is proving to be challenging, and could negatively influence economic growth [214,215].

Energy consumption represents the total amount of energy used for a certain work during certain time periods. In the context of the final energy consumption of countries, or sectors, their yearly energy consumption is described in terms of million tons of oil equivalent (Mtoe). Energy consumption can be reduced by the inference of improving the energy efficiency. Energy efficiency is a measure that represents the amount of energy output for a given energy input, it is strongly connected to the technological advancement. The overall goal is to reduce the amount of inputs and to maintain the same outputs and achieve a more energy efficient state. To evaluate how efficiently an economy consumes energy we use energy intensity. It is a measure that represents the amount of energy used to produce a unit of output, and most commonly refers to the ratio between gross energy consumption and gross domestic product. A reduction of energy intensity can be achieved via structural changes and overall technological advancement. Technological advancement resulting in higher energy efficiency is more important for countries with higher income per capita, while structural change, denoting sectoral shifts within a country, proves to be more important for all income levels [216,217]. Despite the fact that global energy intensity has been constantly decreasing [218], the global energy demand grew by 2% in 2017 [219].

The climate policy is in the EU represented by the "20-20-20" climate and energy targets [220], by the Directive 2010/31/EU [221] on the energy performance of buildings and by Directive 2012/27/EU [222] on energy efficiency. The Directive 2010/31/EU set a strict goal of *"all new buildings being nearly zero energy buildings by 31 December 2020 (public buildings by 31 December 2018)"* [221]. To enable the delivery of the Paris Agreement commitments [223], the European Commission has presented a new package of measures, with milestones to ensure *"a sustainable, competitive, secure and decarbonised energy system by 2050"*, which introduces a necessity for long-term renovation strategies, necessity for skills development and education in the construction and energy efficiency sectors, promotion of smart technologies, and promotion of health and well-being of building users, defined in the revised energy performance of buildings Directive 2018/844/EU [224]. However, even if these existing policies and announced intentions would come to full force, the International Energy Agency (IEA) predicts that the global energy need would expand by 30% between today and 2040. This is due to the fact that even though the energy efficiency improves, the global floor area continues to grow, causing an increase of energy consumption of the building sector of 0.8% a year [225]. Strict regulation of the building sector is necessary to achieve the limit increase in the global average temperature under 2 °C [226].

Energy consumption in the building sector is especially high in developed countries (up to 40% of the total consumption, Table 3) [231]. The high potential of energy savings triggers the development of energy conservation technologies. The global building envelope performance (its capacity to minimize energy demand for heating and cooling) has improved by approximately 1.4% a year from 2010 [232]. The global investment in energy efficiency of buildings reached €108.5 billion in 2016 [233], but still represents only a fraction of the total €4 trillion spent on buildings and construction globally [225]. In the EU the renewables cover, approximately 15% of energy demand for residential buildings, and 4% for non-residential buildings [234]. Through the announced "Smart financing for smart buildings" the EU plans to unlock €10 billion between 2018 and 2020 to fund energy efficiency and renewables in buildings [235], to support the development of technologies to lower the final energy consumption of buildings (Figure 8). However, to reach the 2 °C scenario [212,236], according to IEA, an investment of €25 trillion in the building sector in the following 40 years (until 2050) is needed in technology development [237], p. 43. Investment in digitalization of buildings, such as the introduction of smart controls could lead to lowering of buildings energy consumption by as much as 10% globally [225,238].

Table 3. Energy consumption of the building sector.

Final Energy Consumption	Year	Commercial in Mtoe	Residential in Mtoe	Total in Mtoe	% Share of the Building Sector
EU – 28 [1]	2016	150	285	1108	39
USA [2]	2017	456	504	2467	39
China [3]	2017	84	379	3052	15
India [4]	2015	7.4	52	519	11.4

[1]: Relevant statistics for EU in [227]; [2]: Relevant statistics for USA can be found in [228]; [3]: Relevant statistics for China can be found in [229]; [4]: Relevant statistics for India can be found in [230].

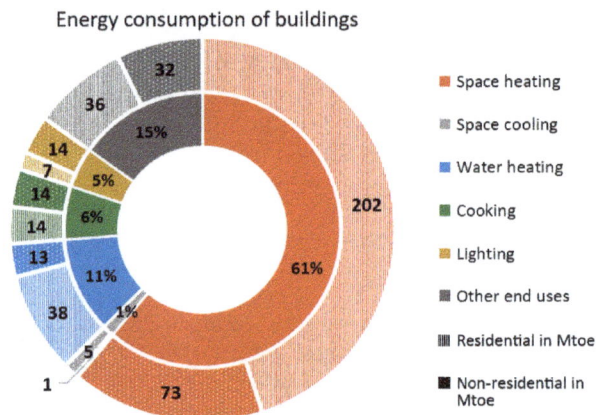

Figure 8. Final energy consumption in buildings for EU28 in 2013. Data according to [230].

Energy use is considered to be the most reliable indicator of the successful implementation of a sustainable building design [239]. Contemporary rating systems for newly built high-performance buildings base their certification on predicted energy performance, acquired with the use of simulation tools based on proposed design [240]. However, studies show that there is a gap between actual and predicted energy consumption (simulated energy consumption) [105,239,241–243]. The use of actual energy performance in certification could further stimulate the development of the building and construction sector and result in higher energy efficiency and lower energy consumption. Among the identified socio-technical factors influencing the gap between the predicted and actual energy performance there are:

- occupants, when there is more occupants than planned, and/or based on their energy-conscious behavior [26,27,244],
- technology, failure or bad design and installation affecting energy efficiency [239,241,242],
- maintenance and management, bad management practices of the building management [245–247].

5.3. Building Design Level

Systematic approaches to building design, which account for interactions between different building functions, have the potential to increase energy efficiency and result in an optimized building performance [248]. This is achieved by an integrated building design process, based on the cooperation of all the actors involved in the project, that agree on far-reaching decisions jointly, to ensure building performance [249]. The challenges of reducing the fragmentation of the design process, the increased complexity and necessity for the involvement of various professions during building design, construction and operation, has resulted in the innovation of collaborative techniques for storing, sharing and analysis of project data [250,251]. The most prominent is Building information modeling (BIM), which in recent years has seen one of the highest increases in the engagement of architecture, engineering, and construction (AEC) professionals, and has undergone one of the most promising developments in the industry. BIM modeling supports the cooperation of professionals involved in project drafting and realization and supports the development of sustainable building facilities. The final model is data rich, representing a full virtual 3D model of the developed construction, storing parametric data including dimensions, design, costs, physical properties of elements, description of relationships between objects, as well as life-cycle information. BIM modeling has the potential to minimize the fragmentation of the building design process, enabling inclusive involvement of large groups of stakeholders ranging from design, construction, operation, and maintenance, having a beneficial impact on clash detection, cost-effectiveness and overall sustainability [252,253].

Buildings' energy efficiency can be achieved through the application of passive or active design strategies. Passive building design, refers to the use of the potential available in the local climate and site conditions, benefiting from passive use of solar radiation and natural ventilation [44,57,58], whereas the design solution leads to energy savings [59]. Active technology solutions refer to onsite energy generation, heat-pumps, and automated control of the indoor environment [54,254]. Maximization of passive features such as orientation, daylighting, cooling and heating, enhanced by location-specific systems covering the remaining gap in energy loads (such as heat pumps and solar panels) lead to more energy and emission conscious designs with higher indoor environmental quality [39]. Further, implementation of passive design measures increases the resiliency of buildings, where in case of a prolonged power outage during extreme weather conditions, the buildings remain habitable and provide shelter passively [255].

The development of all these designing methods is also supported by the Energy Performance of Building Directive (EPBD) [221] with the nZEB requirement. In the EPBD, nZEB is defined as follows:

" ... *a building that has a very high energy performance, ... The nearly zero or very low amount of energy required should be covered, to a very significant extent, by energy from renewable sources, including energy from renewable sources produced on/site or nearby.*"

Theoretical background on nZEB can be found in [42,256,257]. In general, passive building design and active technology solutions can be used in an nZEB concept. A comparison of different building design strategies and their final energy savings can be found in [254].

5.4. Building Technology Level

Building technology development is focused on increasing energy efficiency, which is done not only by developing the technology per se, but by making the technologies more responsive to immediate conditions. This includes local weather conditions, requirements on functionality and IEQ as well as adaptation to individual occupant needs. Implementation of energy efficient technologies, in particular, focused on the retrofitting of buildings, is considered to be one of the main strategies to address energy poverty in households. Worldwide 13% of the population lacks access to modern electricity [258,259], and in the EU only, nearly 11% of citizens are considered energy-poor, and not able to keep their homes adequately warm [260]. The main reasons for energy poverty are considered the rise in prices, low income and poor energy performance of households [260–262]. To energy-efficient building technologies belong (1) high-performance building envelopes (HPBE) [263], (2) efficient appliances such as lighting, heating, ventilation and air-conditioning [264], (3) building automation and control systems (BACS) [265] and more [266].

It is important to mention that contemporary building design process is strongly dependent on information and communication technologies (ICTs) [267], depending on predictive methods and on building simulation [266,268]. This enables a new systematic approach to building design, with increased added value through integrated building design [248]. Because of this high technologization of building design, the term *"intelligent building"* is used with contemporary building design [49]. The impact of building design on energy consumption and IEQ can be evaluated through simulation programs (e.g., ESPr, EnergyPlus, Comsol). This type of prediction enables quick design optimization to take place before construction [269].

Conventional building constructions are static, and once built the adaptation to surrounding conditions, or to current occupants' requirements, takes place in a very limited form. The adaptation of the whole building performance is based on a simple "on-off" switch with limited variability of the system (e.g., ventilation, blinds and heating system), whereas no system is available to resolve and account for the interdependence of the performance indicators. Advanced building technologies, such as HPBE and BACS, enable real-time response to environmental conditions as they do maintain high adaptation capabilities after construction, expanding the performance abilities of buildings [54,270]. This transformation ability can be used to maintain the IEQ, and therefore can be a part of the building automation and climate control systems [271].

5.4.1. High-Performance Building Envelopes

An HPBE is considered an external layer of building, that is able to change its properties in order to provide comfort, as a reaction to the change of external or internal conditions, or requirements. HPEBs can be found under multiple names in literature, including responsive, adaptive, kinetic, intelligent and similar. The core concept is based on the imitation of processes native to the organism, that enable adaptation to conditions in order to provide comfort and equilibrium, whereas stress is put on the reversibility and the ability to repeat the process of adaptation [54,263]. The changes can take place at different scales, ranging from molecular changes to structural changes in the whole system. The adaptation is enabled through a system of sensors, processors, and actuators, or based on the physical properties of the materials used where changes occur at the molecular and composition level (e.g., phase change materials) [8].

5.4.2. Efficient Appliances

Energy policies support the uptake of improved energy conservation practices as well as the adoption of energy efficient appliances and technologies (e.g., heating systems, household appliances, and office equipment). The changeover to energy efficient appliances leads to lower energy consumption (reduction of up to 23% in households) and lower instantaneous electric demand [272]. Appliances equipped with motion sensors, or connected to an automatic centralized control system based on occupancy or utilization, bring higher reduction than conventional appliances and can result in a 10% reduction in energy consumption compared to appliances without any power control system, or equipped with a standby mode [273,274].

5.4.3. Building Automation and Climate Control Systems

When referring to IEQ control a building can be free-running (without automated control), with climate control, or use a mixed strategy. The IEQ of a free-running building is solely dependent on occupant operation. While this approach is the most innate, when comparing buildings with and without climate control, the latter may have higher energy loses [275].

Buildings with integrated room automation (IRA) represent a technical counterpart to free-running buildings. IRA is based on data collection and controls lighting, blinds, heating, cooling, and ventilation [276]. This includes the use of advanced IEQ control strategies using non-predictive control or predictive control, set-back plans and automated adaptation according to occupancy information. The efficiency of an adaptation depends on the processed relevant data and the prediction of the adaptation outcome. The internal microclimate of a building is then a result of the *"computation of multicriterial optimization, taking into account technical and psychological comfort and energy savings"* [108]. For this, systems for the collection of data and processing have to be applied. A non-predictive control system can lead to energy savings of 1–15% while predictive control has an energy saving potential of 16–41% [277].

Current advanced IEQ control systems use different approaches: learning-based methods, rule-based control (RBC), model-based predictive control (MPC) and agent-based control systems [50,51]. Selection of building climate control strategy is highly dependent on the available computational capacity. Intelligent control systems, especially MPC, involve high initial investments in building modeling, simulation, data collection, and processing.

RBC is a *"condition-action"* based system and belongs to the non-predictive systems. It relays on databases of pre-set numerical thresholds of indoor environmental factors (for example temperature threshold for cooling/heating) and corresponding reaction of building climate control technologies (heating system, ventilation, lighting …) [278–280]. Currently it is probably the most commonly used control practice. The system reacts on ad hoc conditions and is dependable on pre-set parameters.

Progressive building climate control techniques are based on artificial intelligence, using probabilistic models to predict the optimum control strategy [281]. They solve probabilistic relations between variable processes represented by weather, building operation and occupant interference and compute the probabilistic outcome of their relations. The MPC approach solves over a finite horizon a number of optimization problems to determine optimal control outputs [265]. The process is continuously repeated over a pre-set time period (ca. 10 minutes sampling rate). A collection of literature references can be found in the databases of the project Opti-control [282]. The MPC can be used to increase energy efficiency in building simulation [281,283–286], to create weather forecasts [265] and for occupancy modeling [276,287,288]. All these studies show that predictive strategies are more efficient than non-predictive conventional strategies when it comes to building climate control, especially for thermal regulation. In a study comparing predictive control with conventional control [289] savings of 15%, 24% and 35% were obtained for cold, mild and warm days respectfully. It has to be mentioned that, currently, only a small fraction of buildings have implemented this kind of building climate control.

To gather information used to operate these smart environments, sensors are used to collect data on environmental conditions [53]. These are distributed through the monitored spaces to collect relevant data. Currently they have to be installed in a fixed position and connected directly to the operating computer. However, wireless sensor networks (WSNs) are a promising technology [290]. This means that currently, static nodes for data collection will become mobile. The sensors are coupled with actuators which are responsible for the system operation (blinds, lightings and heaters).

Indoor environmental climate controls can be linked to immediate occupancy information. There is a big potential in using occupancy information to achieve a more energy efficient and healthy building climate control [63,288]. IRA has a savings potential of up to 34% that can be reached by adjusting lighting and especially ventilation to instantaneous measurements [288]. Scheduled temperature plans can lead to energy savings of up to 18% [277]. Further, artificial intelligence systems are able to learn user preferences and also recognize and make use of space properties (time needed to heat, energy needed) [287].

5.5. Interactions Between Occupant and Building

The occupant preferences determine the ergonomics of designed environments, so it can be said that user is the scale of design and the success of each design is dependable on occupant acceptance and successful operation [61,291]. To main factors that influence energy consumption on the occupant level scale, besides behavioral constraints discussed in Section 4.1, belong:

- the income level of building occupants [260,262],
- the age composition of occupants and their educational level [292,293],
- the available and perceived control of IEQ [21,294,295],
- the effectiveness of training on building system operation control [9,84,246,296,297].

There is a high dependency between energy consumption and criteria set for the IEQ and building design. However, it has to be noted that reducing energy consumption while compromising the IEQ will have an adverse impact on health and will lead to a decrease in productivity.

5.5.1. Income Level of Building Occupants

Independent surveys agree that socioeconomic characteristics such as income level and ownership are significant drivers for energy consumption [298,299]. Low-income level is one of the factors negatively influencing the overall energy efficiency of consumers, and these consumers are also at risk of energy poverty [260]. Studies investigating the probability of investment of consumers in energy efficient technologies have shown that owners of households with a high-income level are more likely to invest in these technologies. However, for these households, other factors, such as age composition and education, are more determining [299–301].

5.5.2. Age Composition and Education of Building Occupants

The age of building occupants is considered to be a good indicator for energy-conscious behavior, such as the adoption of renewable energy practices [301], energy conservation and adoption of energy efficient technologies [292,300]. The studies suggest that younger and middle-aged building occupants are more likely to adopt energy- and climate-conscious behavior, while older occupants place more importance on financial savings. Elderly occupants are further less informed about energy-saving measures available. Lastly, the level of education seems to play a significant role for energy-conscious behavior, where better-educated occupants understand better the environmental problems and have more knowledge how to engage in energy-conscious behaviors [292,302,303].

5.5.3. Available and Perceived Control

While occupant behavior can negatively affect energy consumption [304] and automated control leads to significant energy savings as reported in several studies [39,51,265,283,288,305,306], the difference between available a perceived control can have a major impact on user satisfaction. The lack to effective IEQ control leads to dissatisfaction [15,84,154,295]. However, several studies report higher forgiveness for IEQ factors if occupants have more access to the building controls [28,159,165,295,307]. In line with this, a study made by de Dear and Bragger [308] suggests that occupants in naturally-ventilated buildings are more responsive to indoor building climate (e.g., clothing adjustment) and more satisfied than occupants of buildings with centralized HVAC. Karjalainen and Lappalainen stated that [287]: *"the automated systems should provide conditions that satisfy the average person but also give occupants the opportunity to alter the conditions based on personal wishes"*.

5.5.4. Training on Building System Control Operation

Personal control generally refers to the possibility to manipulate blinds, lighting, open a window or use a fan. It can also refer to the possibility to adjust the automated control. Through sensor data collection the user has access to a vast amount of information. This can lead to frustration and dissatisfaction. The *"rule of simplicity"* should be applied in the user interface [287].

While the conventional design is based on manual operation and most adjustments are intuitive and backed up by experience, contemporary design is highly technical, and thus demanding on user operation. In research by Day and Gunderson [296] importance of effective training on building system control operation (system overrule, impact on energy consumption) was studied. Users that received effective training were more likely satisfied with their indoor environment. On the contrary users with no, or ineffective training, reported discomfort with their working environment. This can have an impact on their overall productivity and health performance.

Further, the technical limitations of indoor environmental controls have to be taken into consideration. If the effective change does not take place in a notable time-period, this will also lead to increased dissatisfaction [154]. This must be taken into consideration mainly for the heating system [309].

One of the indicators of a good indoor environment is the perceived productivity of the occupants. The improvement of productivity through successful design is a subject to a number of studies [157,191,310]. This only confirms the conclusion of [311] that *"the economic benefits of energy-efficient design may be significantly greater than just the energy cost savings"*. This brings the argument of the economic feasibility of improved IEQ [312].

6. Discussion and Conclusions

A comprehensive description of the limits and adaptation opportunities regarding the relationship between an occupant and the built environment was presented in this paper. We have defined the factors that limit the interactions between occupants, IEQ and building design. It is crucial to respect the interdependencies in a human-built environment system for a successful design of healthy, energy efficient and comfortable indoor environment.

The review paper was drafted based on an analysis of more than 300 scientific publications, published between 1960 and 2019, that dealt with the topics concerning IEQ, energy efficiency, occupant comfort, and health, sustainability and adaptability of the built environment. A holistic analysis of selected review papers was presented, showing an increase in publications focusing on the technologization of the built environment, whereas in recently published papers the occurrence of terms such as smart home and model-based predictive control is more frequent.

In this paper, we have reviewed the current requirements on IEQ performance codified in contemporary standards. In conclusion, it can be said that in all respective fields of IEQ standardization, there is a lack of guidelines for the design of non-static building environments. The future of the building design is in the ability of the built environment to adapt to instant performance requirements and environmental conditions, and this is an upcoming challenge for future standardization. A necessity to shift the focus from measurable physical parameters used to evaluate the IEQ performance (thermal comfort, visual comfort, indoor air quality and acoustical quality), was highlighted within this review paper, as scientific evidence proves that other performance parameters such as type of space, building design, and working conditions have an impact on the overall satisfaction of occupants with the building environment.

The review highlights the necessity of a human-centric design of the built environment, where the efficiency of technology can be measured only if it is successfully implemented and used by the building occupants. In here the limiting human factors, defining the boundaries of occupants' physical actions, behavioral adjustment, and physiological adaptation, while interacting with the built environment have to be taken into account.

Energy consumption, conservation, and the impact of climate change are the upcoming challenges for the future IEQ research. The importance of flexibility and adaptability is highlighted by these issues, both of which can be achieved through an automated adaptive building climate control. However, substantial uncertainties when it comes to the set-up, operation, ease of use, maintenance and security still exist. There is a need to create a structure enabling continuous training in the field of building technology and operation for users. The future building environment will be highly automated, based on artificial intelligence and prediction methods. To ensure occupant comfort the basic operational knowledge, opportunity to overrule the system has to be provided, and attention has to be drawn to the user interface and simplicity.

The development of the built environment towards sustainability with the development of energy efficient and healthy technologies has to be broadly recognized, while cost-effectiveness of these technologies has to be preserved. This review can be used as a guide while preparing campaigns to improve the information provided to stakeholders, to raise awareness about energy-efficient building technologies and popularize conservation practices among occupants. We believe that these recommendations will be relevant, especially concerning the emerging markets when it comes to building automatization and that they will contribute to more human-centric and environmentally-conscious building initiatives.

Author Contributions: Conceptualization P.S.; methodology and formal analysis, P.S. and A.H.; writing—original draft preparation, P.S. and A.H.; writing—review and editing, M.R., T.S.M. and A.H.; supervision, M.R. and T.S.M.; funding acquisition, M.R.

Funding: This work was carried out in the framework of the "PaPaBuild" project under the action H2020-MSCA-RISE-2015–grant 690970, COST Action TU1303: Novel Structural Skins and VEGA 1/0050/18: Photovoltaic facades of nearly zero energy buildings.

Conflicts of Interest: The authors declare no conflict of interest.

Abbreviations

CABS	climate adaptive building shell
IEQ	indoor environment quality
HOPE	Health Optimization Protocol for Energy-efficient building
WHO	World Health Organization
LARES	Large Analysis and Review of Housing and Health
QS	questionnaire survey
SBS	sick building syndrome
UNFCCC	United Nations Framework Convention on Climate Change
GHG	greenhouse gas
IEA	International Energy Agency
HPBE	high-performance building envelopes
BACS	building automation and control systems
ICT	information and communication technologies
IRA	integrated room automation
RBC	rule-based control
MPC	model-based predictive control
WSN	wireless sensor networks

References

1. Klepeis, N.E.; Nelson, W.C.; Ott, W.R.; Robinson, J.P.; Tsang, A.M.; Switzer, P.; Behar, J.V.; Hern, S.C. The National Human Activity Pattern Survey (NHAPS): A resource for assessing exposure to environmental pollutants. *J. Expos. Anal. Environ. Epidemiol.* **2001**, *11*, 231–252. [CrossRef] [PubMed]
2. Lucon, O.; Ürge-Vorsatz, D.; Ahmed, A.Z.; Akhbari, H.; Bertoldi, P.; Cabeza, L.F.; Eyre, N.; Gadgil, A.; Harvey, L.D.D.; Jiang, Y.; et al. Buildings. In *Climate Change 2014: Mitigation of Climate Change. Contribution of*

Working Group III to the Fifth Assessment Report of the Intergovernmental Panel on Climate Change; Cambridge University Press: Cambridge, UK, 2014.

3. Ürge-Vorsatz, D.; Eyre, N.; Graham, P.; Harvey, L.D.D.; Hertwich, E.; Jiang, Y.; Kornevall, C.; Majumdar, M.; McMahon, J.E.; Mirasgedis, S.; et al. Energy End-Use: Buildings. In *Global Energy Assessment—Toward a Sustain. Future*; Cambridge University Press: Cambridge, UK; New York, NY, USA, 2012; pp. 649–760.

4. United Nations. *Sustainable Development Goals Report 2018*; United Nations: New York, NY, USA, 2018.

5. Lamb, W.F.; Creutzig, F.; Callaghan, M.W.; Minx, J.C. Learning about urban climate solutions. *Nat. Clim. Chang.* **2018**, *9*, 279–287. [CrossRef]

6. Anderson, J.E.; Wulfhorst, G.; Lang, W. Energy analysis of the built environment—A review and outlook. *Renew. Sustain. Energy Rev.* **2015**, *44*, 149–158. [CrossRef]

7. Paone, A.; Bacher, J. The impact of building occupant behavior on energy efficiency and methods to influence it: A review of the state of the art. *Energies* **2018**, *11*, 953. [CrossRef]

8. Taveres-Cachat, E.; Grynning, S.; Thomsen, J.; Selkowitz, S. Responsive building envelope concepts in zero emission neighborhoods and smart cities—A roadmap to implementation. *Build. Environ.* **2019**, *149*, 446–457. [CrossRef]

9. Hauge, Å.L.; Thomsen, J.; Berker, T. User evaluations of energy efficient buildings: Literature review and further research. *Adv. Build. Energy Res.* **2011**, *5*, 109–127. [CrossRef]

10. Al Horr, Y.; Arif, M.; Katafygiotou, M.; Mazroei, A.; Kaushik, A.; Elsarrag, E. Impact of indoor environmental quality on occupant well-being and comfort: A review of the literature. *Int. J. Sustain. Built Environ.* **2016**, *5*, 1–11. [CrossRef]

11. Al Horr, Y.; Arif, M.; Kaushik, A.; Mazroei, A.; Katafygiotou, M.; Elsarrag, E. Occupant productivity and office indoor environment quality: A review of the literature. *Build. Environ.* **2016**, *105*, 369–389. [CrossRef]

12. Mitchell, C.S.; Zhang, J.; Sigsgaard, T.; Jantunen, M.; Lioy, P.J.; Samson, R.; Karol, M.H. Current State of the Science: Health Effects and Indoor Environmental Quality. *Environ. Health Pers.* **2007**, *115*, 958–964. [CrossRef]

13. Bluyssen, P.M.; Janssen, S.; Van den Brink, L.H.; De Kluizenaar, Y. Assessment of wellbeing in an indoor office environment. *Build. Environ.* **2011**, *46*, 2632–2640. [CrossRef]

14. Seppänen, O.; Fisk, W.J. Some Quantitative Relations between Indoor Environmental Quality and Work Performance or Health. *ASHRAE Res. J.* **2006**, *12*, 957–973. [CrossRef]

15. Frontczak, M.; Wargocki, P. Literature survey on how different factors influence human comfort in indoor environments. *Build. Environ.* **2011**, *46*, 922–937. [CrossRef]

16. Ortiz, M.A.; Kurvers, S.R.; Bluyssen, P.M. A review of comfort, health, and energy use: Understanding daily energy use and wellbeing for the development of a new approach to study comfort. *Energy Build.* **2017**, *152*, 323–335. [CrossRef]

17. Dodge, R.; Daly, A.P.; Huyton, J.; Sanders, L.D. The challenge of defining wellbeing. *Int. J. Wellbeing* **2012**, *2*, 222–235. [CrossRef]

18. Rostron, J. Sick building syndrome: A review of causes, consequences and remedies. *J. Retail Leis. Prop.* **2008**, *7*, 291–303. [CrossRef]

19. Coccolo, S.; Kämpf, J.; Scartezzini, J.-L.; Pearlmutter, D. Outdoor human comfort and thermal stress: A comprehensive review on models and standards. *Urban Clim.* **2016**, *18*, 33–57. [CrossRef]

20. Rashid, M.; Zimring, C. A Review of the Empirical Literature on the Relationships Between Indoor Environment and Stress in Health Care and Office Settings. *Environ. Behav.* **2008**, *40*, 151–190. [CrossRef]

21. O'Brien, W.; Gunay, H.B. The contextual factors contributing to occupants' adaptive comfort behaviors in offices—A review and proposed modeling framework. *Build. Environ.* **2014**, *77*, 77–88. [CrossRef]

22. Stazi, F.; Naspi, F.; D'Orazio, M. A literature review on driving factors and contextual events influencing occupants' behaviours in buildings. *Build. Environ.* **2017**, *118*, 40–66. [CrossRef]

23. Hong, T.; Yan, D.; D'Oca, S.; Chen, C. Ten questions concerning occupant behavior in buildings: The big picture. *Build. Environ.* **2017**, *114*, 518–530. [CrossRef]

24. Balta-Ozkan, N.; Davidson, R.; Bicket, M.; Whitmarsh, L. Social barriers to the adoption of smart homes. *Energy Policy* **2013**, *63*, 363–374. [CrossRef]

25. Rasheed, E.O.; Byrd, H. Can self-evaluation measure the effect of IEQ on productivity? A review of literature. *Facilities* **2017**, *35*, 601–621. [CrossRef]

26. D'Oca, S.; Hong, T.; Langevin, J. The human dimensions of energy use in buildings: A review. *Renew. Sustain. Energy Rev.* **2018**, *81*, 731–742. [CrossRef]

27. Delzendeh, E.; Wu, S.; Lee, A.; Zhou, Y. The impact of occupants' behaviours on building energy analysis: A research review. *Renew. Sustain. Energy Rev.* **2017**, *80*, 1061–1071. [CrossRef]

28. Keyvanfar, A.; Shafaghat, A.; Abd Majid, M.Z.; Bin Lamit, H.; Warid Hussin, M.; Binti Ali, K.N.; Dhafer Saad, A. User satisfaction adaptive behaviors for assessing energy efficient building indoor cooling and lighting environment. *Renew. Sustain. Energy Rev.* **2014**, *39*, 277–295. [CrossRef]

29. Brager, G.S.; De Dear, R.J. Thermal adaptation in the built environment: A literature review. *Energy Build.* **1998**, *27*, 83–96. [CrossRef]

30. Soares, N.; Bastos, J.; Pereira, L.D.; Soares, A.; Amaral, A.R.; Asadi, E.; Rodrigues, E.; Lamas, F.B.; Monteiro, H.; Lopes, M.A.R.; et al. A review on current advances in the energy and environmental performance of buildings towards a more sustainable built environment. *Renew. Sustain. Energy Rev.* **2017**, *77*, 845–860. [CrossRef]

31. Lu, Y.; Wang, S.; Shan, K. Design optimization and optimal control of grid-connected and standalone nearly/net zero energy buildings. *Appl. Energy* **2015**, *155*, 463–477. [CrossRef]

32. Zuo, J.; Zhao, Z.Y. Green building research-current status and future agenda: A review. *Renew. Sustain. Energy Rev.* **2014**, *30*, 271–281. [CrossRef]

33. Martínez-Molina, A.; Tort-Ausina, I.; Cho, S.; Vivancos, J.L. Energy efficiency and thermal comfort in historic buildings: A review. *Renew. Sustain. Energy Rev.* **2016**, *61*, 70–85. [CrossRef]

34. Vásquez-Hernández, A.; Restrepo Álvarez, M.F. Evaluation of buildings in real conditions of use: Current situation. *J. Build. Eng.* **2017**, *12*, 26–36. [CrossRef]

35. Raw, G.J.; Littleford, C.; Clery, L. Saving energy with a better indoor environment. *Arch. Sci. Rev.* **2017**, *60*, 239–248. [CrossRef]

36. Sorrell, S. Reducing energy demand: A review of issues, challenges and approaches. *Renew. Sustain. Energy Rev.* **2015**, *47*, 74–82. [CrossRef]

37. Pérez-Lombard, L.; Ortiz, J.; Pout, C. A review on buildings energy consumption information. *Energy Build.* **2008**, *40*, 394–398. [CrossRef]

38. Marjaba, G.E.; Chidiac, S.E. Sustainability and resiliency metrics for buildings—Critical review. *Build. Environ.* **2016**, *101*, 116–125. [CrossRef]

39. Pacheco, R.; Ordóñez, J.; Martínez, G. Energy efficient design of building: A review. *Renew. Sustain. Energy Rev.* **2012**, *16*, 3559–3573. [CrossRef]

40. Sintov, N.D.; Schultz, P.W. Adjustable green defaults can help make smart homes more sustainable. *Sustainability* **2017**, *9*, 622. [CrossRef]

41. Marikyan, D.; Papagiannidis, S.; Alamanos, E. A systematic review of the smart home literature: A user perspective. *Technol. Forecast. Soc. Chang.* **2019**, *138*, 139–154. [CrossRef]

42. Kylili, A.; Fokaides, P.A. European Smart Cities: The Role of Zero Energy Buildings. *Sustain. Cities Soc.* **2015**, *15*, 86–95. [CrossRef]

43. GhaffarianHoseini, A.; Dahlan, N.D.; Berardi, U.; GhaffarianHoseini, A.; Makaremi, N.; GhaffarianHoseini, M. Sustainable energy performances of green buildings: A review of current theories, implementations and challenges. *Renew. Sustain. Energy Rev.* **2013**, *25*, 1–17. [CrossRef]

44. Stevanović, S. Optimization of passive solar design strategies: A review. *Renew. Sustain. Energy Rev.* **2013**, *25*, 177–196. [CrossRef]

45. Zhao, X.; Zuo, J.; Wu, G.; Huang, C. A bibliometric review of green building research 2000–2016. *Arch. Sci. Rev.* **2018**, *62*, 74–88. [CrossRef]

46. Song, Y.; Wu, S.; Yan, Y.Y. Control strategies for indoor environment quality and energy efficiency-a review. *Int. J. Low Carbon Technol.* **2013**, *10*, 305–312. [CrossRef]

47. Wang, Y.; Kuckelkorn, J.; Liu, Y. A state of art review on methodologies for control strategies in low energy buildings in the period from 2006 to 2016. *Energy Build.* **2017**, *147*, 27–40. [CrossRef]

48. Rockett, P.; Hathway, E.A. Model-predictive control for non-domestic buildings: A critical review and prospects. *Build. Res. Inform.* **2017**, *45*, 556–571. [CrossRef]

49. Wong, J.K.W.; Li, H.; Wang, S.W. Intelligent building research: A review. *Autom. Construct.* **2005**, *14*, 143–159. [CrossRef]

50. Dounis, A.I.; Caraiscos, C. Advanced control systems engineering for energy and comfort management in a building environment—A review. *Renew. Sustain. Energy Rev.* **2009**, *13*, 1246–1261. [CrossRef]

51. Shaikh, P.H.; Nor, N.B.M.; Nallagownden, P.; Elamvazuthi, I.; Ibrahim, T. A review on optimized control systems for building energy and comfort management of smart sustainable buildings. *Renew. Sustain. Energy Rev.* **2014**, *34*, 409–429. [CrossRef]

52. Kolokotsa, D. Artificial intelligence in buildings: A review of the application of fuzzy logic. *Adv. Build. Energy Res.* **2007**, *1*, 29–54. [CrossRef]

53. Singh, A.; Gaur, A.; Kumar, A.; Singh, M.K.; Kapoor, K.; Mahanta, P.; Kumar, A.; Mukhopadhyay, S.C. Sensing Technologies for Monitoring Intelligent Buildings: A Review. *IEEE Sens. J.* **2018**, *18*, 4847–4860. [CrossRef]

54. Loonen, R.C.G.M.; Trčka, M.; Cóstola, D.; Hensen, J.L.M. Climate adaptive building shells: State-of-the-art and future challenges. *Renew. Sustain. Energy Rev.* **2013**, *25*, 483–493. [CrossRef]

55. Rezaei, S.D.; Shannigrahi, S.; Ramakrishna, S. A review of conventional, advanced, and smart glazing technologies and materials for improving indoor environment. *Sol. Energy Mater. Sol. Cells* **2017**, *159*, 26–51. [CrossRef]

56. Casini, M. Active dynamic windows for buildings: A review. *Renew. Energy* **2018**, *119*, 923–934. [CrossRef]

57. Quesada, G.; Rousse, D.; Dutil, Y.; Badache, M.; Hallé, S. A comprehensive review of solar facades. Opaque solar facades. *Renew. Sustain. Energy Rev.* **2012**, *16*, 2820–2832. [CrossRef]

58. Quesada, G.; Rousse, D.; Dutil, Y.; Badache, M.; Hallé, S. A comprehensive review of solar facades. Transparent and translucent solar facades. *Renew. Sustain. Energy Rev.* **2012**, *16*, 2643–2651. [CrossRef]

59. Sadineni, S.B.; Madala, S.; Boehm, R.F. Passive building energy savings: A review of building envelope components. *Renew. Sustain. Energy Rev.* **2011**, *15*, 3617–3631. [CrossRef]

60. Fisk, W.J. Review of some effects of climate change on indoor environmental quality and health and associated no-regrets mitigation measures. *Build. Environ.* **2015**, *86*, 70–80. [CrossRef]

61. Gunay, H.B.; O'Brien, W.; Beausoleil-Morrison, I. A critical review of observation studies, modeling, and simulation of adaptive occupant behaviors in offices. *Build. Environ.* **2013**, *70*, 31–47. [CrossRef]

62. Yan, D.; O'Brien, W.; Hong, T.; Feng, X.; Burak Gunay, H.; Tahmasebi, F.; Mahdavi, A. Occupant behavior modeling for building performance simulation: Current state and future challenges. *Energy Build.* **2015**, *107*, 264–278. [CrossRef]

63. Mirakhorli, A.; Dong, B. Occupancy behavior based model predictive control for building indoor climate—A critical review. *Energy Build.* **2016**, *129*, 499–513. [CrossRef]

64. Foucquier, A.; Robert, S.; Suard, F.; Stéphan, L.; Jay, A. State of the art in building modelling and energy performances prediction: A review. *Renew. Sustain. Energy Rev.* **2013**, *23*, 272–288. [CrossRef]

65. Sakhare, V.V.; Ralegaonkar, R.V. Indoor environmental quality: Review of parameters and assessment models. *Arch. Sci. Rev.* **2014**, *57*, 147–154. [CrossRef]

66. Peretti, C.; Schiavon, S. Indoor environmental quality surveys. A brief literature review. In *Indoor Air 2011*; Center for the Built Environment: Dallas, TX, USA, 2011.

67. Heinzerling, D.; Schiavon, S.; Webster, T.; Arens, E. Indoor environmental quality assessment models: A literature review and a proposed weighting and classification scheme. *Build. Environ.* **2013**, *70*, 210–222. [CrossRef]

68. Guan, Y.D.; Hosni, M.H.; Jones, B.W.; Gielda, T. Literature Review of the Advances in Thermal Comfort Modeling. *ASHRAE Trans.* **2003**, *109*, 908–916.

69. Djongyang, N.; Tchinda, R.; Njomo, D. Thermal comfort: A review paper. *Renew. Sustain. Energy Rev.* **2010**, *14*, 2626–2640. [CrossRef]

70. Park, J.Y.; Nagy, Z. Comprehensive analysis of the relationship between thermal comfort and building control research—A data-driven literature review. *Renew. Sustain. Energy Rev.* **2018**, *82*, 2664–2679. [CrossRef]

71. Rupp, R.F.; Vásquez, N.G.; Lamberts, R. A review of human thermal comfort in the built environment. *Energy Build.* **2015**, *105*, 178–205. [CrossRef]

72. Zhang, H.; Arens, E.; Zhai, Y. A review of the corrective power of personal comfort systems in non-neutral ambient environments. *Build. Environ.* **2015**, *91*, 15–41. [CrossRef]

73. Yang, L.; Yan, H.; Lam, J.C. Thermal comfort and building energy consumption implications—A review. *Appl. Energy* **2014**, *115*, 164–173. [CrossRef]

74. Peffer, T.; Pritoni, M.; Meier, A.; Aragon, C.; Perry, D. How people use thermostats in homes: A review. *Build. Environ.* **2011**, *46*, 2529–2541. [CrossRef]

75. De Dear, R.J.; Akimoto, T.; Arens, E.A.; Brager, G.; Candido, C.; Cheong, K.W.D.; Li, B.; Nishihara, N.; Sekhar, S.C.; Tanabe, S.; et al. Progress in thermal comfort research over the last twenty years. *Indoor Air* **2013**, *23*, 442–461. [CrossRef]

76. Karjalainen, S. Thermal comfort and gender: A literature review. *Indoor Air* **2012**, *22*, 96–109. [CrossRef]

77. Taleghani, M.; Tenpierik, M.; Kurvers, S.; Van Den Dobbelsteen, A. A review into thermal comfort in buildings. *Renew. Sustain. Energy Rev.* **2013**, *26*, 201–215. [CrossRef]

78. Enescu, D. A review of thermal comfort models and indicators for indoor environments. *Renew. Sustain. Energy Rev.* **2017**, *79*, 1353–1379. [CrossRef]

79. Carlucci, S.; Causone, F.; De Rosa, F.; Pagliano, L. A review of indices for assessing visual comfort with a view to their use in optimization processes to support building integrated design. *Renew. Sustain. Energy Rev.* **2015**, *47*, 1016–1033. [CrossRef]

80. Van Den Wymelenberg, K.G. Visual comfort, discomfort glare, and occupant fenestration control: Developing a research agenda. *LEUKOS* **2014**, *10*, 207–221. [CrossRef]

81. Altmonte, S. Daylight for Energy Savings and Psycho-Physiological Well-Being in Sustainable Built Environments. *J. Sustain. Dev.* **2008**, *1*, 3–16. [CrossRef]

82. Todorovic, M.S.; Kim, J.T. Beyond the science and art of the healthy buildings daylighting dynamic control's performance prediction and validation. *Energy Build.* **2012**, *46*, 159–166. [CrossRef]

83. Aries, M.B.C.; Aarts, M.P.J.; Van Hoof, J. Daylight and health: A review of the evidence and consequences for the built environment. *Light. Res. Technol.* **2015**, *47*, 6–27. [CrossRef]

84. Galasiu, A.D.; Veitch, J.A. Occupant preferences and satisfaction with the luminous environment and control systems in daylit offices: A literature review. *Energy Build.* **2006**, *38*, 728–742. [CrossRef]

85. Reinten, J.; Braat-Eggen, P.E.; Hornikx, M.; Kort, H.S.M.; Kohlrausch, A. The indoor sound environment and human task performance: A literature review on the role of room acoustics. *Build. Environ.* **2017**, *123*, 315–332. [CrossRef]

86. Navai, M.; Veitch, J.A. *Acoustic Satisfaction in Open-Plan Offices: Review and Recommendations*; Research Report; NRC Institute for Research in Construction: Ottawa, ON, Canada, 2006.

87. Vardaxis, N.G.; Bard, D.; Persson Waye, K. Review of acoustic comfort evaluation in dwellings—Part I: Associations of acoustic field data to subjective responses from building surveys. *Build. Acoust.* **2018**, *25*, 151–170. [CrossRef]

88. Vardaxis, N.G.; Bard, D. Review of acoustic comfort evaluation in dwellings: Part II—impact sound data associated with subjective responses in laboratory tests. *Build. Acoust.* **2018**, *25*, 171–192. [CrossRef]

89. Chua, K.J.; Chou, S.K.; Yang, W.M.; Yan, J. Achieving better energy-efficient air conditioning—A review of technologies and strategies. *Appl. Energy* **2013**, *104*, 87–104. [CrossRef]

90. Sundell, J.; Levin, H.; Nazaroff, W.W.; Cain, W.S.; Fisk, W.J.; Grimsrud, D.T.; Gyntelberg, F.; Li, Y.; Persily, A.K.; Pickering, A.C.; et al. Ventilation rates and health: Multidisciplinary review of the scientific literature. *Indoor Air* **2011**, *21*, 191–204. [CrossRef]

91. Bornehag, C.G.; Sundell, J.; Bonini, S.; Custovic, A.; Malmberg, P.; Skerfving, S.; Sigsgaard, T.; Verhoeff, A. Dampness in buildings as a risk factor for health effects, EUROEXPO: A multidisciplinary review of the literature (1998–2000) on dampness and mite exposure in buildings and health effects. *Indoor Air* **2004**, *14*, 243–257. [CrossRef]

92. Destaillats, H.; Maddalena, R.L.; Singer, B.C.; Hodgson, A.T.; McKone, T.E. Indoor pollutants emitted by office equipment: A review of reported data and information needs. *Atmos. Environ.* **2008**, *42*, 1371–1388. [CrossRef]

93. Wang, S.; Ma, Z. Supervisory and Optimal Control of Building HVAC Systems: A Review. *HVAC Res.* **2011**, *14*, 3–32. [CrossRef]

94. Lai, C.K. Particle deposition indoors: A review. *Indoor Air* **2002**, *12*, 211–214. [CrossRef]

95. Yu, B.F.; Hu, Z.B.; Liu, M.; Yang, H.L.; Kong, Q.X.; Liu, Y.H. Review of research on air-conditioning systems and indoor air quality control for human health. *Int. J. Refrig.* **2009**, *32*, 3–20. [CrossRef]

96. Mendell, M.J.; Fisk, W.J.; Kreiss, K.; Levin, H.; Alexander, D.; Cain, W.S.; Girman, J.R.; Hines, C.J.; Jensen, P.A.; Milton, D.K.; et al. Improving the health of workers in indoor environments: Priority research needs for a national occupational research agenda. *Am. J. Public Health* **2002**, *92*, 1430–1440. [CrossRef]

97. Persily, A.K.; Emmerich, S.J. Indoor air quality in sustainable, energy efficient buildings. *HVAC Res.* **2012**, *18*, 4–20. [CrossRef]

98. Tham, K.W. Indoor air quality and its effects on humans—A review of challenges and developments in the last 30 years. *Energy Build.* **2016**, *130*, 637–650. [CrossRef]

99. Van Eck, N.J.; Waltman, L. Software survey: VOSviewer, a computer program for bibliometric mapping. *Scientometrics* **2010**, *84*, 523–538. [CrossRef]

100. Prüss-Üstün, A.; Corvalán, C. *Preventing Disease Through Healthy Environs*; World Health Organization: Genève, Switzerland, 2006.

101. Redlich, C.A.; Sparer, J.; Cullen, M.R. Sick-building syndrome. *Lancet* **1997**, *349*, 1013–1016. [CrossRef]

102. Bottalico, P.; Astolfi, A. Investigations into vocal doses and parameters pertaining to primary school teachers in classrooms. *J. Acoust. Soc. Am.* **2012**, *131*, 2817–2827. [CrossRef]

103. Srinivasan, S.; O'Fallon, L.R.; Dearry, A. Creating Healthy Communities, Healthy Homes, Healthy People: Initiating a Research Agenda on the Built Environment and Public Health. *Am. J. Public Health* **2003**, *93*, 1446–1450. [CrossRef]

104. World Health Organization (WHO); Regional Office for Europe. *Declaration: fourth Ministerial Conference on Environment and Health, Budapest, Hungary, 23–25 June 2004*; WHO Regional Office for Europe: Copenhagen, Denmark, 2004.

105. Berardi, U. Sustainability Assessment in the Construction Sector: Rating Systems and Rated Buildings. *Sustain. Dev.* **2012**, *20*, 411–424. [CrossRef]

106. Bluyssen, P.M. Towards new methods and ways to create healthy and comfortable buildings. *Build. Environ.* **2010**, *45*, 808–818. [CrossRef]

107. Cox, C. *Health Optimisation Protocol for Energy-Efficient Buildings Pre-Normative and Socio-Economic Res. to Create Healthy and Energy-Efficient Buildings*; TNO: Delft, The Netherlands, 2005.

108. Sarbu, I.; Sebarchievici, C. Aspects of indoor environmental quality assessment in buildings. *Energy Build.* **2013**, *60*, 410–419. [CrossRef]

109. Bronsema, B.; Bjorck, M.; Clausen, G.; Firzner, K.; Flatheim, G.; Follin, T.; Haverinen, U.; Jamriska, M.; Kurnistki, J.; Maroni, M.; et al. *Performance Criteria of Buildings for Health and Comfort*; CIB General Secretariat: Rotterdam, The Netherlands, 2004; ISBN 90-6363-038-7.

110. Bluyssen, P.M. *The Indoor Environ. Handbook: How to Make Buildings Healthy and Comfortable*; Earthscan: London, UK, 2009.

111. ASHRAE. *Thermal Environmental Conditions for Human Occupancy*; ANSI/ASHRAE Standard 55-2017; American Society of Heating, Refrigerating and Air-Conditioning Engineers: Atlanta, GA, USA, 2017.

112. ISO 17772-1. *Energy Performance of Builds—Indoor Environmental Quality—Part 1: Indoor Environmental Input Parameters for the Design and Assessment of Energy Performance of Builds*; ISO: Genève, Switzerland, 2017.

113. ISO 7730. *Ergonomics of the Thermal Environ.—Analytiacl Determination and Interpretation of Thermal Comfort Using Calculation of the PMV and PPD Indices and Local Thermal Comfort Criteria*; ISO: Genève, Switzerland, 2005.

114. CEN. *Light and Lighting—Basic Terms and Criteria for Specifying Lighting Requirements*; European Standard EN 12665:2011; European Committee for Standardisation: Brussels, Belgium, 2011.

115. CEN. *Dayligh in Buildings*; European Standard EN 17037:2018; European Committee for Standardisation: Brussels, Belgium, 2018.

116. Bellia, L.; Bisegna, F.; Spada, G. Lighting in indoor environments: Visual and non-visual effects of light sources with different spectral power distributions. *Build. Environ.* **2011**, *46*, 1984–1992. [CrossRef]

117. Hraška, J. Chronobiological aspects of green buildings daylighting. *Renew. Energy* **2014**, *73*, 109–114. [CrossRef]

118. Van Bommel, W.J.M. Non-visual biological effect of lighting and the practical meaning for lighting for work. *Appl. Ergon.* **2006**, *37*, 461–466. [CrossRef]

119. Alrubaih, M.S.; Zain, M.F.M.; Alghoul, M.A.; Ibrahim, N.L.N.; Shameri, M.A.; Elayeb, O. Research and development on aspects of daylighting fundamentals. *Renew. Sustain. Energy Rev.* **2013**, *21*, 494–505. [CrossRef]

120. IES. 100 Significant Papers. Available online: http://www.ies.org/edoppts/100papers.cfm (accessed on 28 January 2018).

121. IESNA. *The IESNA Lighting Handbook*, 9th ed.; IESNA: New York, NY, USA, 2000.

122. U.S. Department of Energy. *Energy Efficiency Trends in Residential and Commercial Buildings*; United States Department of Energy: Washington, DC, USA, 2008.

123. VonNeida, B.; Maniccia, D.; Tweed, A. An analysis of the energy and cost savings potential of occupancy sensors for commercial lighting systems. *J. Illum. Eng. Soc.* **2000**, *30*, 11–125. [CrossRef]

124. WHO. *WHO Guildelines for Indoor Air Quality: Selected Polutants*; WHO Regional Office for Europe: Bonn, Switzerland, 2010.

125. CEN. *Energy performance of buildings – Ventilation for buildings – Part 3: For non-residential buildings - performance requirements for ventilation and room-conditioning systems (Modules M5-1. M5-4)*; European Standard EN 16798-3:2017; European Committee for Standardisation: Brussels, Belgium, 2017.

126. ASHRAE. *Ventilation for Acceptable Indoor Air Quality*; ANSI/ASHRAE Standard 62.1-2016; American Society of Heating, Refrigerating and Air-Conditioning Engineers: Atlanta, GA, USA, 2016.

127. ASHRAE. *Ventilation and Acceptable Indoor Air Quality in Low-Rise Residential Buildings*; ANSI/ASHRAE Standard 62.2-2016; American Society of Heating, Refrigerating and Air-Conditioning Engineers: Atlanta, GA, USA, 2016.

128. CEN. *Ventilation for buildings—Design Criteria for the Indoor Environment*; European Prestandard CR 1752:1998; European Committee for Standardisation: Brussels, Belgium, 1998.

129. ASHRAE. *Indoor Air Quality Guide*; American Society of Heating, Refrigerating and Air-Conditioning Engineers: Atlanta, GA, USA, 2009; ISBN 978-1-933742-59-5.

130. Brown, A.L.; Kang, J.; Gjestland, T. Towards standardization in soundscape preference assessment. *Appl. Acoust.* **2011**, *72*, 387–392. [CrossRef]

131. DIN. *Akustik—Begriffe*; German National Standard DIN 1320:2009; Deutsches Institut Fur Normung E.V.: Berglin, Germany, 2009.

132. Genuit, K. Objective Evaluation of Acoustic Quality Based on a Realtive Approach. In Proceedings of the Inter-Noise'96, 25th Anniversary Congress Liverpool, Liverpool, UK, 30 July–2 August 1996; pp. 1061 p1–1061 p6.

133. Gramez, A.; Boubenider, F. Acoustic comfort evaluation for a conference room: A case study. *Appl. Acoust.* **2017**, *118*, 39–49. [CrossRef]

134. ISO 10140. Acoustics: Building Elements Sound Insulation Measurements. Sound and Vibration Standards. Available online: http://www.acoustic-standards.co.uk/bs-10140.htm (accessed on 8 February 2015).

135. ISO 717. *Acoustics—Rating of Sound Insulation in Buildings and of Build. Elements*; ISO: Genève, Switzerland, 2013.

136. ISO 3382. *Acoustics—Measurement of Room Acoustic Parameters*; ISO: Genève, Switzerland, 2012.

137. ISO 12354. *Build. Acoustics—Estimation of Acoustic Performance of Buildings from the Performance of Elements*; ISO: Genève, Switzerland, 2017.

138. World Health Organization (WHO). *Constitution of the World Health Organization*; WHO: Geneva, Switzerland, 2006.

139. Jamrozik, A.; Ramos, C.; Zhao, J.; Bernau, J.; Clements, N.; Vetting Wolf, T.; Bauer, B. A novel methodology to realistically monitor office occupant reactions and environmental conditions using a living lab. *Build. Environ.* **2018**, *130*, 190–199. [CrossRef]

140. Lai, A.C.K.; Mui, K.W.; Wong, L.T.; Law, L.Y. An evaluation model for indoor environmental quality (IEQ) acceptance in residential buildings. *Energy Build.* **2009**, *41*, 930–936. [CrossRef]

141. Chiang, C.M.; Chou, P.C.; Lai, C.M.; Li, Y.Y. A methodology to assess the indoor environment in care centers for senior citizens. *Build. Environ.* **2001**, *36*, 561–568. [CrossRef]

142. Zalejska-Jonsson, A.; Wilhelmsson, M. Impact of perceived indoor environment quality on overall satisfaction in Swedish dwellings. *Build. Environ.* **2013**, *63*, 134–144. [CrossRef]

143. Frontczak, M.; Schiavon, S.; Goins, J.; Arens, E.; Zhang, H.; Wargocki, P. Quantitative relationships between occupant satisfaction and satisfaction aspects of indoor environmental quality and building design. *Indoor Air* **2012**, *22*, 119–131. [CrossRef]

144. Wargocki, P.; Frontczak, M.; Schiavon, S.; Goins, J.; Arens, E.; Zhang, H. Satisfaction and self-estimated performance in relation to indoor environmental parameters and building features. In Proceedings of the 10th International Conference on Healthy Builds 2012, Brisbane, Australia, 8–12 July 2012.

145. Humphreys, M.A. Quantifying occupant comfort: Are combined indices of the indoor environment practicable? *Build. Res. Inform.* **2005**, *33*, 317–325. [CrossRef]

146. Chiang, C.M.; Lai, C.M. A study on the comprehensive indicator of indoor environment assessment for occupants' health in Taiwan. *Build. Environ.* **2002**, *37*, 387–392. [CrossRef]

147. Wong, L.T.; Mui, K.W.; Hui, P.S. A multivariate-logistic model for acceptance of indoor environmental quality (IEQ) in offices. *Build. Environ.* **2008**, *43*, 1–6. [CrossRef]

148. Astolfi, A.; Pellerey, F. Subjective and objective assessment of acoustical and overall environmental quality in secondary school classrooms. *J. Acoust. Soc. Am.* **2008**, *123*, 163–173. [CrossRef]

149. Lai, J.H.K.; Yik, F.W.H. Perception of importance and performance of the indoor environmental quality of high-rise residential buildings. *Build. Environ.* **2009**, *44*, 352–360. [CrossRef]

150. Cao, B.; Ouyang, Q.; Zhu, Y.; Huang, L.; Hu, H.; Deng, G. Development of a multivariate regression model for overall satisfaction in public buildings based on field studies in Beijing and Shanghai. *Build. Environ.* **2012**, *47*, 394–399. [CrossRef]

151. Marino, C.; Nucara, A.; Pietrafesa, M. Proposal of comfort classification indexes suitable for both single environments and whole buildings. *Build. Environ.* **2012**, *57*, 58–67. [CrossRef]

152. Ncube, M.; Riffat, S. Developing an indoor environment quality tool for assessment of mechanically ventilated office buildings in the UK—A preliminary study. *Build. Environ.* **2012**, *53*, 26–33. [CrossRef]

153. Gayathri, L.; Perera, B.A.K.S.; Sumanarathna, D.M.G.A.N.M. Factors Affecting the Indoor Environmental Quality in Sri Lanka: Green vs. Conventional Hotel Buildings. In Proceedings of the 5th World Construction Symposium 2016: Greening Environment, Eco Innovations & Entrepreneurship, Colombo, Sri Lanka, 29–31 July 2016; pp. 210–220.

154. Leaman, A.; Bordass, B. Productivity in buildings: The "killer" variables. *EcoLibrium* **2005**, *4*, 16–20. [CrossRef]

155. Bakker, L.G.; Hoes-van Oeffelen, E.C.M.; Loonen, R.C.G.M.; Hensen, J.L.M. User satisfaction and interaction with automated dynamic facades: A pilot study. *Build. Environ.* **2014**, *78*, 44–52. [CrossRef]

156. Castaldo, V.L.; Pigliautile, I.; Rosso, F.; Cotana, F.; De Giorgio, F.; Pisello, A.L. How subjective and non-physical parameters affect occupants' environmental comfort perception. *Energy Build.* **2018**, *178*, 107–129. [CrossRef]

157. Heerwagen, J. Green buildings, organizational success and occupant productivity. *Build. Res. Inform.* **2000**, *28*, 353–367. [CrossRef]

158. Fisk, W.J. Health and Productivity Gains from Better Indoor Environments and their Relationship with Building Energy Efficiency. *Annu. Rev. Energy Environ.* **2000**, *25*, 537–566. [CrossRef]

159. Bluyssen, P.M.; Aries, M.; Van Dommelen, P. Comfort of workers in office buildings: The European HOPE project. *Build. Environ.* **2011**, *46*, 280–288. [CrossRef]

160. Molina, C.; Pickering, A.C.; Valjbjorn, O.; De Bartoli, M. *Sick Build Syndrome A practical Guide*; Office for Publications of the European Communities: Luxembourg City, Luxembourg, 1989.

161. Berglund, B.; Lindvall, T. Sensory reactions to "sick buildings". *Environ. Int.* **1986**, *12*, 147–159. [CrossRef]

162. Crawford, J.O.; Bolas, S.M. Sick building syndrome, work factors and occupational stress. *Scand. J. Work Environ. Health* **1996**, *22*, 243–250. [CrossRef]

163. Preziosi, P.; Czernichow, S.; Gehanno, P.; Hercberg, S. Workplace air-conditioning and health services attendance among French middle-aged women: A prospective cohort study. *Int. J. Epidemiol.* **2004**, *33*, 1120–1123. [CrossRef]

164. Seppänen, O.; Fisk, W.J. A conceptual model to estimate cost effectiveness of the indoor environment improvements. In Proceedings of the Healthy Builds 2003 Conference, Singapore; Healthy Buildings Inc.: Singapore, 2003; pp. 368–374.

165. Nicol, J.F.; Humphreys, M.A. Adaptive thermal comfort and sustainable thermal standards for buildings. *Energy Build.* **2002**, *34*, 563–572. [CrossRef]

166. Wyon, D. Indoor environmental effects on productivity. *IAQ* **1996**, *96*, 5–15.

167. Heat-Shield. Available online: https://www.heat-shield.eu/ (accessed on 2 November 2018).

168. Meerbeek, B.W.; Van Loenen, E.J.; Te Kulve, M.; Aarts, M. User Experience of Automated Blinds in Offices. In Proceedings of the Experiencing Light 2012 International Conference on the Effects of Light on Wellbeing, Eindhoven, The Netherlands, 12–13 November 2012; De Kort, Y., Aarts, A.W., Beute, M.J., Ijsselsteijn, F., Lakens, W.A., Smolders, D., An, C.H.J., Van Rijswijk, L., Eds.; Technische Universiteit Eindhoven: Eindhoven, The Netherlands, 2012; pp. 1–5.

169. Aries, M.B.C.; Veitch, J.A.; Newsham, G.R. Windows, view, and office characteristics predict physical and psychological discomfort. *J. Environ. Psychol.* **2010**, *30*, 533–541. [CrossRef]

170. Franchi, M.; Carrer, P.; Kotzias, D.; Viegi, G. *Towards Healthy Air in Dwellings in Europe—The THADE Report*; EFA Central Office: Brussels, Belgium, 2004.

171. Franchi, M.; Carrer, P.; Kotzias, D.; Rameckers, E.M.A.L.; Seppänen, O.; Van Bronswijk, J.E.M.H.; Viegi, G.; Gilder, J.A.; Valovirta, E. Working towards healthy air in dwellings in Europe. *Allergy* **2006**, *61*, 864–868. [CrossRef]

172. Autrup, H.; Calow, P.; Dekant, W.; Greim, H.; Hanke, W.; Janssen, C.; Jansson, B.; Komulainen, H.; Ladefoged, O.; Linders, J.; et al. *Opinion on Risk Assessment on Indoor Air Quality*; European Commision: Brussels, Belgium, 2007; ISBN 9789279127564.

173. World Health Organization (WHO). *Preventing Noncommunicable Diseases (NCDs) by Reducing Environmental Risk Factors*; World Health Organisation: Geneva, Switzerland, 2017.

174. Landström, U.; Åkerlund, E.; Kjellberg, A.; Tesarz, M. Exposure levels, tonal components, and noise annoyance in working environments. *Environ. Int.* **1995**, *21*, 265–275. [CrossRef]

175. Jensen, K.L.; Arens, E.; Zagreus, L. Acoustical Quality in Office Workstations, as Assessed by Occupant Surveys. In Proceedings of the Indoor Air 2005, Beijing, China, 4–9 September 2005; University of California: Berkeley, CA, USA, 2005; pp. 2401–2405.

176. Veitch, J.A.; Charles, K.E.; Farley, K.M.J.; Newsham, G.R. A model of satisfaction with open-plan office conditions: COPE field findings. *J. Environ. Psychol.* **2007**, *27*, 177–189. [CrossRef]

177. Niemann, H.; Bonnefoy, X.; Braubach, M.; Hecht, K.; Maschke, C.; Rodrigues, C.; Robbel, N. Noise-induced annoyance and morbidity results from the pan-European LARES study. *Noise Health* **2006**, *8*, 63–79. [CrossRef]

178. Van Kempen, E.; Van Kamp, I.; Lebret, E.; Lammers, J.; Emmen, H.; Stansfeld, S. Neurobehavioral effects of transportation noise in primary schoolchildren: A cross-sectional study. *Environ. Health Glob. Access Sci. Source* **2010**, *9*, 25. [CrossRef]

179. Baker, N.; Nikolopoulou, M.; Steemers, K. Thermal comfort in urban spaces: Different forms of adaptation Proceedings Rebuild. In Proceedings of the 1999 REBUILD International Conference: The Cities of Tomorrow, Barcelona, Spain, 4–6 October 1999.

180. Kenny, G.P.; Yardley, J.E.; Martineau, L.; Jay, O. Physical work capacity in older adults: Implications for the aging worker. *Am. J. Ind. Med.* **2008**, *51*, 610–625. [CrossRef]

181. Lorenzoni, I.; Nicholson-Cole, S.; Whitmarsh, L. Barriers perceived to engaging with climate change among the UK public and their policy implications. *Glob. Environ. Chang.* **2007**, *17*, 445–459. [CrossRef]

182. Blanchard, B.; Blyler, J.E. Human-Factors Engineering. In *System Engineering Management*; John Wiley & Sons: Hoboken, NJ, USA, 2016.

183. Bamberg, S.; Möser, G. Twenty years after Hines, Hungerford, and Tomera: A new meta-analysis of psycho-social determinants of pro-environmental behaviour. *J. Environ. Psychol.* **2007**, *27*, 14–25. [CrossRef]

184. Wohlwill, J.F. Human Adaptation to Levels of Environmental Stimulation. *Hum. Ecol.* **1974**, *2*, 127–147. [CrossRef]

185. Gifford, R. The Dragons of Inaction: Psychological Barriers That Limit Climate Change Mitigation and Adaptation. *Am. Psychol.* **2011**, *66*, 290–302. [CrossRef]

186. Gilman, S. Joint position sense and vibration sense: Anatomical organisation and assessment. *J. Neurol. Neurosurg. Psychiatry* **2002**, *73*, 473–477. [CrossRef]

187. Patapoutian, A.; Peier, A.M.; Story, G.M.; Viswanath, V. Thermotrp channels and beyond: Mechanisms of temperature sensation. *Nat. Rev. Neurosci.* **2003**, *4*, 529–539. [CrossRef]

188. EN. Directive 2006/25/EC of the European Parliament and of the Council of 5 April 2006 on the Minimum Health and Safety Requirements Regarding the Exposure of Workers to Risks Arising from Physical Agents (Artificial Optical Radiation) (19th Individual Direct. 2006. ISBN 9789279122552). Available online: http://eur-lex.europa.eu/legal-content/EN/TXT/?uri=CELEX%3A02006L0025-20140101 (accessed on 6 February 2018).

189. Bernstein, J.A.; Alexis, N.; Bacchus, H.; Bernstein, I.L.; Fritz, P.; Horner, E.; Li, N.; Mason, S.; Nel, A.; Oullette, J.; et al. The health effects of nonindustrial indoor air pollution. *J. Allergy Clin. Immunol.* **2008**, *121*, 585–591. [CrossRef]

190. Noonan, B.; Bancroft, R.W.; Dines, J.S.; Bedi, A. Heat- and Cold-induced Injuries in Athletes: Evaluation and Management. *J. Am. Acad. Orthop. Surg.* **2012**, *20*, 744–754. [CrossRef]

191. Toftum, J.; Andersen, R.V.; Jensen, K.L. Occupant performance and building energy consumption with different philosophies of determining acceptable thermal conditions. *Build. Environ.* **2009**, *44*, 2009–2016. [CrossRef]

192. Gross, H.; Blechinger, F.; Achtner, B. Human Eye. In *Handbook of Optical Systems, Volume 4, Survey of Optical Instruments*; Herbert, G., Ed.; Wiley: Hoboken, NJ, USA, 2008; pp. 1–88.

193. Curry, D.G.; Martinsen, G.L.; Hopper, D.G. Capability of the human visual system. *Cockpit Disp. X* **2003**, *5080*, 58–69. [CrossRef]

194. Boff, K.R.; Lincoln, J.E. *Eng. Data Compendium. Human Perception and Performance*; Wiley and Sons: New York, NY, USA, 1988.

195. Davis, J.; Hsieh, Y.H.; Lee, H.C. Humans perceive flicker artifacts at 500 Hz. *Sci. Rep.* **2015**, *5*, 7861. [CrossRef]

196. De Valois, R.L.; De Valois, K.K. *Spatial Vision*; Oxford University Press: Oxford, UK, 1990.

197. Ruch, T.C.; Fulton, J.F. *Medical Physiology and Biophysics*; W.B. Saunders: Philadelphia, PA, USA, 1960.

198. Lind, B.; Sihlbom, H.; Nordwall, A.; Malchau, H. Normal range of motion of the cervical spine. *Arch. Phys. Med. Rehab.* **1989**, *70*, 692–695.

199. Xie, B. Virtual Auditory Display and Spatial Hearing. In *Head-Related Transfer Function and Virtual Auditory Display*, 2nd ed.; Ross Publishing, Inc.: Richmond, VA, USA, 2013; pp. 1–42.

200. Hartmann, W.M. How we localize sound. *Phys. Today* **1999**, *52*, 24–29. [CrossRef]

201. McQueen, C.A.; Bond, J.; Ramos, K.; Lamb, J.; Guengerich, F.P.; Lawrence, D.; Walker, M.; Campen, M.; Schnellmann, R.; Yost, G.S.; et al. *Comprehensive Toxicology, Volumes 1-14*, 2nd ed.; Elsevier: Amsterdam, The Netherlands, 2010.

202. Butcher, K.J. Health Issues. In *CIBSE Guide A—Environ.al Design*; CIBSE: Norwich, UK, 2006.

203. Bushdid, C.; Magnasco, M.O.; Vosshall, L.B.; Keller, A.; Mixture, M. 1 Trillion Olfactory Stimuli. *Science* **2016**, *343*, 1370–1373. [CrossRef]

204. Feron, V.; Arts, J.; Kuper, C.; Slootweg, P.; Woutersen, R. Health risks associated with inhaled nasal toxicants. *Crit. Rev. Toxicol.* **2001**, *31*, 313–347. [CrossRef]

205. World Health Organization (WHO). *Indoor Air Quality Guidelines: Household Fuel Combustion*; World Health Organization: Geneva, Switzerland, 2014; ISBN 978 92 4 154887 8.

206. World Health Organization (WHO). *Ambient Air Pollution: A Global Assessment of Exposure and Burden of Disease*; World Health Organization: Geneva, Switzerland, 2016; ISBN 978 92 4 151135 3.

207. Nakano, S.; Washizu, A. Acceptance of energy efficient homes in large Japanese cities: Understanding the inner process of home choice and residence satisfaction. *J. Environ. Manag.* **2018**, *225*, 84–92. [CrossRef]

208. Altenburg, T.; Assmann, C. *Green Industrial Policy: Concept, Policies, Country Experiences*; UN Environment; German Development Institute: Bonn, Germany, 2014.

209. Hafer, M.; Howley, W.; Chang, M.; Ho, K.; Tsau, J.; Razavi, H. Occupant engagement leads to substantial energy savings for plug loads. In Proceedings of the 2017 IEEE Conference on Technologies for Sustainability, SusTech 2017, Phoenix, AZ, USA, 12–14 November 2017.

210. Kahma, N.; Matschoss, K. The rejection of innovations? Rethinking technology diffusion and the non-use of smart energy services in Finland. *Energy Res. Soc. Sci.* **2017**, *34*, 27–36. [CrossRef]

211. WCED. *Report of the World Commision on Environ. and Dev.: Our Common Future*; Oxford University Press: New York, NY, USA, 1987.

212. United Nations. *Paris Agreement*; United Nations: Paris, France, 2015.

213. Climate Action Tracker. Rating Countries. Available online: http://climateactiontracker.org/countries.html (accessed on 29 March 2018).

214. Marques, L.M.; Fuinhas, J.A.; Marques, A.C. Augmented energy-growth nexus: Economic, political and social globalization impacts. *Energy Procedia* **2017**, *136*, 97–101. [CrossRef]

215. Shahbaz, M.; Zakaria, M.; Shahzad, S.J.H.; Mahalik, M.K. The energy consumption and economic growth nexus in top ten energy-consuming countries: Fresh evidence from using the quantile-on-quantile approach. *Energy Econ.* **2018**, *71*, 282–301. [CrossRef]

216. Deichmann, U.; Reuter, A.; Vollmer, S.; Zhang, F. Relationship between Energy Intensity and Economic Growth New Evidence from a Multi—Country Multi—Sector Data Set. *World Bank Work. Pap.* **2018**. [CrossRef]

217. Croner, D.; Frankovic, I. A Structural Decomposition Analysis of Global and National Energy Intensity Trends. *Energy J.* **2018**, *39*, 219–231. [CrossRef]

218. The World Bank. Energy intensity level of primary energy (MJ/$2011 PPP GDP) Data. Sustain. energy for all. Available online: https://data.worldbank.org/indicator/EG.EGY.PRIM.PP.KD?view=chart (accessed on 6 February 2018).

219. IEA. Energy Efficiency 2018—Analysis and outlooks to 2040. *Market Rep. Ser.* **2017**, 1–143. [CrossRef]

220. European Commission. *A Roadmap for Moving to a Competitive Low Carbon Economy in 2050*; European Commission: Brussels, Belgium, 2011.

221. EN. Directive 2010/31/EU of the European Parliament and of the Council of 19 May 2010 on the energy performance of buildings. *Off. J. Eur. Union* **2010**, *153*, 13–35.

222. EN. Directive 2012/27/EU of the European Parliament and of the Councilof 25 October 2012 on energy efficiency. *Off. J. Eur. Union* **2012**, *315*, 1–56.

223. European Union. *Intended Nationally Determined Contribution of the EU and its Member States*; European Union: Brussels, Belgium, 2015.

224. En. Directive (EU) 2018/844 of the European Parliament an of the Council of 30 May 2018 amending Directive 2010/31/EU on the energy performance of buildings and Directive 2012/27/EU on energy efficiency. *Off. J. Eur. Union* **2018**, *156*, 17.

225. UN Environment. *Towards a Zero-Emission, Efficient, and Resilient Buildings and Construction Sector*; UN Environment: Nairobi, Kenya, 2017.

226. Fekete, H.; Luna, L.; Sterl, S.; Hans, F.; Gonzales, S.; Hohne, N.; Wong, L.; Deng, Y.; Ur, G.U.R.; Berg, T.; et al. *Improvement in Warming Outlook as India and China Move Ahead, but Paris Agreement Gap Still Looms Large*; Climate Action Tracker: Cologne/Berlin, Germany, 2017.

227. Eurostat. Energy Balances. Energy Balances in the MS Excel File Format (2018 Edition). Available online: http://ec.europa.eu/eurostat/web/energy/data/energy-balances (accessed on 29 March 2018).

228. U.S. Department of Energy. Use of Energy in the United States Explained. U.S. Energy Information Administration. Available online: https://www.eia.gov/energyexplained/index.php?page=us_energy_use (accessed on 30 January 2019).

229. National Bureau of Statistics of China. China Statistical Yearbook 2016. Available online: http://www.stats. gov.cn/tjsj/ndsj/2018/indexeh.htm (accessed on 30 January 2019).

230. Ministry of Statistics and Programme Implementation Government of India. ENERGY—Statistical Year Book India 2018. Statistical Year Book India 2018. Available online: http://www.mospi.gov.in/statistical-year-book-india/2018/185 (accessed on 30 January 2019).

231. United Nations Environment Programme. Why Buildings. Sustain. Buildings and Climate Initiative. 2015. Available online: http://www.unep.org/sbci/AboutSBCI/Background.asp (accessed on 12 February 2015).

232. International Energy Agency. *Tracking Clean Energy Progress 2017*; International Energy Agency: Paris, France, 2017.

233. International Energy Agency. *Energy Efficiency 2017*; International Energy Agency: Paris, France, 2017.

234. European Commission. Energy. EU Building Database. Available online: https://ec.europa.eu/energy/en/eu-buildings-database (accessed on 31 January 2019).

235. European Commission. *Annex. Accelerating Clean Energy in Buildings to the Communication from the Commision to the European Parliament, the Council, the European Economic and Social Committee, the Committee of the Regions and the European Investment Bank. Clean Energy For All*; European Commission: Brussels, Belgium, 2016.

236. RIJSBERMAN, F.R.; SWART, R.J. (Eds.) *Targets and Indicators of Climatic Change*; The Stockholm Environment Institute: Stockholm, Sweden, 1990; ISBN 9188116212.

237. International Energy Agency. *Transition to Sustain. Buildings. Strategies and Opportunities to 2050*; OECD/IEA: Paris, France, 2013.

238. Anda, M.; Temmen, J. Smart metering for residential energy efficiency: The use of community based social marketing for behavioural change and smart grid introduction. *Renew. Energy* **2014**, *67*, 119–127. [CrossRef]

239. Li, C.; Hong, T.; Yan, D. An insight into actual energy use and its drivers in high-performance buildings. *Appl. Energy* **2014**, *131*, 394–410. [CrossRef]

240. Vierra, S. Green Building Standards and Certification Systems|WBDG—Whole Building Design Guide. WHole Building Design Guide. 2016. Available online: https://www.wbdg.org/resources/green-building-standards-and-certification-systems (accessed on 6 February 2019).

241. Liang, J.; Qiu, Y.; Hu, M. Mind the energy performance gap: Evidence from green commercial buildings. *Resour. Conserv. Recycl.* **2019**, *141*, 364–377. [CrossRef]

242. Shrubsole, C.; Hamilton, I.G.; Zimmermann, N.; Papachristos, G.; Broyd, T.; Burman, E.; Mumovic, D.; Zhu, Y.; Lin, B.; Davies, M. Bridging the gap: The need for a systems thinking approach in understanding and addressing energy and environmental performance in buildings. *Indoor Built Environ.* **2018**, *28*, 100–117. [CrossRef]

243. Mcelroy, D.J.; Rosenow, J. Policy implications for the performance gap of low-carbon building technologies. *Build. Res. Inform.* **2018**, *47*, 611–623. [CrossRef]

244. Gill, Z.M.; Tierney, M.J.; Pegg, I.M.; Allan, N. Low-energy dwellings: The contribution of behaviors to actual performance. *Build. Res. Inform.* **2010**, *38*, 491–508. [CrossRef]

245. Gerarden, T.; Newell, R.; Stavins, R. Assessing the energy-efficiency ga. *J. Econ. Lit.* **2017**, *55*, 1486–1525. [CrossRef]

246. Borgstein, E.H.; Lamberts, R.; Hensen, J.L.M. Mapping failures in energy and environmental performance of buildings. *Energy Build.* **2018**, *158*, 476–485. [CrossRef]

247. Costa, A.; Keane, M.M.; Torrens, J.I.; Corry, E. Building operation and energy performance: Monitoring, analysis and optimisation toolkit. *Appl. Energy* **2013**, *101*, 310–316. [CrossRef]

248. Baudains, P.; Bishop, S.; Duffour, P.; Marjanovic-Halburd, L.; Psarra, S.; Spataru, C. A systems paradigm for integrated building design. *Intel. Build. Int.* **2014**, *6*, 201–214. [CrossRef]

249. Reed, W.G.; Gordon, E.B. Integrated design and building process: What research and methodologies are needed? *Build. Res. Inform.* **2000**, *28*, 325–337. [CrossRef]

250. Eastman, C.; Teicholz, P.; Sacks, R.; Liston, K. *BIM Handbook: A Guide to Build. Information Modeling for Owners, Managers, Designers, Enginners, and Contractors*; John Wiley & Sons, Inc.: New Jersey, NJ, USA, 2008.

251. Hetherington, R.; Laney, R.; Peak, S.; Oldham, D. Integrated building design, information and simulation modelling: The need for a new hierarchy. In *Build. Simulation 2011*; IBPSA: Sydney, Australia, 2011.

252. Ghaffarianhoseini, A.; Tookey, J.; Ghaffarianhoseini, A.; Naismith, N.; Azhar, S.; Efimova, O.; Raahemifar, K. Building Information Modelling (BIM) uptake: Clear benefits, understanding its implementation, risks and challenges. *Renew. Sustain. Energy Rev.* **2017**, *75*, 1046–1053. [CrossRef]

253. Reizgevicius, M.; Kutut, V.; Cibulskiene, D.; Nazarko, L. Promoting Sustainability through Investment in Building Information Modeling (BIM) Technologies: A Design Company Perspective. *Sustainability* **2018**, *10*, 21. [CrossRef]

254. Hernandez, P.; Kenny, P. From net energy to zero energy buildings: Defining life cycle zero energy buildings (LC-ZEB). *Energy Build.* **2010**, *42*, 815–821. [CrossRef]

255. Ozkan, A.; Kesik, T.; Yilmaz, A.Z.; O'Brien, W. Development and visualization of time-based building energy performance metrics. *Build. Res. Inform.* **2019**, *47*, 493–517. [CrossRef]

256. Boermans, T.; Hermelink, A.; Schischar, S.; Grözinger, J.; Offermann, M.; Thomsen, K.E.; Jorsen, R.; Aggerholm, S.O. *Principles for Nearly Zero-Energy Buildings. Paving the Way for Effective Implementation of Policy Requirements*; Buildings Performance Institute Europe: Brussels, Belgium, 2011.

257. Torcellini, P.; Pless, S.; Deru, M.; Crawley, D. Zero Energy Buildings: A critical Look at the Definition. In *ACEEE Summer Study*; American Council for and Energy-Efficient Economy: Pacific Grove, CA, USA, 2006.

258. United Nations. *Transforming Our World: The 2030 Agenda for Sustainable Develeopment*; United Nations: New York, NY, USA, 2015.

259. United Nations. Ensure Access to Affordable, Reliable, Sustainable and Modern Energy. Sustainable Development Goals. Available online: https://www.un.org/sustainabledevelopment/energy/ (accessed on 26 February 2019).

260. Pye, S.; Dobbins, A.; Baffert, C.; Brajković, J.; Grgurev, I.; De Miglio, R.; Deane, P. Energy poverty and vulnerable consumers in the energy sector across the EU: Analysis of policies and measures. *Policy Rep.* **2015**, *2*, 91. [CrossRef]

261. Scarpellini, S.; Alexia Sanz Hernández, M.; Moneva, J.M.; Portillo-Tarragona, P.; Rodríguez, M.E.L. Measurement of spatial socioeconomic impact of energy poverty. *Energy Policy* **2019**, *124*, 320–331. [CrossRef]

262. Schleich, J. Energy efficient technology adoption in low-income households in the European Union—What is the evidence? *Energy Policy* **2019**, *125*, 196–206. [CrossRef]

263. Capeluto, G.; Ochoa, C. Intelligent Envelopes for High-Performance Buildings: Design and Strategy. In *Green Energy and Technology*; Springer International Publishing: Cham, Germany, 2017; pp. 1–134.

264. Bilton, M.; Woolf, M.; Djapic, P.; Aunedi, M.; Carmichael, R.; Strbac, G.; Woolf, M.; Djapic, P.; Aunedi, M.; Carmichael, R.; et al. *Impact of Energy Efficient Appliances on Network Utilisation*; Imperial College London: London, UK, 2014.

265. Oldewurtel, F.; Parisio, A.; Jones, C.N.; Gyalistras, D.; Gwerder, M.; Stauch, V.; Lehmann, B.; Morari, M. Use of model predictive control and weather forecasts for energy efficient building climate control. *Energy Build.* **2012**, *45*, 15–27. [CrossRef]

266. Loonen, R.C.G.M.; Singaravel, S.; Trčka, M.; Cóstola, D.; Hensen, J.L.M. Simulation-based support for product development of innovative building envelope components. *Autom. Constr.* **2014**, *45*, 86–95. [CrossRef]

267. Kramers, A.; Svane, Ö. *ICT Applications for Energy Efficiency in Buildings*; Centre for Sustainable Communications: Stockholm, Sweden, 2011.

268. Crawley, D.B.; Hand, J.W.; Kummert, M.; Griffith, B.T. Contrasting the Capabilities of Build. Energy Performance Simulation Programs. *Build. Environ.* **2008**, *43*, 661–673. [CrossRef]

269. Xu, J.; Kim, J.-H.; Hong, H.; Koo, J. A systematic approach for energy efficient building design factors optimization. *Energy Build.* **2015**, *89*, 87–96. [CrossRef]

270. Upadhyay, K.; Ansari, A.A. Intelligent and Adaptive Facade System—The Impact on the Performance and Energy Efficiency of Buildings. *J. Civil Eng. Environ. Technol.* **2017**, *4*, 295–300.

271. Kasinalis, C.; Loonen, R.C.G.M.; Cóstola, D.; Hensen, J.L.M. Framework for assessing the performance potential of seasonally adaptable facades using multi-objective optimization. *Energy Build.* **2014**, *79*, 106–113. [CrossRef]

272. Borg, S.P.; Kelly, N.J. The effect of appliance energy efficiency improvements on domestic electric loads in European households. *Energy Build.* **2011**, *43*, 2240–2250. [CrossRef]

273. Hakim, H.; Martini; Hindarto, D.; Riyanto, I.; Margatama, L. Motion Sensor Application on Building Lighting Installation for Energy Saving and Carbon Reduction Joint Crediting Mechanism. *Appl. Syst. Innov.* **2018**, *1*, 23. [CrossRef]

274. Garg, V.; Bansal, N.K. Smart occupancy sensors to reduce energy consumption. *Energy Build.* **2000**, *32*, 81–87. [CrossRef]

275. De Groote, M.; Lefever, M. *Reaching the Untapped Potential Driving Transformational Change in the Construction Value Chain*; Building Performance Institute Europe: Brussels, Belgium, 2015.

276. Lehmann, B.; Gyalistras, D.; Gwerder, M.; Wirth, K.; Carl, S. Intermediate complexity model for Model Predictive Control of Integrated Room Automation. *Energy Build.* **2013**, *58*, 250–262. [CrossRef]

277. Gyalistras, D.; Gwerder, M.; Oldewurtel, F.; Jones, C.N.; Morari, M.; Lehmann, B.; Wirth, K.; Stauch, V. Analysis of Energy Savings Potentials for Integrated Room Automation. In Proceedings of the 10th REHVA World Congr. Clima, Antalya, Turkey, 9–12 May 2010; p. 8.

278. Gwerder, M.; Gyalistras, D.; Oldewurtel, F.; Lehmann, B.; Stauch, V.; Tödtli, J. Potential Assessment of Rule-Based Control for Integrated Room Automation. In Proceedings of the 10th REHVA World Congress, Sustain. Energy Use in Buildings-CLIMA 2010, Antalya, Turkey, 9–12 May 2010; pp. 9–12.

279. Tamani, N.; Ahvar, S.; Santos, G.; Istasse, B.; Praca, I.; Brun, P.E.; Ghamri, Y.; Crespi, N.; Becue, A. Rule-based model for smart building supervision and management. In Proceedings of the 2018 IEEE Int. Conference on Services Computing, SCC 2018—Part of the 2018 IEEE World Congress on Services, San Francisco, CA, USA, 2–7 July 2018; pp. 9–16.

280. Mainetti, L.; Mighali, V.; Patrono, L.; Rametta, P. A novel rule-based semantic architecture for IoT building automation systems. In Proceedings of the 2015 23rd Int. Conference on Software, Telecommunications and Computer Networks, SoftCOM 2015, Split, Croatia, 16–18 September 2015; pp. 124–131.

281. Kolokotsa, D.; Pouliezos, A.; Stavrakakis, G.; Lazos, C. Predictive control techniques for energy and indoor environmental quality management in buildings. *Build. Environ.* **2009**, *44*, 1850–1863. [CrossRef]

282. OptiControl. OptiControl Project—Literature. Available online: http://www.opticontrol.ethz.ch/07E-Literature.html#top (accessed on 2 February 2015).

283. Ma, Y. *Model Predictive Control for Energy Efficient Buildings*; UC Berkeley: Berkeley, CA, USA, 2012.

284. Široký, J.; Oldewurtel, F.; Cigler, J.; Prívara, S. Experimental analysis of model predictive control for an energy efficient building heating system. *Appl. Energy* **2011**, *88*, 3079–3087. [CrossRef]

285. Oldewurtel, F.; Sturzenegger, D.; Andersson, G.; Morari, M.; Smith, R.S. Towards a standardized building assessment for demand response. In *52nd IEEE Conference on Decision and Control*; IEEE: Florence, Italy, 2013; pp. 7083–7088, ISBN 0743-1546.

286. Tanaskovic, M.; Sturzenegger, D.; Smith, R.; Morari, M. Robust Adaptive Model Predictive Building Climate Control. *IFAC Pap.* **2017**, *50*, 1871–1876. [CrossRef]

287. Karjalainen, S.; Lappalainen, V. Integrated control and user interfaces for a space. *Build. Environ.* **2011**, *46*, 938–944. [CrossRef]

288. Oldewurtel, F.; Sturzenegger, D.; Morari, M. Importance of occupancy information for building climate control. *Appl. Energy* **2013**, *101*, 521–532. [CrossRef]

289. Cho, S.; Zaheer-uddin, M. Predictive control of intermittently operated radiant floor heating systems. *Energy Conv. Manag.* **2003**, *44*, 1333–1342. [CrossRef]

290. Rawi, M.; Al-Anbuky, A. Passive House sensor networks: Human centric thermal comfort concept. In Proceedings of the 2009 5th Int. Conference Intelligent Sensors, Sensor Networks and Information Processing (ISSNIP), Melbourne, Australia, 7–10 December 2009; pp. 255–260.

291. Owens, S.; Driffill, L. How to change attitudes and behaviours in the context of energy. *Energy Policy* **2008**, *36*, 4412–4418. [CrossRef]

292. Mills, B.; Schleich, J. Residential energy-efficient technology adoption, energy conservation, knowledge, and attitudes: An analysis of European countries. *Energy Policy* **2012**, *49*, 616–628. [CrossRef]

293. Niamir, L.; Filatova, T.; Voinov, A.; Bressers, H. Transition to low-carbon economy: Assessing cumulative impacts of individual behavioral changes. *Energy Policy* **2018**, *118*, 325–345. [CrossRef]

294. Lee, S.Y.; Brand, J.L. Effects of control over office workspace on perceptions of the work environment and work outcomes. *J. Environ. Psychol.* **2005**, *25*, 323–333. [CrossRef]

295. Boerstra, A.; Beuker, T.; Loomans, M.; Hensen, J. Impact of available and perceived control on comfort and health in European offices. *Arch. Sci. Rev.* **2013**, *56*, 30–41. [CrossRef]

296. Day, J.K.; Gunderson, D.E. Understanding high performance buildings: The link between occupant knowledge of passive design systems, corresponding behaviors, occupant comfort and environmental satisfaction. *Build. Environ.* **2014**, *84*, 114–124. [CrossRef]

297. Lazowski, B.; Parker, P.; Rowlands, I.H. Towards a smart and sustainable residential energy culture: Assessing participant feedback from a long-term smart grid pilot project. *Energy Sustain. Soc.* **2018**, *8*, 27. [CrossRef]

298. Girod, B.; Stucki, T.; Woerter, M. How do policies for efficient energy use in the household sector induce energy-efficiency innovation? An evaluation of European countries. *Energy Policy* **2017**, *103*, 223–237. [CrossRef]

299. Baldini, M.; Trivella, A.; Wente, J.W. The impact of socioeconomic and behavioural factors for purchasing energy efficient household appliances: A case study for Denmark. *Energy Policy* **2018**, *120*, 503–513. [CrossRef]

300. Ameli, N.; Brandt, N. Determinants of households' investment in energy efficiency and renewables: Evidence from the OECD survey on household environmental behaviour and attitudes. *Environ. Res. Lett.* **2015**, *10*, 044015. [CrossRef]

301. Sardianou, E.; Genoudi, P. Which factors affect the willingness of consumers to adopt renewable energies? *Renew. Energy* **2013**, *57*, 1–4. [CrossRef]

302. Dieu-Hang, T.; Grafton, R.Q.; Martínez-Espiñeira, R.; Garcia-Valiñas, M. Household adoption of energy and water-efficient appliances: An analysis of attitudes, labelling and complementary green behaviours in selected OECD countries. *J. Environ. Manag.* **2017**, *197*, 140–150. [CrossRef]

303. Thøgersen, J. Frugal or green? Basic drivers of energy saving in European households. *J. Clean. Prod.* **2018**, *197*, 1521–1530. [CrossRef]

304. Masoso, O.T.; Grobler, L.J. The dark side of occupants' behaviour on building energy use. *Energy Build.* **2010**, *42*, 173–177. [CrossRef]

305. David, D. *On the Adaptation of Build. Controls to the Envelope and the Occupants*; École Polytechnique fédérale de Laussane: Laussane, Switzerland, 2010.

306. Papantoniou, S.; Kolokotsa, D.; Kalaitzakis, K. Building optimization and control algorithms implemented in existing BEMS using a web based energy management and control system. *Energy Build.* **2015**, *98*, 44–55. [CrossRef]

307. Ackerly, K.; Brager, G. Window signalling systems: Control strategies and occupant behaviour. *Build. Res. Inform.* **2013**, *41*, 342–360. [CrossRef]

308. De Dear, R.J.; Brager, G.S. The adaptive model of thermal comfort and energy conservation in the built environment. *Int. J. Biometeorol.* **2001**, *45*, 100–108. [CrossRef]

309. Luo, M.; De Dear, R.J.; Ji, W.; Lin, B.; Ouyang, Q.; Zhu, Y. The Dynamics of Thermal Comfort Expectations. *Build. Environ.* **2016**, *95*, 322–329. [CrossRef]

310. Jensen, K.L.; Toftum, J.; Friis-Hansen, P. A Bayesian Network approach to the evaluation of building design and its consequences for employee performance and operational costs. *Build. Environ.* **2009**, *44*, 456–462. [CrossRef]

311. Romm, J.J.; Browning, W.D. Greening the Building and the Bottom Line. In *Proceedings of the Second International Green Buildings Conference and Exposition*; Whitter, M., Cohn, T.B., Eds.; Rocky Mountain Institute: Snowmass Village, CO, USA, 1995.

312. Hanie, O.; Aryan, A.; MohammadReza, L.; Elham, L. Understanding the Importance of Sustainable Buildings in Occupants Environmental Health and Comfort. *J. Sustain. Dev.* **2010**, *3*, 194–200. [CrossRef]

![energies logo] *energies*

MDPI

Article

Virtual Organization Structure for Agent-Based Local Electricity Trading

Amin Shokri Gazafroudi [1,*], Javier Prieto [1,2] and Juan Manuel Corchado [1,2,3,4]

[1] BISITE Research Group, University of Salamanca, Edificio I+D+i, 37008 Salamanca, Spain;
javierp@usal.es (J.P); corchado@usal.es (J.M.C)

[2] Air Institute, IoT Digital Innovation Hub (Spain), Carbajosa de la Sagrada, 37188 Salamanca, Spain

[3] Department of Electronics, Information and Communication, Faculty of Engineering, Osaka Institute of Technology, Osaka 535-8585, Japan

[4] Pusat Komputeran dan Informatik, Universiti Malaysia Kelantan, Karung Berkunci 36, Pengkaan Chepa, Kota Bharu 16100, Kelantan, Malaysia

* Correspondence: shokri@usal.es

Received: 10 March 2019; Accepted: 17 April 2019; Published: 22 April 2019

Abstract: End-users are more active because of demand response programs and the penetration of distributed energy resources in the bottom-layer of the power systems. This paper presents a virtual organization of agents of the power distribution grid for local energy trade. An iterative algorithm is proposed; it enables interaction between end-users and the Distribution Company (DisCo). Then, the performance of the proposed algorithm is evaluated in a 33-bus distribution network; its effectiveness is measured in terms of its impact on the energy trading scenarios and, thus, of its contribution to the energy management problem. According to the simulation results, although aggregators do not play the role of decision makers in the proposed model, our iterative algorithm is profitable for them.

Keywords: decentralized energy management system; local energy trading; multi-agent system; optimization; smart grid

---

1. Introduction

Smart grids are based on connected IoT and embedded devices that communicate with each other in the power network. Thus, improving the functionality of smart grids, smart buildings, and their IoT devices (e.g., energy management) has become a major research concern [1]. According to the infrastructure provided by smart grids, Demand Response (DR) programs introduce active players into the power distribution system. Hence, end-users wish to participate as bidirectional energy customers, which are called *prosumers*, in the distribution network [2]. Therefore, new market structures are needed to provide energy based on decentralized approaches. Here, there are several studies in the literature that have worked on the energy transaction approach in power distribution grids.

Pratt et al. [3] proposed energy transaction nodes that connect buildings and the local electricity market. Jokic et al. [4] proposed a price-based method for energy management. In [5–7], a multi-agent-based transactive energy market was designed to decentralize decisions. Shafie-khah et al. [8] proposed a price-based method for solving the energy management problem locally based on supervision of the central price controller.

In addition, there are several research papers that have discussed the interplay between agents in the distribution grid based on demand response programs. In [9], the DR program was performed considering several suppliers and consumers. Deng et al. [10] presented a distributed framework based on a dual decomposition technique, which regulates the demand of end-users. In [11], a distributed model was described to determine optimal power flow in radial networks. Bahrami et al. [12]

proposed centralized energy trading as a bi-level model. In [13], a decentralized DR framework was presented. The local electricity market defined in [14] gave independence to market agents, enabling them perform energy transactions freely among each other. In [15], a trading mechanism was designed among micro-grids. Zhang et al. [16] proposed a hierarchical structure for energy exchange in distribution grids. In [17], the energy trading problem was addressed among the agents in the power distribution system where the authors modeled the energy flexibility by the Ising-based model. In [18] and [19], the authors presented decentralized approaches from the perspective of end-users and other relevant decision makers to manage energy flexibility based on the desired reliability level in the distribution network.

Even though several works in the literature have modeled the bidirectional behavior of players to produce/consume energy in the distribution networks, an interplay model has not been addressed for energy trade management between end-users, aggregators, and the Distribution Company (DisCo). In this paper, a virtual organization structure for agents in the power distribution system is proposed for energy transactions between end-users and the DisCo based on an iterative algorithm. Thus, energy transactions are based on a bottom-up hierarchical structure from end-users to aggregators, from the aggregator to the DisCo, and from the DisCo to the wholesale electricity market, respectively. In this way, the main contributions of this paper can be summarized as follows:

- A new virtual organization of agents' structure in the distribution network.
- A novel iterative algorithm for energy trade between end-users and the DisCo in the power distribution system.
- The evaluation of energy trading scenarios through the proposed model.

In the following, the organization of this paper is described. In Section 2, agents and their corresponding virtual organizations are defined. The problem formulation is described in Section 3. Section 4 discusses our findings on the basis of the simulation results. Finally, the paper is concluded in Section 5.

2. Virtual Organization of Agents in the Power Distribution Grid

After the restructuring of power systems, different players emerged in the system. In this paper, the proposed agent structure in the distribution network is described. Thus, different organizations of agents are defined in the system, which consist of end-users, aggregators, and the DisCo. In the following, each of these agents and their interconnections are described.

2.1. End-Users

End-users are agents in the bottom layer of the power distribution system that act as consumers, producers, or prosumers in the system. In this paper, a bottom-up approach is presented to trade energy through end-users, aggregators, the DisCo, and the wholesale market. Thus, end-users manage their energy production/consumption on the basis of their interactions with the aggregators and the DisCo. Furthermore, the end-users have several agents (e.g., Information Provider (IP), Prediction Engine (PE), and Decision Maker System (DMS)), which make up an organization of agents. Each of these agents are described below.

- The *Information Provider (IP)* records information of all other agents, as well as the environmental conditions. Furthermore, the IP is responsible for sending/receiving information to/from the external agents that correspond to its organization, as shown in Figure 1.
- The *Prediction Engine (PE)* forecasts the uncertain variables (e.g., the energy generated from distributed energy resources, electrical consumption, electricity price, etc.) of end-users based on information provided by the IP. In this way, the values predicted by the PE are the inputs of the DMS.
- The *Decision-Making System (DMS)* is in charge of making optimum decisions for its corresponding organization (e.g., end-user, aggregator, and the DisCo). On the one hand, the inputs of the DMS

are received from the IP and the PE. On the other hand, the outputs of the DMS are sent to the IP, which exchanges them with the external agents from the corresponding organization. Figure 1 shows interactions between agents in the end-user's organization.

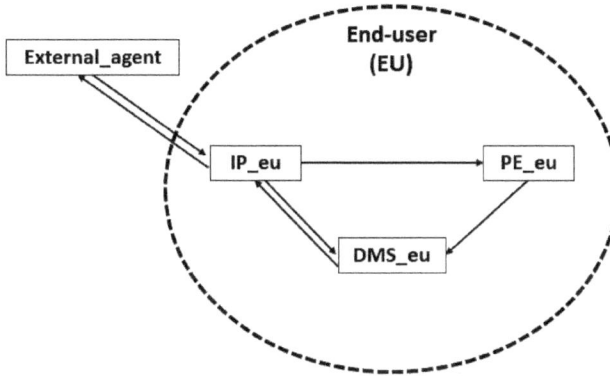

Figure 1. Organization of end-user agents.

2.2. Aggregators

Aggregators (AGG) are one type of reseller player in the restructured power system. In this paper, aggregators are defined as agents that are in charge of trading energy with end-users in their corresponding regions. Furthermore, they are able to conduct energy transactions with the DisCo in this model. In the proposed agent-based structure, aggregators have several agents such as IP and End-Users (EU) for creating agent organizations in each region of the distribution network. Furthermore, according to Figure 2, each aggregator conducts data transactions with the DisCo (as an external agent of its organization) through its IP agent.

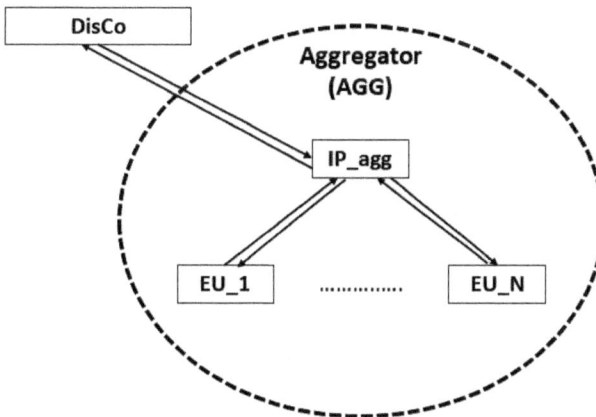

Figure 2. Organization of aggregator agents.

2.3. Distribution Company

The DisCo is the only agent that trades energy with the wholesale market. Moreover, the DisCo has the IP and the DMS agents for data exchange with the aggregators and end-users as external agents and makes optimum decisions, respectively, as shown in Figure 3.

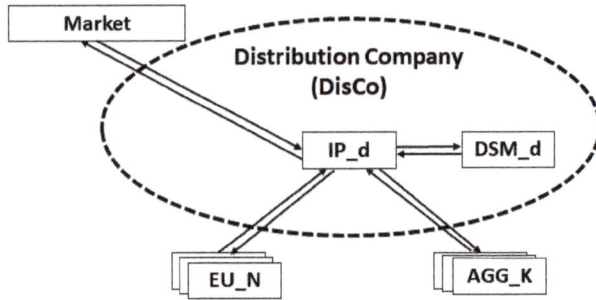

Figure 3. Organization of the DisCo agents.

3. Problem Formulation

In this section, the proposed energy trading problem is described; it is based on the iterative algorithm designed to conduct energy transactions between the end-users and the DisCo as decision makers in the system. In other words, the decision-making problems for the DMSs of end-users and the DisCo are presented in this section.

3.1. Energy Trading Model

In this structure, end-users can trade energy with the DisCo, P_{jt}^{D2L}, and their corresponding aggregators, P_{jt}^{L2A}, at prices λ^{D2L} and λ_{kt}^{L2A}, respectively. Then, aggregators exchange energy, P_{kt}^{A2D}, with the DisCo. However, the wholesale market can only trade with the DisCo, P_t^M, as shown in Figure 4. Equation (1) represents the balancing equation for energy trade between end-user j and the DisCo and its corresponding aggregator. Here, P_{jt} and L_{jt} represent energy production and consumption of end-user j and time step t. End-users play the role of consumers ($L_{jt}^{net} \geq 0$) or producers ($L_{jt}^{net} < 0$) according to (2) and (3). Here, γ_j^P and γ_j^C are defined as coefficients, which present the potential of end-user j as a producer and a consumer, respectively. Furthermore, Equation (4) expresses that the end-user can only buy electricity from the DisCo, and there is a one-way energy transaction between end-users and the DisCo. Equation (5) represents shiftable limits to constrain end-users as active agents in the bottom layer of the distribution network.

$$P_{jt} = L_{jt} + P_{jt}^{L2A} - P_{jt}^{D2L}, \ \forall j, t \tag{1}$$

$$L_{jt}^{net} = L_{jt} - P_{jt}, \ \forall j, t \tag{2}$$

$$-\gamma_j^P L_{jt} \leq L_{jt}^{net} \leq \gamma_j^C L_{jt}, \ \forall j, t \tag{3}$$

$$P_{jt}^{D2L} \geq 0, \ \forall j, t \tag{4}$$

$$\sum_t L_{jt}^{net} = 0, \ \forall j \tag{5}$$

According to our bottom-up energy trading approach, the summation of the energy exchanged between end-users and aggregators is traded with the DisCo as represented in (6). The maximum and minimum constraints for the price of energy traded between aggregators and the DisCo, λ_{kt}^{A2D}, are represented in (7). Here, λ_t^M represents electricity price in the wholesale market, and δ_{kt} is defined as a coefficient to guarantee the profit of the energy transaction for aggregators ($\delta_{kt} \geq 1$). Besides, the balancing equation in the layer of the DisCo for energy exchange between the DisCo and the wholesale market is presented in (8).

$$P_{kt}^{A2D} = \sum_{j \in A_k} P_{jt}^{L2A}, \forall k, t \tag{6}$$

$$\delta_{kt} \lambda_{kt}^{L2A} \le \lambda_{kt}^{A2D} \le \lambda_t^M, \forall t, k \tag{7}$$

$$P_t^M = \sum_j P_{jt}^{D2L} - \sum_k P_{kt}^{A2D}, \forall t \tag{8}$$

Here, the objective functions of end-users, aggregators, and the DisCo are represented in (9), (10), and (11), respectively. In (9), the objective function of end-user *j* consists of two terms and states the end-user's expected cost. The first term represents the expected cost of the energy sold by the DisCo, and the second term expresses the expected profit from the energy sold to the aggregator ($P_{jt}^{L2A} > 0$) or the expected cost of the energy purchased from the aggregator ($P_{jt}^{L2A} < 0$). In (10), OF_k^a consists of two terms, which are the expected cost of energy transactions with the end-users and the expected profit from exchanging energy with the DisCo. In (11), OF^d includes three terms consisting of the expected cost of energy transaction with aggregators, the expected cost of energy traded with the wholesale market, and the expected profit from energy sold to end-users.

$$OF_{j \in A_k}^e = \lambda^{D2L} \sum_t P_{jt}^{D2L} - \sum_t \lambda_{kt}^{L2A} P_{jt}^{L2A} \tag{9}$$

$$OF_k^a = \sum_t \sum_{j \in A_k} \lambda_{kt}^{L2A} P_{jt}^{L2A}$$
$$- \sum_t \lambda_{kt}^{A2D} P_{kt}^{A2D} \forall k \tag{10}$$

$$OF^d = \sum_t \lambda_{kt}^{A2D} P_{kt}^{A2D} + \sum_t \lambda_t^M P_t^M$$
$$- \lambda^{D2L} \sum_t \sum_j P_{jt}^{D2L} \tag{11}$$

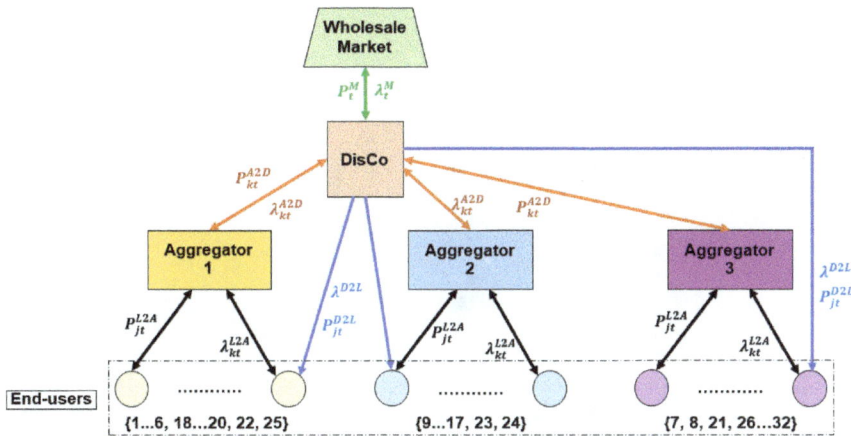

Figure 4. Agents and energy trading framework for the distribution network adapted with permission from [19].

3.2. Proposed Iterative Algorithm

In this section, an iterative algorithm is proposed that models the energy trade through the interaction between end-users and the DisCo. Here, the end-users and the DisCo are defined as agents who manage energy in the power distribution network. In the following, the energy management problems of both end-user and the DisCo are represented:

- End-users energy trading problem (Problem E):

$$\min EC^e = \sum_j OF_j^e$$
$$s.t. : (1)-(3), (5)-(6).$$

Decision making: L_{jt}, P_{jt}, L_{jt}^{net}, P_{jkt}^{L2A}, P_{kt}^{A2D}. Fixed: P_{jt}^{D2L}, λ_{kt}^{A2D}. Passed to problem D: P_{kt}^{A2D}.

- DSO's problem (Problem D):

$$\min EC^d = OF^d$$
$$s.t. : (4), (7)-(8).$$

Decision making: P_{jt}^{D2L}, λ_{kt}^{A2D}, P_t^M. Fixed: P_{kt}^{A2D}. Passed to problem E: P_{jt}^{D2L}, λ_{kt}^{A2D}.

On the one hand, in Problem E, end-users manage their own energy independently and control energy traded through aggregators and the DisCo, P_{kt}^{A2D}, hierarchically. On the other hand, the DisCo determines the price of the energy it trades with the aggregators, λ_{kt}^{A2D}, in Problem D through the proposed algorithm, which has been presented in Figure 5. Therefore, P_{kt}^{A2D} is a fixed variable in Problem D, and P_{jt}^{D2L} and λ_{kt}^{A2D} are fixed variables in Problem E. EC^e and EC^d represent total expected costs of end-users and the DisCo, respectively. Note that the proposed energy trading problem is not the Mathematical Program with Equilibrium Constraints (MPEC) problem and Mixed Complementarity Problem (MCP). Thus, no complementarity has been defined between equations and variables in the proposed problem. The price of energy traded between the DSO and aggregators, λ_{kt}^{A2D}, is just limited to Equation (7), and it is not a dual variable of the balancing equation. Furthermore, λ_{kt}^{A2D} is determined by the DSO.

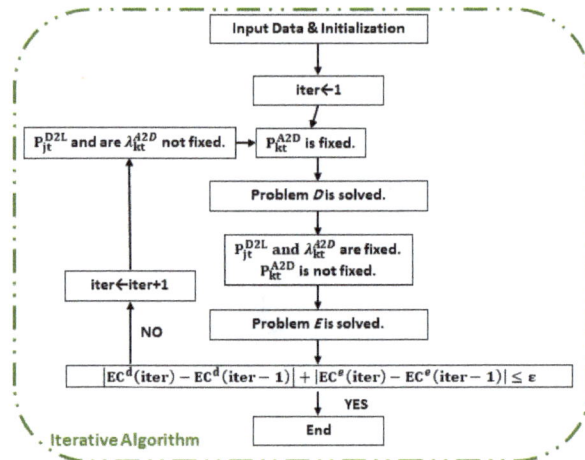

Figure 5. Proposed iterative algorithm for energy trade between end-users and the Distribution Company (DisCo).

4. Simulation Results

4.1. Case Study

In this paper, a 33-bus test system was used [19,20] to assess the proposed energy trading problem as shown in Figure 6. As shown in Figure 6, three regions have been considered, which are managed by their corresponding aggregators. A1–A3 represent Aggregator 1–Aggregator 3 as shown in Figure 4. The energy price that was traded in each of those regions was different as shown in Table 1. Furthermore, we assumed that $\lambda^{D2L} = 0.6$ (€/kWh) and $\delta_{kt} = 1.1$ according to [16–19]. Furthermore, it was assumed that $\gamma_j^P = \gamma_j^C = 0.1$ to cover the electrical demand of end-users' electrical demand from 90%–110% by shaving their demand in the peak-time and shifting them (or not) in the off-peak time via their energy storage systems.

Table 1. Prices of energy traded between consumers and aggregators adapted with permission from [19].

Time (h)	$\lambda_{k=1,t}^{L2A}$ (€/kWh)	$\lambda_{k=2,t}^{L2A}$ (€/kWh)	$\lambda_{k=3,t}^{L2A}$ (€/kWh)	λ_t^M (€/kWh)
1	0.05	0.08	0.06	0.13
2	0.05	0.08	0.07	0.12
3	0.05	0.09	0.07	0.15
4	0.04	0.07	0.05	0.11
5	0.11	0.18	0.15	0.30
6	0.12	0.20	0.16	0.32
7	0.13	0.22	0.17	0.35
8	0.15	0.24	0.19	0.40
9	0.16	0.25	0.20	0.42
10	0.24	0.41	0.33	0.66
11	0.26	0.42	0.36	0.71
12	0.28	0.43	0.37	0.74
13	0.25	0.40	0.32	0.69
14	0.18	0.26	0.21	0.50
15	0.15	0.24	0.20	0.41
16	0.14	0.22	0.18	0.40
17	0.15	0.25	0.19	0.42
18	0.20	0.36	0.30	0.60
19	0.21	0.36	0.29	0.65
20	0.22	0.41	0.30	0.67
21	0.24	0.42	0.33	0.70
22	0.12	0.22	0.16	0.35
23	0.11	0.19	0.15	0.28
24	0.06	0.09	0.07	0.15

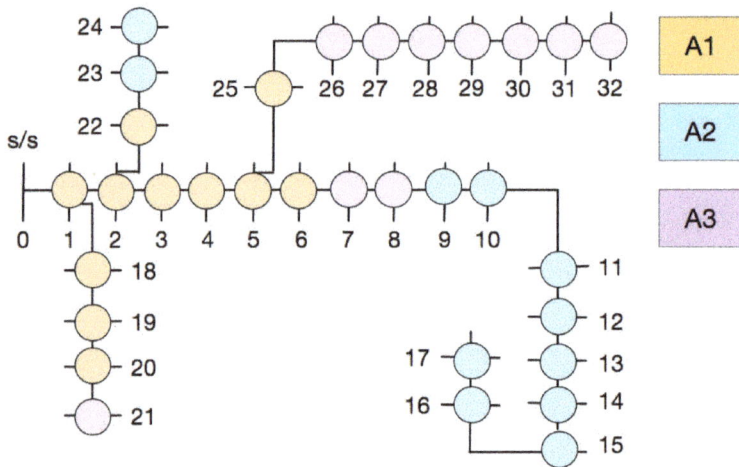

Figure 6. The 33-bus test system and corresponding regions of the aggregators adapted with permission from [19].

4.2. Evaluation of the Proposed Iterative Algorithm

In this section, we assess the performance of the proposed iterative algorithm for energy trade between end-users and the DisCo. Thus, two scenarios were defined for evaluation. In Scenario 1, S_1, Equation (5) is not considered in the problem. In other words, in S_1, the electrical load of end-users is modeled as an interruptible load. Hence, end-users shave their peak load to minimize their cost for energy transaction in S_1. However, Scenario 2, S_2, includes all constraints of the problem. In other words, in S_2, the electrical load of end-users is modeled as a shiftable load. In this way, the total amount of energy shifted by end-users in 24 h should be equal to zero. Therefore, if end-users shave N% of their desired electrical consumption in the peak time, they should shift their shaved consumption (N% of their desired demand) to the off-peak time. In this way, the total expected costs of end-users (EC^e), aggregators ($EC^a = \sum_k OF_k^a$), and the DisCo (EC^d) were compared in two cases with the aim of finding an energy trading solution. In Case 1, the energy trading problem was solved from the perspective of end-users as independent agents. Hence, end-users manage energy in the distribution network without the interplay with the DisCo and aggregators. However, in Case 2, the energy trading problem was solved based on the interaction between end-users and the DisCo by our proposed iterative algorithm.

As seen in Table 2, in Case 1, EC^e, EC^a, and EC^d are negative in S_1. In other words, Case 1 was profitable for all end-users, aggregators, and the DisCo. This is because of the bottom-up energy trading flow from end-users to aggregators, from aggregators to the DisCo, and from the DisCo to the wholesale market. In S_2, the total expected costs of aggregators was positive in Case 1. However, EC^a was negative in S_2 of Case 2.

Table 2. Impact of the proposed iterative algorithm on the expected costs of end-users, aggregators, and the DisCo.

Expected Cost (Case)	S_1	S_2
EC^e (Case 1) (€)	−2394.438	−714.291
EC^a (Case 1) (€)	−239.444	733.548
EC^d (Case 1) (€)	−2273.819	−1461.078
EC^e (Case 2) (€)	159.767	1111.734
EC^a (Case 2) (€)	−239.444	−100.082
EC^d (Case 2) (€)	−8607.231	−5612.034

In other words, Case 2 (proposed iterative algorithm) was a profitable case for aggregators. Moreover, there are bidirectional energy transactions between end-users and aggregators, aggregators and the DisCo, and the DisCo and the market in both Cases of S_2 as seen in Figures 7 and 8. As shown in Figure 7, there is no energy trade between end-users and the DisCo in Case 1, because λ^{D2L} is greater than λ^{L2A}_{kt} in all time steps. In this way, end-users did not purchase energy from the DisCo because the energy trading problem was solved from the perspective of end-users in Case 1. However, S_2 was not profitable for end-users in Case 2. In this way, although aggregators were not decision makers in our proposed iterative algorithm, Case 2 (iterative algorithm) was profitable for aggregators in both scenarios.

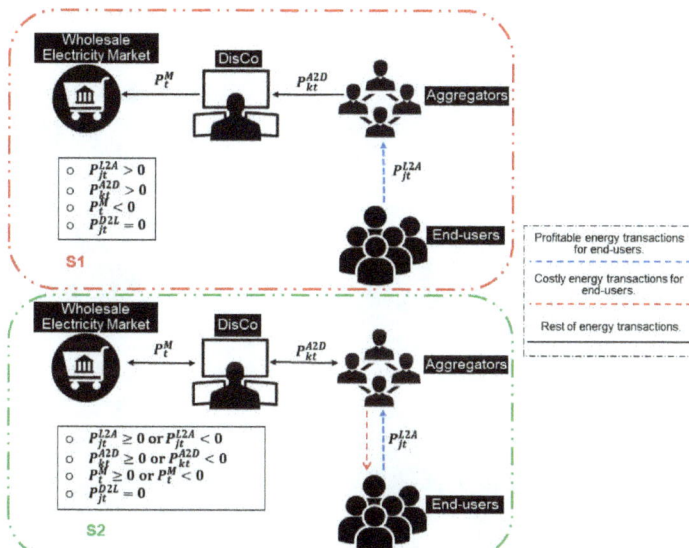

Figure 7. Energy trading flow between agents in the distribution network in Case 1.

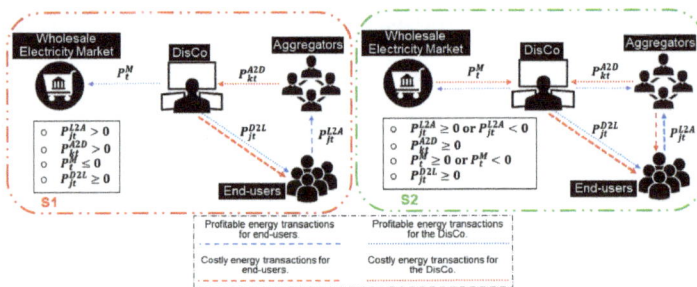

Figure 8. Energy trading flow between agents in the distribution network in Case 2 (proposed iterative algorithm).

5. Conclusions

This paper has proposed a virtual organization structure for energy trade between the agents of the distribution network. Furthermore, an iterative algorithm has been proposed for energy transaction management between end-users and the DisCo. The proposed algorithm has been evaluated in terms of its impact on the energy trade scenarios. According to the simulation results, it has been found that:

- If all end-users participate as interruptible loads in the distribution network, the energy trade was more profitable for all the agents.
- Our proposed algorithm was profitable for aggregators and the DisCo, who are policy makers in the power distribution system.
- The proposed algorithm was costly for all end-users in comparison with the decentralized approach (which is not practical in current power systems) to manage energy by end-users in the distribution network, because the DisCo is in charge of determining the amount of energy that can be traded between the DisCo and end-users.

In future work, we are going to discuss how to model the uncertainty of distributed energy resources that are decentralized and how a distributed energy management system can be modeled considering peer-to-peer energy trading among end-users and aggregators based on a mathematical program with equilibrium constraints and mixed complementarity problems.

Author Contributions: A.S.G. developed the proposed iterative algorithm for trading among agents. The remaining co-authors addressed key ideas for the project development.

Funding: This work has been supported by the Salamanca Ciudad de Cultura y Saberes Foundation under the Atracción del Talento program (CHROMOSOM project).

Acknowledgments: Amin Shokri Gazafroudi acknowledges the support of the Ministry of Education of the Junta de Castilla y Leon and of the European Social Fund through a grant from predoctoral recruitment of research personnel associated with the research project *Arquitectura multiagente para la gestión eficaz de redes de energía a través del uso de tecnicas de inteligencia artificial* of the University of Salamanca.

Conflicts of Interest: The authors declare no conflict of interest regarding the publication of this paper.

References

1. Sayadi, H.; Makrani, H.M.; Randive, O.; PD, S.M.; Rafatirad, S.; Homayoun, H. Customized machine learning-based hardware-assisted malware detection in embedded device. In Proceedings of the 17th IEEE International Conference On Trust, Security And Privacy in Computing And Communications (IEEE TrustCom-18), New York, NY, USA, 1–3 August 2018.
2. Gazafroudi, A.S.; Prieto-Castrillo, F.; Pinto, T.; Corchado, J.M. Organization-Based Multi-Agent System of Local Electricity Market: Bottom-Up Approach. In Proceedings of the 15th International Conference on Practical Applications of Agents and Multi-Agent Systems (PAAMS), Porto, Portugal, 21–23 June 2017.

3. Pratt, A.; Krishnamurthy, D.; Ruth, M.; Wu, H.; Lunacek, M.; Vaynshenk, P. Transactive Home Energy Management Systems. *IEEE Electrif. Mag.* **2016**, *4*, 8–14.

4. Jokic, A.; van den Bosch, P.P.J.; Hermans, R.M. Distributed, Price-based Control Approach to Market-based Operation of Future Power Systems. In Proceedings of the 2009 6th International Conference on the European Energy Market, Leuven, Belgium, 27–29 May, 2009.

5. Sajjadi, S.M.; Mandal, P.; Tseng, T.L.; Velez-Reyes, M. Transactive energy market in distribution systems: A case study of energy trading between transactive nodes. In Proceedings of the 2016 North American Power Symposium (NAPS), Denver, CO, USA, 18–20 September 2016.

6. Shafie-khah, M.; Catalão, J.P.S. A Stochastic Multi-Layer Agent-Based Model to Study Electricity Market Participants Behavior. *IEEE Trans. Power Syst.* **2015**, *30*, 867–881,.

7. Nunna, H.K.; Srinivasan, D. Multiagent-Based Transactive Energy Framework for Distribution Systems With Smart Microgrids. *IEEE Trans. Ind. Inform.* **2017**, *13*, 2241–2250. [CrossRef]

8. Warrington, J.; Mariéthoz, S.; Jones, C.N.; Morari, M. Predictive power dispatch through negotiated locational pricin. In Proceedings of the IEEE PES Innovative Smart Grid Technologies Conference Europe (ISGT Europe), Gothenberg, Sweden, 11–13 October 2010.

9. Chai, B.; Chen, J.; Yang, Z.; Zhang, Y. Demand response management with multiple utility companies: A two-level game approach. *IEEE Trans. Smart Grid* **2014**, *5*, 722–731. [CrossRef]

10. Deng, R.; Yang, Z.; Hou, F.; Chow, M.-Y.; Chen, J. Distributed realtime demand response in multiseller-multibuyer smart distribution grid. *IEEE Trans. Power Syst.* **2015**, *30*, 2364–2374. [CrossRef]

11. Disfani, V.R.; Fan, L.; Miao, Z. Distributed dc optimal power flow for radial networks through partial primal dual algorithm. In Proceedings of the 2015 IEEE Power & Energy Society General Meeting, Denver, CO, USA, 26–30 July 2015, pp. 1–5.

12. Bahrami, S.; Amini, M.H.; Shafie-khah, M.; Catalão, J.P.S. A decentralized renewable generation management and demand response in power distribution networks. *IEEE Trans. Sustain. Energy* **2018**. [CrossRef]

13. Bahrami, S.; Amini, M.H.; Shafie-khah, M.; Catalão, J.P.S. A decentralized electricity market scheme enabling demand response deployment. *IEEE Trans. Power Syst.* **2018**, *33*, 4218–4227. [CrossRef]

14. Mustafa, M.A.; Cleemput, S.; Abidin, A. A local electricity trading market: Security analysis. In Proceedings of the IEEE PES Innovative Smart Grid Technologies Conference Europe (ISGT Europe), Ljubljana, Slovenia, 9–12 October 2016.

15. Park, S.; Lee, J.; Bae, S.; Hwang, G.; Choi, J.K. Contribution-Based Energy-Trading Mechanism in Microgrids for Future Smart Grid: A Game Theoretic Approach. *IEEE Trans. Ind. Electron.* **2016**, *63*, 4255–4265. [CrossRef]

16. Zhang, C.; Wang, Q.; Wang, J.; Pinson, P.; Morales, J.M.; Ostergaard, J. Real-time procurement strategies of a proactive distribution company with aggregator-based demand response. *IEEE Trans. Smart Grid* **2016**, *9*, 766–776. [CrossRef]

17. Prieto-Castrillo, F.; Gazafroudi, A.S.; Prieto, J.; Corchado, J.M. An Ising Spin-Based Model to Explore Efficient Flexibility in Distributed Power Systems. *Complexity* **2018**, *2018*, 5905932. [CrossRef]

18. Gazafroudi, A.S.; Corchado, J.M.; Keane, A.; Soroudi, A. Decentralised flexibility management for EVs. *IET Renew. Power Gener.* **2019**, *13*, 952–960. [CrossRef]

19. Gazafroudi, A.S.; Prieto-Castrillo, F.; Pinto, T.; Corchado, J.M. Energy Flexibility Management in Power Distribution Systems: Decentralized Approach. In Proceedings of the IEEE International Conference on Smart Energy Systems and Technologies (SEST), Sevilla, Spain, 10–12 September 2018; pp. 1–6.

20. Mithulananthan, N.; Hung, Q.D.; Kwang, Y. *Intelligent Network Integration of Distributed Renewable Generation*; Springer International Publishing: Berlin, Germany, 2017.

Article

Demand Response Optimization Using Particle Swarm Algorithm Considering Optimum Battery Energy Storage Schedule in a Residential House

Ricardo Faia, Pedro Faria *, Zita Vale and João Spinola

Polytechnic of Porto, Rua DR. Antonio Bernardino de Almeida, 431, 4200-072 Porto, Portugal;
rfmfa@isep.ipp.pt (R.F.); zav@isep.ipp.pt (Z.V.); jafps@isep.ipp.pt (J.S.)
* Correspondence: pnf@isep.ipp.pt; Tel.: +351-228-340-511; Fax: +351-228-321-159

Received: 4 March 2019; Accepted: 25 April 2019; Published: 30 April 2019

Abstract: Demand response as a distributed resource has proved its significant potential for power systems. It is capable of providing flexibility that, in some cases, can be an advantage to suppress the unpredictability of distributed generation. The ability for participating in demand response programs for small or medium facilities has been limited; with the new policy regulations this limitation might be overstated. The prosumers are a new entity that is considered both as producers and consumers of electricity, which can provide excess production to the grid. Moreover, the decision-making in facilities with different generation resources, energy storage systems, and demand flexibility becomes more complex according to the number of considered variables. This paper proposes a demand response optimization methodology for application in a generic residential house. In this model, the users are able to perform actions of demand response in their facilities without any contracts with demand response service providers. The model considers the facilities that have the required devices to carry out the demand response actions. The photovoltaic generation, the available storage capacity, and the flexibility of the loads are used as the resources to find the optimal scheduling of minimal operating costs. The presented results are obtained using a particle swarm optimization and compared with a deterministic resolution in order to prove the performance of the model. The results show that the use of demand response can reduce the operational daily cost.

Keywords: demand response; distributed generation; particle swarm optimization; prosumer

1. Introduction

The future of power systems has been guided of a new structure where consumers (end-users) are considered as a central entity. This vision is presented in the Strategic Energy Technology (SET) plan of the European Union [1]. The transformation of end-users' roles allows these entities to have an active contribution in electric power systems. The prosumer is a new concept that has its origin in the proliferation of Distributed Generation (DG) in end-user facilities. The Prosumer definition is presented in Reference [2], where prosumers are considered agents that can either consume or produce energy. The integration of renewable energy sources (RESs) and energy storage systems results in the increase the complexity of energy management. In Reference [3], some methods to optimize renewable energy systems management are revised.

Regarding demand response (DR) programs, the potential for participation in facilities is significantly increased by the distributed energy resources and especially the energy storage systems. With the participation in DR programs, the roles of the consumers change from a passive entity to an active entity that manages both local consumption and generation resources [4]. DR constitutes a modification of load profile in response to monetary or price signals, and thus provides flexibility and aims to help power systems during peak hours of demand or contingencies cases [5]. As the DR programs are able to reschedule part of the load, the use of these programs is a way to increase the flexibility of the grid management, avoiding the need to invest in more capacity [6].

Categorizing DR programs, it can be divided into two main categories: incentive-based DR programs and price-based DR programs. The incentive-based DR programs are referred to as the first category for DR programs, where the consumers can offer an incentive to change their consumption patterns. Direct load control programs, load curtailment programs, demand bidding programs, and emergency demand reduction programs are examples of incentive-based DR programs. The "price-based DR programs" are the second category of DR programs, where the consumers are charged with different rates at different consumptions times. Therefore, the retail electricity tariff is affected by the cost of electricity supply. The price-based DR programs types are a time of use pricing, critical peak pricing, real-time pricing, and inclining block rate [7]. Advanced infrastructure metering is needed to implement DR programs at the residential, commercial, or industrial level. Such infrastructure (i.e., smart meters) is able to measure and store energy utilization at different times and also obtain the current usage information remotely.

The European Union has shown significant interest in the concept of smart metering. According to [8], it is expected by 2020 to invest ~45 million euros for 200 million smart electricity meters and 45 million smart meters of natural gas. This facilitates the application of DR programs in most electrical facilities.

Regarding the formulation of DR optimization problems, linear programming (LP) or nonlinear programming (NLP) can be used. Frequently the DR problems are able to use binary decision variables for determining the status (ON or OFF) of various consumers or appliances; in these situations, mixed-integer linear programming (MILP) or mixed-integer nonlinear programming (MINLP) may be used. In Reference [9], the authors use MILP to optimize DR and generate scheduling in a residential community grid using renewable energies, batteries, and electric vehicles. In this optimization, a minimization problem of purchased energy costs of the residential community has been solved. In Reference [10] a cost minimization in smart building microgrid considering DR optimization and day-ahead operation is implemented using MILP. This case study is composed of two different smart buildings with 30 and 90 houses. During the optimization process, the optimal schedule of house appliances is found. Another MILP approach is applied in Reference [11], showing how strategies like DR can achieve suitability in any region considering the presence of high penetration of renewable-based generation.

An example of NLP applied for DR optimization is presented in Reference [12], where the unit commitment problem for a microgrid is solved. The optimization problem finds the amount of load reduction and paid incentives for each time interval. Another example of MINLP has been presented in Reference [13], which considers the minimization of purchase gas and electricity from the grid by including the consumption of different loads at different periods. The optimal day-ahead scheduling of resources in energy hubs is determined.

The DR application in end consumers has been over time applied through an aggregator. It works as a service provider, and the DR services must be paid to this provider. In Reference [14], an aggregation of thermostatically controlled loads for performing DR is presented. In this case, the air conditioning consumption is considered as the load. The aggregation services are not restricted to the application of DR programs, in Reference [15] an aggregation of generalized energy storage can be found. The aggregator storage is used to participate in the energy and regulation market. DR programs targeting independent users, without the need of contracts or service providers, are also possible [16,17]. These applications are considered independent because the user is not connected to any aggregator. Usually, when the application is independent the user has a device installed in its house to control the loads. In Reference [16], the controller is a PV inverter, while in Reference [17], a home energy management system is used. The controllable loads can be divided into passive (i.e., air conditioning, fridges, washing machine) and active (i.e., DG, ESS, vehicle-to-grid, PV) loads [18]. In References [16,19] the DR is applied on discrete loads, which only have two states: on or off.

With focus on artificial intelligence (AI), its application in power systems has increased in the past years. The metaheuristics are a very popular part of AI for solving optimization problems. These techniques have acceptable performance in order to solve engineering problems by finding a near-optimal solution with a limited computation burden. Metaheuristics can be applied in problems with a large number of decision variables and easily adapted to a problem that has several constraints [20]. A PSO variant is used in Reference [17] for finding the optimal operation of price-driven demand response with a load shifting dispatch strategy for photovoltaic, storage battery, and power grid systems. The optimization algorithm is implemented on Home Energy Manage System. In Reference [21], the PSO algorithm is also used. The DR is optimized considering the variation of electricity price imposed by DSO to provoke a consumption reduction. In the microgrid environment [22], a PSO is used for solve the DR optimization problem. In this case a dynamic pricing model is considered for increase the profit of costumers. In Reference [23] a PSO algorithm is proposed to optimize the performance of a smart microgrid in a short term to minimize operating costs and emissions. Other algorithms like genetic algorithm [24], simulated annealing [25], and differential evolution [26] are frequently used algorithms to solve DR optimization problems.

The present paper proposes DR optimization considering the optimal battery schedule in a residential house with Photovoltaic (PV) generation. A PSO approach is implemented to solve the optimization problem (MILP), and the results are compared with a deterministic resolution (CPLEX solver). The consumer (residential house) is provided with independent management that approaches the several resources capabilities and contributions for the minimization of energy bought from the grid. The main contributions of this paper are as follows.

(1) To perform DR without any contract with the DR service provider—this presented methodology allows the user to perform DR actions without any connection with DR services provider. The consumer is provided with independent management that approaches the several resources capabilities and contributions for the minimization of bought energy from the grid.

(2) The implementation of PSO which is a very simple metaheuristic to implement, open access, multiplatform (Windows, MacOS, Linux, etc.), executable from an Arduino/Raspberry and also is the cheapest implementation option. Referring to the presented solution in [16], which uses a CPLEX solver for MATLAB/TOMLAB platform, the implementation of the PSO is a much affordable solution, once that MATLAB and TOMLAB are non-open access. PSO can be implemented in an open access environment and can be executed in free simple platforms, such as Python.

(3) The proposed methodology represents an optimization problem that can considerably improve the consumer's energy savings—the combined use of resources (PV production, storage capacity, and loads flexibility) allows for a significant reduction in daily operation costs. The optimal solution obtained by PSO has a daily cost of 3.28 €, while an operation without PV production, storage capacity and loads flexibility has a cost of 16.83 € per day, which is five times higher than PSO result for best scenario. If one considers a base scenario that was obtained by using a simple management mechanism considering the PV production and storage capacity, the daily cost is 9.33 €, which is three times higher than PSO result for the best scenario. The assessment of PSO can be verified in the comparison of the base scenario and the optimized base scenario with the PSO. The daily costs with PSO decreases 1.38 €.

The paper is structured into seven sections: In Section 1 an introduction about DR and how to solve DR problems is presented. Section 2 presents the proposed methodology; in Section 3 the problem formulation is presented. Section 4 presents the algorithm (PSO) and its adaptation to the problem formulation. In Section 5, the case study is presented as well as all input variables and PSO parameters. Section 6 presents the results, and the conclusions are presented in Section 7.

2. Proposed Methodology

With the goal to reduce the electricity bill of the end consumers is introduced the presented methodology. This methodology aims to minimize the operation costs considering the batteries and flexibility provided by the DR actions. The costs minimization considered the grid, the PV systems, energy storage batteries, and consumption flexibility through load scheduling. The end consumer is connected to the grid, and has a tariff contract that allows selling energy in the grid in exchange for monetary payment. This methodology is able to be expanded to other consumers with different conditions and with different numbers of resources. Figure 1 presents the context scheme of the idea proposed. This scheme is typical for a household prosumer. The scheme of Figure 1 has a unit generation (PV), energy storage system (ESS) (battery), one inverter module, the controllable and noncontrollable loads, and a smart meter.

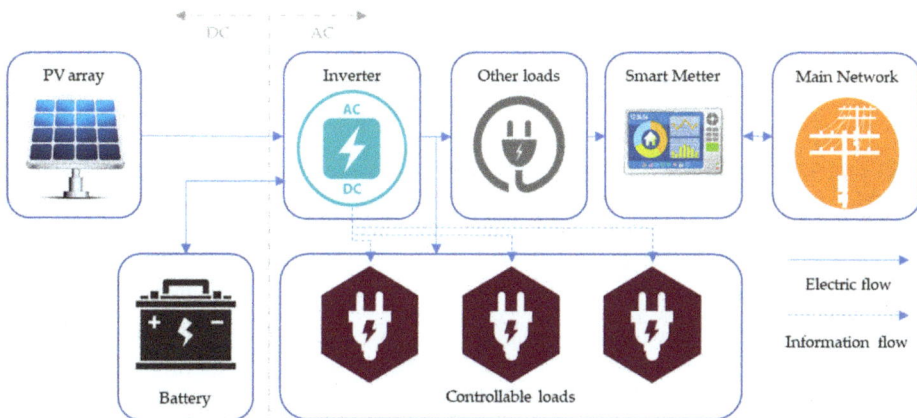

Figure 1. Implementation scheme of the proposed work.

For household, the use of PV generation is considered free (the generation unit is household property). In this paper, the PV generation is considered priority above all others, meaning that when it is available it will always be used either by load's necessities, battery charge, or injection in the grid. The connection with the grid is considered bidirectional. The PV rated power is usually limited by a contract between retailer and household. This limitation occurs because it can be a source of problems for the physical grid. In this way, it is difficult to reach a situation in which, as limit case if no injection to the grid is allowed, the PV is higher than the load plus the energy that can be used to charge the battery. However, if it happens, the inverter will disconnect the PV in order to avoid overvoltage. In Figure 1 one can see power flows and information flows. The information flows are connected to the inverter and controllable loads. In this case, the inverter is enabled with a control and management system that allows controlling loads, adding DR actions in household installation.

In general, the consumer can take advantage of the use of PV generation, ESS, and DR actions to minimize the cost of consumption from the grid. The consumer can look for the periods where electricity is cheaper to satisfy the consumption and charge the ESS, and the periods where the electricity price is most expensive to sell the excess electricity from the facility. Thus, it can be considered as a management system for the consumer to improve his energy bill.

Figure 2 is a representative illustration of the load's control using relays. The controller, in this case, is a component of the inverter. Each controllable load must have one relay associated with it, which allows for its control. So, when the controller sends the signal to the relay, the load is connected or disconnected from the electrical circuit. In this case, this control is considered a DR cut (direct load control). The scheme in Figure 2 considers only one relay for simplification; however, the proposed methodology is able to consider several relays, one for each load in the facility.

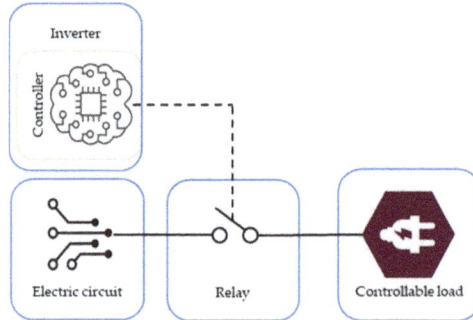

Figure 2. The control scheme for the demand response (DR) cut with one load.

3. Problem Formulation

The mathematical formulation is presented throughout this section. With the formulation presented it is intended to simulate the interaction of a consumer with the grid. The main goal is to minimize the operation costs, considering that the user has storage units and is also enabled to do DR in specific loads. The presented optimization model is considered a mixed-integer linear problem. Equation (1) presents the objective function.

$$Minimize \ f = Energy \ Bill + DR \ Curtailment \tag{1}$$

Equation (1) is comprised of the sum of two different parcels: the energy bill present in Equation (2) and the DR curtailment present in Equation (3). The Energy Bill represents the cost of buying and selling energy, and the DR curtailment refers to cost weighting associated with kWh curtailment.

In Equation (2) the variable P_t^{grid} represents the flow of energy between household and grid, $I_t^{grid \ in}$ is an indicator variable for power flow into the grid and control the energy buy $\left(I_t^{grid \ in} = 1\right)$ and energy sell $\left(I_t^{grid \ in} = 0\right)$, $C_t^{grid \ in}$ represents the cost of buying electricity and $C_t^{grid \ out}$ represents the cost of selling electricity. The Energy bill in Equation (2), consider the costs of buying electricity $\left(I_t^{grid \ in} \times P_t^{grid}\right) \times C_t^{grid \ in}$ and the revenues of selling electricity $\left(\left(1 - I_t^{grid \ in}\right) \times P_t^{grid}\right) \times C_t^{grid \ out}$. In each period ($t$) the user can make a single operation (buy or sell).

$$Energy \ Bill = \sum_{t=1}^{T}\left[\left(\left(I_t^{grid \ in} \times P_t^{grid}\right) \times C_t^{grid \ in} - \left(\left(1 - I_t^{grid \ in}\right) \times P_t^{grid}\right) \times C_t^{grid \ out}\right) \times \frac{1}{\Delta t}\right] + DCP \tag{2}$$

$$I_t^{grid \ in} = \begin{cases} 1, \ if \ P_t^{grid} > 0 \\ 0, \ otherwise \end{cases} \qquad \forall t \in \{1, \dots, T\}$$

Also, in Equation (2) the term $\left(\left(1 - I_t^{grid \ in}\right) \times P_t^{grid}\right)$ represents the power sent to the network. The term Δt is used for to adjust the consumption to the tariff price because normally the tariff is available in €/kWh and the optimization can be scheduled at different time intervals (e.g., 15 min). *DCP* represents the daily contracted power cost. If the term P_t^{grid} has a positive value during optimization it means that there is electricity consumption from the network. However, if it has a negative value it means that there is a sale of electricity to the network. Equation (3) presents the DR curtailment.

$$DR\ Curtailment = \sum_{t=1}^{T}\left(\sum_{l=1}^{L} P_{l,t}^{cut} \times X_{l,t}^{cut} \times W_{l,t}^{cut}\right) \tag{3}$$

If the DR curtailment equation is implemented the cost of load is cut with the use of weights, and in fact does not have cost for the user. The variable $P_{l,t}^{cut}$ represents the cut energy of load (l) in period (t), the $X_{l,t}^{cut}$ represents the decision binary variable to active the cut of load (l) in period (t), and $W_{l,t}^{cut}$ represents the cut weight of load (l) in period (t). The term $\left(P_{l,t}^{cut} \times X_{l,t}^{cut} \times W_{l,t}^{cut}\right)$ shows the interest of the user to perform cut in load (l) in period (t).

Equation (4) represents the balance between load and generation, $P_{b,t}^{bat}$ represents the energy charged or discharged by baterry (b) in period (t). If the value of $P_{b,t}^{bat}$ is less than 0 the battery is discharging, otherwise, if the value of $P_{b,t}^{bat}$ is greater than 0, the battery is charging. The variable $P_{p,t}^{PV}$ represents the photovoltaic production of unit p at period t, and P_t^{load} corresponds to the value of load at period t.

$$P_t^{grid} = P_t^{load} + \sum_{b=1}^{B} P_{b,t}^{bat} - \sum_{l=1}^{L} P_{l,t}^{cut} \times X_{l,t}^{cut} - \sum_{p=1}^{P} P_{p,t}^{PV}, \forall t \in \{1,\dots,T\} \tag{4}$$

The Equation (5) shows the balance of battery systems.

$$E_{b,t}^{stor} = E_{b,t-1}^{stor} + P_{b,t}^{bat} \times \frac{1}{\Delta t}, \forall t \in \{2,\dots T\}, \forall b \in \{1,\dots,B\} \tag{5}$$

Variable $E_{b,t}^{stor}$ represents the state of the battery b in period t, in other words, it represents the amount of energy it has available. So, by Equation (5) the current battery state is obtained by adding the previous state $E_{b,t-1}^{stor}$ to the value of the variable $P_{b,t}^{bat}$. The power term $P_{b,t}^{bat}$ in Equation (5) is multiplied by $\frac{1}{\Delta t}$ to convert power into energy units. The system is governed by the following constraints (Equations (6)–(10)).

$$-P_t^{grid\ min} \leq P_t^{grid} \leq P_t^{grid\ max} \ \forall t \in \{1,\dots,T\} \tag{6}$$

$$P_{l,t}^{cut} = P_{l,t}^{cut\ max} \ \forall l \in \{1,\dots,L\}, \ \forall t \in \{1,\dots,T\} \tag{7}$$

$$0 \leq E_{b,t}^{stor} \leq E_{b,t}^{stor\ max} \ \forall b \in \{1,\dots,B\}, \ \forall t \in \{1,\dots,T\} \tag{8}$$

$$-P_{b,t}^{dch\ max} \leq P_{b,t}^{bat} \leq P_{b,t}^{ch\ max} \ \forall b \in \{1,\dots,B\}, \ \forall t \in \{1,\dots,T\} \tag{9}$$

$$X_{l,t}^{cut} = \begin{cases} 1 \\ 0 \end{cases} \ \forall l \in \{1,\dots,L\}, \ \forall t \in \{1,\dots,T\}. \tag{10}$$

In Equation (6), the variable $P_t^{grid\ min}$ and $P_t^{grid\ max}$ represent the limit values for variable P_t^{grid}. Equation (7) identifies that $P_{l,t}^{cut}$ can only take the maximum value $P_{l,t}^{cut\ max}$. The $P_{b,t}^{bat}$ variables can take a value between $-P_{b,t}^{dch\ max}$ and $P_{b,t}^{ch\ max}$; if the value of $P_{b,t}^{bat}$ is less than zero it represents a discharge and if the value is greater than zero it represents a charge. The variable $X_{l,t}^{cut}$ is a binary variable and represents a decision variable. When $X_{l,t}^{cut}$ is equal to 1 the cut of load (l) at period (t) is active.

4. Particle Swarm Optimization

PSO was proposed by Kennedy and Eberhart in 1995, and it is a random search algorithm that simulates the foraging and flocking of birds in nature [27]. When birds look randomly for food in a given area, each bird can be associated with a single solution and can be considered as a particle in the swarm.

For PSO implementations assume that it has j particles in the n-dimensional search space and each particle represent a solution in the search space. Equation (11) presents the position vector of particle j and in Equation (12) the velocity vector for particle j.

$$\vec{x}_i^j = \left(x_{i,1}^j, x_{i,2}^j, \ldots, x_{i,n}^j\right) \tag{11}$$

$$\vec{v}_i^j = \left(v_{i,1}^j, v_{i,2}^j, \ldots, v_{i,n}^j\right) \tag{12}$$

where, $\vec{x}_i^j$ represents the position vector of particle j for n. variables at iteration i. The $\vec{v}_i^j$ represents the velocity vector of particle j for n variables. When the search process starts, both vectors are generated randomly between the respective limits of the n variables.

Equation (13) represents the velocity update equation. This equation is composed of three different components: the $w_i^j \vec{v}_i^j$ component represents the previous positions in memory search, $c1_i^j r1_i^j\left(P_{best}^j - \vec{x}_i^j\right)$ corresponds to the cognitive learning component, and $c2_i^j r2_i^j\left(G_{best} - \vec{x}_i^j\right)$ is a global learning component. Equation (14) represents the position update.

$$\vec{v}_{i+1}^j = w_i^j \times \vec{v}_i^j + c1_i^j \times r1_i^j \times \left(P_{best}^j - \vec{x}_i^j\right) + c2_i^j \times r2_i^j \times \left(G_{best} - \vec{x}_i^j\right) \tag{13}$$

$$\vec{x}_{i+1}^j = \vec{v}_{i+1}^j + \vec{x}_i^j \tag{14}$$

where, $\vec{v}_{i+1}^j$ is the velocity vector at iteration $i+1$; w_i^j represents the inertia weight obtained through Equation (15); $c1_i^j$ and $c2_i^j$ are acceleration coefficients, which are obtained by Equations (16) and (17), respectively; and $r1_i^j$ and $r2_i^j$ are two uniformly distributed random numbers independently generated within [0,1] for the n-dimensional search space. $P_{best}^j = \left(x_{pbest,1}^j, x_{pbest,2}^j, \ldots, x_{pbest,n}^j\right)$ denotes the historical best position and $G_{best} = \left(x_{gbest,1}, x_{gbest,2}, \ldots, x_{gbest,n}\right)$ denotes the population historical best position. Equation (15) presents an inertia weight.

$$w_i^j = w^{max} - \left(\frac{w^{max} - w^{min}}{i^{max}}\right) \times i \tag{15}$$

where, w^{max} is the maximum value for inertia weight, w^{min} is the minimum value for inertia weight, and i^{max} represents the maximum value of iterations. The inertia weight present in Equation (15) is a linear decreasing method during the search process. The inertia weight reduction ensures strong global exploration properties in the initial phase and strong local exploitation properties in the advanced phase. The inertia weight is calculated at each iteration and is the same for the set of particles at each iteration [28]. Equations (16) and (17) present the acceleration coefficients calculation:

$$c1_i^j = c_1^{max} - \left(\frac{c_1^{max} - c_1^{min}}{i^{max}}\right) \times i \tag{16}$$

$$c2_i^j = c_2^{min} + \left(\frac{c_2^{max} - c_2^{min}}{i^{max}}\right) \times i \tag{17}$$

where, c_1^{max} and c_1^{min} are the maximum and minimum values for the personal acceleration coefficient, respectively. $c1_i^j$ decreases over the iterations, which means that the acceleration component for the personal position at the beginning of the search is high allowing exploration. The parameters c_2^{min} and c_2^{max} represent the minimum and maximum values for the global acceleration coefficient. $c2_i^j$ increases over the iterations, which means that the acceleration component for the global position at the end of the search is high allowing exploitation. The encoding of the solutions is crucial for the success of the algorithm. Equation (18) shows the encoded vector used for solving the problem present in Section 2.

$$\vec{x}_i^j = \left[\left\{P_{1.1}^{bat}, \ldots, P_{B,T}^{bat}\right\}, \left\{X_{1.1}^{cut}, \ldots, X_{L,T}^{cut}\right\}\right] \tag{18}$$

where, $\{P_{1.1}^{bat}, \ldots, P_{B,T}^{bat}\}$ is a group of continuous variables representing the electricity amount of charge or discharge in each battery (b) at period (t) and $\{X_{1.1}^{cut}, \ldots, X_{L,T}^{cut}\}$ are binary variables to enable the possibility of performed cut action in load (l) at period (t). Therefore, particle $\overrightarrow{x}$ has dimensions of $n = B \times T + L \times T$. This encoding allows a direct evaluation in Equation (1).

The PSO implementation starts by defining the search space limits by setting the lower and upper bounds of each variable. In Equation (19), xlb^j represents the lower limits for the solution of j particle and xub^j in Equation (20) represent the upper limit for j particle.

$$xlb^j = \left[\{-P_{1.1}^{dch\ max}, \ldots, -P_{B,T}^{dch\ max}\}, \{X_{1.1}^{cut\ min}, \ldots, X_{L,T}^{cut\ min}\}\right] \tag{19}$$

$$xub^j = \left[\{P_{1.1}^{ch\ max}, \ldots, P_{B,T}^{ch\ max}\}, \{X_{1.1}^{cut\ max}, \ldots, X_{L,T}^{cut\ min}\}\right] \tag{20}$$

$$\overrightarrow{x}_1^j = rand\left[xlb^j, xub^j\right] \tag{21}$$

Equation (21) presents the process of initialization where the initial solution was created. In this case, a random process into allowed bounds is executed. $rand\left[xlb^j, xub^j\right]$ is a random number within the lower xlb^j and the upper xub^j bounds of j particle for n variables.

Equation (22) presents the boundary constrains method. The search process over the iterations will generate new solutions that may not be within the initially stipulated limits. To address this issue the boundary control strategies are used to repair infeasible individuals. In this paper is used a boundary control technique known as bounce-back [20].

$$\overrightarrow{x}_i^j = \begin{cases} rand\left(xlb^j, \overrightarrow{x}_i^j\right) & if\ \overrightarrow{x}_i^j < xlb^j \\ rand\left(\overrightarrow{x}_i^j, xub^j\right) & if\ \overrightarrow{x}_i^j > xub^j \\ \overrightarrow{x}_i^j & otherwise \end{cases} \tag{22}$$

In contrast to random reinitialization (the most used control technique), bounce-back uses the information on the progress towards the optimum region by reinitialized the variable value between the base variable value and the bound being violated. Making use of domain knowledge about the problem, the Equations (23) and (24) is proposed as a direct repair equation. The Equation (23) concerns the direct repair of $E_{b,t}^{stor}$.

$$E_{b,t}^{stor} = \begin{cases} 0 & if & E_{b,t}^{stor} < 0 \\ P_{b,t}^{ch\ max} & if\ E_{b,t}^{stor} > E_{b,t}^{stor\ max} & \forall b \in \{1, \ldots, B\},\ \forall t \in \{1, \ldots, T\} \\ E_{b,t}^{stor} & otherwise \end{cases} \tag{23}$$

Although boundary control is used it can only control the variables P^{bat} and X^{cut}, the variable E^{stor} is a variable of control and balance, and when it is repaired other variables are necessarily changed. For the repair process E^{stor} is needed to test two different conditions, $E_{b,t}^{stor} < 0$ represents a greater discharge than the allowed one, being that it fixes the variable to the minimum value. $E_{b,t}^{stor} > E_{b,t}^{stor\ max}$ means that the battery has a charge greater than the allowed, the value of maximum energy in the battery is fixed in maximum that can accumulate. Equation (24) presents the direct repair for P^{bat} variable.

$$P_{b,t}^{bat} = \begin{cases} E_{b,t}^{stor} - E_{b,t-1}^{stor} & if & E_{b,t}^{stor} < 0 \\ E_{b,t}^{stor} - E_{b,t-1}^{stor} & if\ E_{b,t}^{stor} > E_{b,t}^{stor\ max} & \forall b \in \{1, \ldots, B\},\ \forall t \in \{2, \ldots, T\} \\ P_{b,t}^{bat} & otherwise \end{cases} \tag{24}$$

P^{bat} is repaired in Equation (22), but with the direct repair used in Equation (23) the variable P^{bat} may not be correct, and it is necessary to perform direct repair on it. So, a battery power level test is performed, if $E_{b,t}^{stor} < 0$ the value for $P_{b,t}^{bat}$ is equal to the difference between the battery power level in the previous period $E_{b,t-1}^{stor}$ and the current period $E_{b,t}^{stor}$. The same rule is applied when the battery power level is greater than the allowed maximum $E_{b,t}^{stor} > E_{b,t}^{stor\ max}$.

The particles should be evaluated according to a fitness function $f'(\vec{x})$, Equation (25), including objective function $f(\vec{x})$ Equation (1) and constrains violation $pf(\vec{x})$.

$$f'(\vec{x}) = f(\vec{x}) + pf(\vec{x}) \tag{25}$$

$$pf(\vec{x}) = \begin{cases} \sum_{t=1}^{T} t \times \rho & if\ P_t^{grid} \le P_t^{grid\ min} \cap P_t^{grid} \ge P_t^{grid\ max} \\ 0 & otherwise \end{cases} \qquad \forall t \in \{1, \dots, T\} \tag{26}$$

where, $pf(\vec{x})$ in Equation (26) represents the penalty value for a solution $\vec{x}$. Despite the application of bounce-back method Equation (22) and direct repair methods (23) and (24), the solution may still be infeasible. The penalty value is obtained checking the limits of variable P_t^{grid} for every period. In each period that the variable is out of limit is counted and multiplied by a penalty amount ρ, the sum of all individual (per period) penalties represents the total penalties per each solution.

Pseudocode of the PSO algorithm is presented in Algorithm 1.

Algorithm 1. PSO pseudocode.

INITIALIZE
Set control parameters $w^{max}, w^{min}, c_1^{max}, c_1^{min}, c_2^{max}, c_2^{min}, j^{max}$, and i^{max}.
Create an initial Pop (Equation (21)) and initial velocities.
IF Direct repair is used **THEN**
    Apply direct repair to unfeasible individuals
END IF
Evaluate the fitness of Pop (Equation (25)).
Create a P_{best} vector for every particle.
Create a G_{best} vector of the swarm.
FOR $i = 1$ to i^{max}
    FOR $j = 1$ to j^{max}
        Velocity update (Equation (13))
        Position update (Equation (14))
        Update w_i, $c1_i$ and $c2_i$ (Equations (15)–(17))
        Verify boundary constraints for P^{bat} (Equation (9))and X^{cut} (Equation (10))
        IF Boundary constraints are violated **THEN**
            Apply boundary control (Equation (22))
        END IF
        Verify boundary constraints for E^{stor} (Equation (8)) and P^{bat} (Equation (9))
        IF Boundary constraints are violated **THEN**
            Apply direct repair (Equations (23) and (24))
        END IF
        Evaluate fitness of $\vec{x}$ (Equation (25)).
        Verify boundary constraints for P^{grid} (Equation (6))
        IF P^{grid} is out of limits **THEN**
            Apply penalty function (Equation (26))
            Update fitness value (Equation (25))
        END IF
        Update P_{best} vector for i particle.
    END FOR
    Update G_{best} vector of the swarm.
END FOR

Basically, if in the evaluation process constraints violations are identified, the individual is randomly repaired using the initialization process from Equation (22). The pseudocode of Algorithm 1 is displayed step-by-step, starts with the definition of the parameters related to the PSO. The search begins with the creation of the initial population. After being evaluated, the best position of each particle and the best position of the population are defined. The main cycle starts, and at each iteration of the main cycle, another cycle is performed for each particle. For each particle a new velocity is generated, updated, verified, and evaluated. When all particles repeat the process, the value of the best personal position of each particle and the best overall position of the population is updated.

5. Case Study

This section presents the case study. The optimization problem was solved using PSO metaheuristic and compared to a solution obtained by a CPLEX solver in MATLAB™/TOMSYM™ environment to compare the results.

The proposed methodology addresses a Portuguese consumer and complies with actual Portuguese legislation, which allows small producers (consumers with local generation) to use the energy produced to satisfy the own load necessities and sell it to the grid. The consumer has a supply power contract of 10.35 KVA with the retailer, and it is characterized by three different periods: peak, intermediate, and off-peak [29]. The prices applied to a consumer operation are present in Table 1. The input prices in Table 1 are real values of a Portuguese retailer (https://www.edp.pt/particulares/energia/tarifarios), which provides a realistic case study. The prosumer can inject his excess production into grid, but a limit is imposed by the retailer. The maximum value injected into grid is half of its contracted power, approximately 5.1 kW. The real prices and real condition inclusion in this problem contribute to more accurate in this study and prove the real value of the methodology application.

Table 1. Prices of the different periods and contracted power.

Parameter	Energy (€/kWh)			Contracted Power (€/Day)
	Peak	**Intermediate**	**Off-Peak**	
Buy from grid	0.2738	0.1572	0.1038	0.5258
Periods	10.30 h–13 h 19.30 h–21 h	08 h–10.30 h 13 h–19.30 h, 21 h–22 h	22 h–02 h 02 h–08 h	
Sell to grid		0.1659 *		—
DR weight	0	0.2	0.4	

* is used for all periods.

The DR weights present in Table 1 are defined by the consumer taking into account the energy price variation within the day, adapted from [16]. The use of DR is more appreciated when the energy is cheaper, so the weight of 0 is given in peak periods (highest price). With this weight distribution, the DR actions are expected to be executed during peak periods. Equation (3) gives the amount of DR actions contributing to the objective function. It does not represent costs for the consumer, but is rather a consumer's preference that influences the scheduling. In Table 2 are presented the problem input variables adapted from [16].

The system has two PV panels with different production, one has a maximum production of 7.5 kW and other has a maximum production of 2.5 kW. This PV panels and the battery storage unit are connected to the inverter. The battery can receive power from the PV production or the grid. In this case study, the inverter has two functionalities: the first is to convert the power from DC to AC and vice versa; the other functionality is to give the signal to manage certain loads. In this study, three different loads are considered: a dishwasher, an air conditioner, and a water heater. Figure 3 shows the disaggregated consumption and PV generation forecasts. In this case study, the forecast is performed for the next 24 h. In real-time operation, the forecast can be updated at every instant. Each time that user update the forecast can perform a new optimization. Regarding the influence of the forecasting results on optimization, in the case that the presented day-ahead forecasting strategies in References [30,31] are considered, the forecasting error, using Supporter Vector Machine algorithms to predict the values for the next 24 h, will be 9.11%.

Table 2. Problem input variables.

Parameters	Symbol	Value	Units
Maximum power injected to grid	$-P_t^{grid\ min}$	−5.1	kW
Maximum power required from grid	$P_t^{grid\ max}$	1000	kW
Maximum power accumulated in battery	$E_{b,t}^{stor\ max}$	12	kW
Maximum energy of battery discharge	$-P_{b,t}^{dch\ max}$	−6/4	kWh
Maximum energy of battery charge	$P_{b,t}^{ch\ max}$	6/4	kWh
Total Periods	T	96	–
Total of controllable loads	L	3	–
Total of batteries	B	1	–
Total of PV units	P	2	–
Adjust parameter	Δt	4 *	–

* The factor of 4 comes from the fact that there are four 15-min periods in an hour.

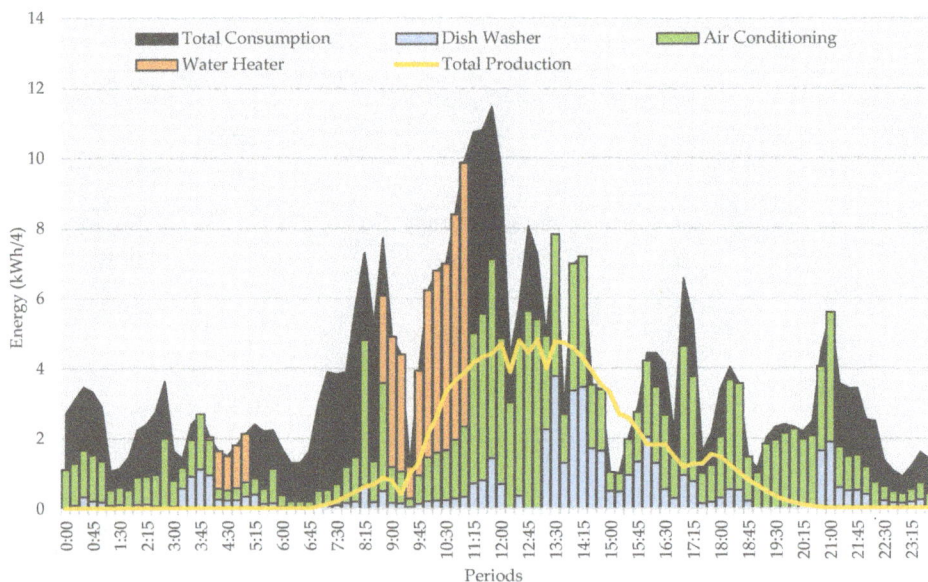

Figure 3. Disaggregated consumption by appliance and photovoltaic (PV) generation.

Figure 3 presents a typical load profile with a peak of 11.5 kW at ~11.45 h. The consumption per controllable load is present in Figure 3 with different colors. The total consumption includes the sum of all loads and the same situation for PV but is the sum of two PV units. The peak of production is forecasted between the 12.00 h and 13.30 h with 6 kW. In some periods, such as 10:30, the sum of the controllable loads corresponds to the total consumption. Table 3 presents the parameters for PSO; they were obtained from a previous study.

Table 3. Particle swarm optimization (PSO) parameters used.

Parameters	Symbol	Value
Population size	j^{max}	500
Maximum numbers of iterations	i^{max}	500
Maximum inertia weight	w^{max}	0.4
Minimum inertia weight	w^{min}	0.9
Maximum cognitive weight	c_1^{max}	1.5
Minimum cognitive weight	c_1^{min}	0.5
Maximum global weight	c_2^{max}	1.5
Minimum global weight	c_2^{min}	0.5
Number of evaluations	–	250,000
Number of trials	–	30

The member of evaluation is equal to $j^{max} \times i^{max}$ and presents the number of fitness function is evaluated during the search process. Considering that the PSO is an algorithm of a random nature, a group of 30 trials is performed. With a sample of 30 results, it is possible to extract a more robust conclusion from the application of the PSO to the problem in question.

6. Results

This section presents the results and analysis obtained from the implementation of the proposed methodology and respective case study. Table 4 presents the results for Equation (1) in both the CPLEX (deterministic) obtaining the optimal value, and PSO obtained an approximate resolution. Four different scenarios were created considering the resources combination: the scenario "PV + Bat + DR" combine the all available resources (PV production, the storage capacity and loads flexibility), scenario "PV + Bat" combines the PV production and storage capacity resources and "PV" scenario only considers the PV production resource. The nonoptimized value is used as a base case scenario and was obtained by using a simple management mechanism; the scenario "PV + Bat" considers PV production and storage capacity, and the "Without resources" scenario does not consider any resource. Analyzing the results of CPLEX for the set of scenarios can conclude that "PV + Bat + DR" presents the smallest fitness function. It can be said with resources combinations brings benefits for household management.

Table 4. Results for Equation (2) (€/day).

Resources Combination Scenarios		CPLEX	PSO		
			Min	Mean	STD
Values optimized	PV + Bat + DR	3.1874	3.2771	3.3381	0.0469
	PV + Bat	7.8652	7.9454	8.0595	0.1169
	PV	8.8478	8.8478	8.8478	0
Nonoptimized values	PV + Bat		9.3298		
	Without resources		16.8570		

The analysis of results is performed for the "PV + Bat + DR" scenario. The results present in Table 4 of PSO correspond to 30 trials. The minimum value that the PSO reached is 2.8% higher when compared with CPLEX value. Analyzing the standard deviation (std) value for the sample of PSO results is possible to conclude that it is relatively small and the values of the 30 trials should be relatively close to the mean value. The STD analysis is important because it is a measure that expresses the degree of dispersion of 30 trials solutions. Figure 4 presented the results related to the DR actions applied to the profile shown in Figure 3.

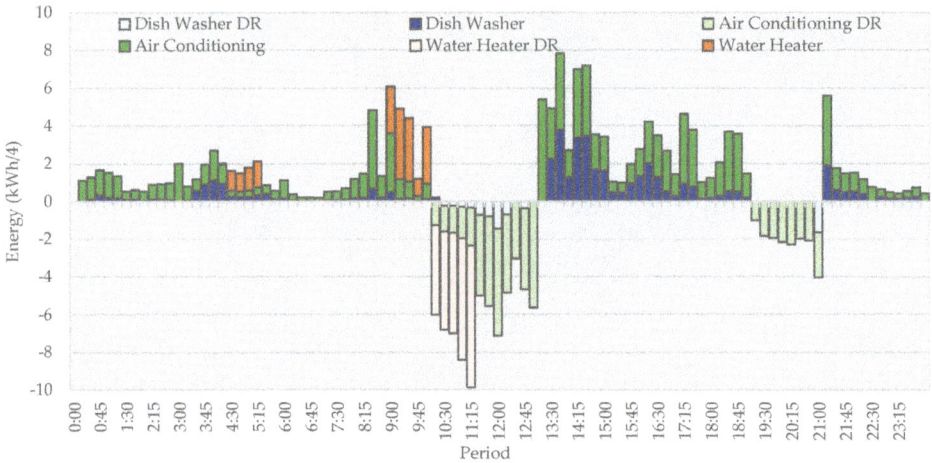

Figure 4. DR result regarding initial profile.

In Figure 4, the positive values correspond to the consumption of appliances that had no changes with the application of the methodology. Negative values are energy that has been reduced due to cut of loads. With the loads cut, reduction of 63% in the total consumption of three loads (dish washer, air conditioner, and water heater) was obtained. The DR actions are performed during 10.00 h to 13.00 h and 19.00 to 21.00. Crossing this information with Table 1, one can see that these periods correspond to a peak hour, precisely when energy is more expensive. During peak hours the consumption with the present optimization methodology is 44.8 kW, without its application and not considering PV generation and energy storage systems, the consumption will be 115.4 kW. This reduction represents 20% reduction of total daily consumption. In this way, it is concluded that the present methodology has an impact on the consumption of peak hours. In Figure 5 are presented the total load consumption (controllable and noncontrollable loads), the battery actions (charge and discharge), and the final load (load consumption plus battery charge).

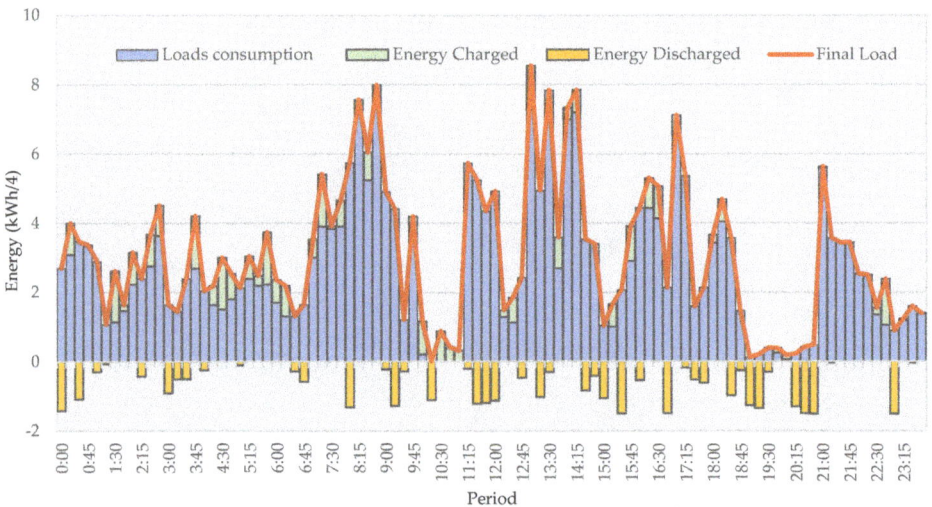

Figure 5. Load consumptions, battery actions, and final load scheduling.

Figure 5 shows that due to this condition, the generation (see Figure 3) exceeds the consumption needs, and in this case, the energy surplus will either be used to charge the battery or sell to the grid. In this way, the user avoids buying energy from the grid to charge the battery and to meet consumption necessities. The battery discharge cycles are mostly represented between 11.00 and 21.00 periods that correspond to a peak and intermediate hour. Table 5 presents a summary of the results obtained by both methods applied.

Table 5. Summary of results.

Scenario	Method	Equation (1)	Equation (2)	Equation (3)	Daily Costs (€)	Daily Revenues (€)	Monthly Costs (€)
PV + Bat + DR	CPLEX	3.1874	3.1874	0	6.9380	3.7505	95.6233
PV + Bat + DR	PSO *	3.2771	3.2771	0	6.0565	2.7794	98.3140
PV + Bat	PSO *	7.9922	7.9922	0	8.5136	0.5683	239.7661
PV + Bat	Nonoptimized	9.3298	9.3298	0	9.3298	0	279.8928

* represent the values of trial with the minimum fitness value.

With the proposed methodology, the daily cost of operation for CPLEX is 3.18 (€) and 3.28 (€) for PSO, but if the PV system, battery and DR do not exist and the daily costs are 16.83 (€). When compared the results of Table 5 is possible to observe that daily cost for CPLEX is larger compared to PSO daily cost, but the value of revenues in CPLEX are also large than PSO values. With the case study present in Section 5, the value of Equation (1) is equal to Equation (2) in both of methods, which means that the value of Equation (3) is zero because Equation (1) is the sum of Equation (2) and (3). When Equation (3) has the value zero represents that the DR is performed on periods with weight equal to zero and do not have a contribution to Equation (1). Table 5 also presents the monthly costs, which are calculated considering that the profile present in Figure 3 is repeated for the 30 days of the month. The value obtained for PSO is 2.96 (€) higher.

7. Conclusions

The present work addresses a methodology for resource scheduling (PV battery, storage capacity, and load flexibility) in a residential house that has not any contract with a DR service provider. Usually, the DR services for residential consumers are available using a DR service provider. In contrast, in the presented methodology the user is independent of applying his preferences in decision-making. In this case, the PV inverter, installed to convert the PV production into DC to AC, can control the charge or discharge of the battery system and the interruption of the loads. The optimization results for $P_{b,t}^{bat}$ and $P_{l,t}^{cut}$ are the inputs for the PV inverter control to act on the battery system and controllable loads.

The optimization problem was solved using a stochastic method (PSO) and a deterministic method (CPLEX). The results obtained by PSO have a close approximation to the deterministic results. The simple implementation and open access possibility of programming PSO over different platforms are factors that potentiate its use in this type of problems. In fact, in the present work, it was possible to demonstrate the results of running a PSO-based algorithm on a connection with the inverter of the PV system for control of the connected loads and the charge or discharge of the battery storage system.

The numerical results presented demonstrate that it is possible to obtain advantages by using the optimal combination of available resources. Table 4 presents the fitness function value for different resources combination, showing that the scenario that combines the all available resources is the best. Although PSO can obtain near-optimal solutions, its solution using the best combination resource scenario is better than the normal operating solution. With the comparison between the base scenario and the same scenario with PSO optimization, it is possible to make the assessment of the PSO approach. The daily cost optimized by PSO for the base scenario is 14% lower compared with the obtained in the nonoptimized base scenario.

As the presented methodology was built for been applied in an independent agent, the agent facility (residential house) needs to be prepared with equipment to perform the actions that the presented method imposes. This condition may be a weakness of the methodology, as it will increase the initial investment in equipment. Assuming that the DR program is implemented efficiently, such investment can be recovered over time, as the user does not need to pay fees to any service provider to use the service. The use of PSO instead of CPLEX can make the initial investment more appealing, for reasons already discussed in the introduction.

For future work, an analysis incorporating more DR actions (e.g., reduction and shifting capabilities) in the presented methodology can be done. Also, robust optimization considering the forecast error in PV production and domestic consumption can also be made to analyze the impact of forecasts errors in the electricity bill.

Author Contributions: Investigation, R.F.; Methodology, P.F. and Z.V.; Resources, Z.V.; Software, R.F. and J.S.; Writing—original draft, R.F.; Writing—review & editing, P.F. and Z.V.

Funding: The present work was done and funded in the scope of the following projects: SIMOCE Project (P2020-23575) and UID/EEA/00760/2019 funded by FEDER Funds through COMPETE program and by National Funds through FCT. Ricardo Faia is supported by national funds through Fundação para a Ciência e a Tecnologia (FCT) with PhD grant reference SFRH/BD/133086/2017.

Conflicts of Interest: The authors declare no conflicts of interest.

Glossary/Nomenclature

Abbreviations

AI	Artificial Intelligence
DR	Demand Response
DG	Distributed Generation
ESS	Energy Storage System
LP	Linear Programming
MATLAB	Matrix Laboratory
MILP	Mixed-integer Linear Programming
MINLP	Mixed-integer Nonlinear Programming
NLP	Nonlinear Programing
PSO	Particle Swarm Optimization
PV	Photovoltaic
RESs	Renewable Energy Sources
SET	Strategic Energy Technology

Indices

b	Battery unit
n	Dimension
i	Iteration
l	Load unit
j	Particle
t	Period
p	Photovoltaic unit

Parameters

$C_t^{grid\ in}$	Cost of buying electricity to the grid
$C_t^{grid\ out}$	Cost of selling electricity to the grid
$W_{l,t}^{cut}$	Cut weight of load
DCP	Daily contracted power cost
xlb^j	Lower bond for $\vec{x}^j$
$P_t^{grid\ max}$	Maximum limit for P_t^{grid}
i^{max}	Maximum number of iterations

j^{max}	Maximum numbers of particles
$p_{l,t}^{cut\ max}$	Maximum value for cut load
$p_{b,t}^{ch\ max}$	Maximum value for energy charge
$p_{b,t}^{dch\ max}$	Maximum value for energy discharge
c_2^{max}	Maximum value for global acceleration coefficient
w^{max}	Maximum value for inertia weight
c_1^{max}	Maximum value for personal acceleration coefficient
$E_{b,t}^{stor\ max}$	Maximum value of accumulated energy in battery
$P_t^{grid\ min}$	Minimum limit for P_t^{grid}
c_2^{min}	Minimum value for global acceleration coefficient
w^{min}	Minimum value for inertia weight
c_1^{min}	Minimum value for personal acceleration coefficient
Δt	Multiplicative factor related with the time to calculate energy
B	Number of batteries
L	Number of controllable loads
T	Number of Periods
ρ	Penalty value
$P_{p,t}^{PV}$	Photovoltaic production
xub^j	Upper bond for $\vec{x}^j$
P_t^{load}	Value of load

Variables

$I_t^{grid\ in}$	Binary variable for control the flow direction
$P_{l,t}^{cut}$	Cut power of load
$X_{l,t}^{cut}$	Decision binary variable to active the cut of loads
$P_{b,t}^{bat}$	Energy charged or discharged by battery
$f(\vec{x})$	Fitness function
$f'(\vec{x})$	Fitness function with penalty
P_t^{grid}	Flow of energy between household and grid
p_{best}^j	Historical best position
w_i^j	Inertia weight
$pf(\vec{x})$	Penalty function
$c1_i^j$ and $c2_i^j$	Personal and global acceleration coefficients
G_{best}	Population historical best position
$\vec{x}_i^j$	Position vector
$E_{b,t}^{stor}$	State of the battery
$r1_i^j$ and $r2_i^j$	Uniform distribution random numbers
$\vec{v}_i^j$	Velocity vector

References

1. European Union. *The Strategic Energy Technology (SET) Plan*; European Union: Brussels, Belgium, 2017.
2. Parag, Y.; Sovacool, B.K. Electricity market design for the prosumer era. *Nat. Energy* **2016**, *1*, 16032. [CrossRef]
3. Bhandari, B.; Lee, K.-T.; Lee, G.-Y.; Cho, Y.-M.; Ahn, S.-H. Optimization of hybrid renewable energy power systems: A review. *Int. J. Precis. Eng. Manuf. Technol.* **2015**, *2*, 99–112. [CrossRef]
4. Roldán-Blay, C.; Escrivá-Escrivá, G.; Roldán-Porta, C. Improving the benefits of demand response participation in facilities with distributed energy resources. *Energy* **2019**, *169*, 710–718. [CrossRef]
5. Paterakis, N.G.; Erdinç, O.; Catalão, J.P.S. An overview of Demand Response: Key-elements and international experience. *Renew. Sustain. Energy Rev.* **2017**, *69*, 871–891. [CrossRef]
6. Neves, D.; Silva, C.A. Optimal electricity dispatch on isolated mini-grids using a demand response strategy for thermal storage backup with genetic algorithms. *Energy* **2015**, *82*, 436–445. [CrossRef]

7. Jordehi, A.R. Optimisation of demand response in electric power systems, a review. *Renew. Sustain. Energy Rev.* **2019**, *103*, 308–319. [CrossRef]

8. European Commission. *Benchmarking Smart Metering Deployment in the EU-27 with a Focus on Electricity*; European Commission: Brussels, Belgium, 2014.

9. Nan, S.; Zhou, M.; Li, G. Optimal residential community demand response scheduling in smart grid. *Appl. Energy* **2018**, *210*, 1280–1289. [CrossRef]

10. Zhang, D.; Shah, N.; Papageorgiou, L.G. Efficient energy consumption and operation management in a smart building with microgrid. *Energy Convers. Manag.* **2013**, *74*, 209–222. [CrossRef]

11. Pina, A.; Silva, C.; Ferrão, P. The impact of demand side management strategies in the penetration of renewable electricity. *Energy* **2012**, *41*, 128–137. [CrossRef]

12. Nwulu, N.I.; Xia, X. Optimal dispatch for a microgrid incorporating renewables and demand response. *Renew. Energy* **2017**, *101*, 16–28. [CrossRef]

13. Alipour, M.; Zare, K.; Abapour, M. MINLP Probabilistic Scheduling Model for Demand Response Programs Integrated Energy Hubs. *IEEE Trans. Ind. Inform.* **2018**, *14*, 79–88. [CrossRef]

14. Zhou, X.; Shi, J.; Tang, Y.; Li, Y.; Li, S.; Gong, K. Aggregate Control Strategy for Thermostatically Controlled Loads with Demand Response. *Energies* **2019**, *12*, 683. [CrossRef]

15. Yao, Y.; Zhang, P.; Chen, S. Aggregating Large-Scale Generalized Energy Storages to Participate in the Energy and Regulation Market. *Energies* **2019**, *12*, 1024. [CrossRef]

16. Spínola, J.; Faria, P.; Vale, Z. Photovoltaic inverter scheduler with the support of storage unit to minimize electricity bill. *Adv. Intell. Syst. Comput.* **2017**, *619*, 63–71.

17. Hussain, B.; Khan, A.; Javaid, N.; Hasan, Q.; Malik, S.A.; Ahmad, O.; Dar, A.; Kazmi, A. A Weighted-Sum PSO Algorithm for HEMS: A New Approach for the Design and Diversified Performance Analysis. *Electronics* **2019**, *8*, 180. [CrossRef]

18. Shen, J.; Jiang, C.; Li, B. Controllable Load Management Approaches in Smart Grids. *Energies* **2015**, *8*, 11187–11202. [CrossRef]

19. Kong, D.-Y.; Bao, Y.-Q.; Hong, Y.-Y.; Wang, B.-B.; Huang, H.-B.; Liu, L.; Jiang, H.-H. Distributed Control Strategy for Smart Home Appliances Considering the Discrete Response Characteristics of the On/Off Loads. *Appl. Sci.* **2019**, *9*, 457. [CrossRef]

20. Jordehi, A.R. A review on constraint handling strategies in particle swarm optimisation. *Neural Comput. Appl.* **2015**, *26*, 1265–1275. [CrossRef]

21. Faria, P.; Vale, Z.; Soares, J.; Ferreira, J. Demand Response Management in Power Systems Using Particle Swarm Optimization. *IEEE Intell. Syst.* **2013**, *28*, 43–51. [CrossRef]

22. Shehzad Hassan, M.A.; Chen, M.; Lin, H.; Ahmed, M.H.; Khan, M.Z.; Chughtai, G.R. Optimization Modeling for Dynamic Price Based Demand Response in Microgrids. *J. Clean. Prod.* **2019**, *222*, 231–241. [CrossRef]

23. Aghajani, G.R.; Shayanfar, H.A.; Shayeghi, H. Demand side management in a smart micro-grid in the presence of renewable generation and demand response. *Energy* **2017**, *126*, 622–637. [CrossRef]

24. Hu, M.; Xiao, F. Price-responsive model-based optimal demand response control of inverter air conditioners using genetic algorithm. *Appl. Energy* **2018**, *219*, 151–164. [CrossRef]

25. Qian, L.P.; Wu, Y.; Zhang, Y.J.A.; Huang, J. Demand response management via real-time electricity price control in smart grids. *Smart Grid Netw. Data Manag. Bus. Model.* **2017**, 169–192.

26. Lezama, F.; Sucar, L.E.; de Cote, E.M.; Soares, J.; Vale, Z. Differential evolution strategies for large-scale energy resource management in smart grids. In Proceedings of the Genetic and Evolutionary Computation Conference Companion on GECCO '17, Berlin, Germany, 15–19 July 2017; ACM Press: New York, New York, USA, 2017; pp. 1279–1286.

27. Eberhart, R.; Kennedy, J. A new optimizer using particle swarm theory. In Proceedings of the Sixth International Symposium on Micro Machine and Human Science (MHS'95), Nagoya, Japan, 4–6 October 1995; pp. 39–43.

28. Faia, R.; Pinto, T.; Vale, Z.; Corchado, J.M. Strategic Particle Swarm Inertia Selection for the Electricity Markets Participation Portfolio Optimization Problem. *Appl. Artif. Intell.* **2018**, *32*, 1–23. [CrossRef]

29. ERSE Tarifas e Precos Para a Energia Eletrica e Outros Servicos em 2019. Available online: http://www.erse.pt/pt/electricidade/tarifaseprecos/2019/Documents/DiretivaERSE13-2018(TarifasePreçosEE2019).pdf (accessed on 6 February 2019).

30. Jozi, A.; Pinto, T.; Praca, I.; Vale, Z. Day-ahead forecasting approach for energy consumption of an office building using support vector machines. In Proceedings of the 2018 IEEE Symposium Series on Computational Intelligence (SSCI), Bangalore, India, 18–21 November 2018; pp. 1620–1625.
31. Jozi, A.; Pinto, T.; Praça, I.; Vale, Z. Decision Support Application for Energy Consumption Forecasting. *Appl. Sci.* **2019**, *9*, 699. [CrossRef]

MDPI

St. Alban-Anlage 66

4052 Basel

Switzerland

Tel. +41 61 683 77 34

Fax +41 61 302 89 18

www.mdpi.com

Energies Editorial Office

E-mail: energies@mdpi.com

www.mdpi.com/journal/energies

www.ingramcontent.com/pod-product-compliance
Lightning Source LLC
Chambersburg PA
CBHW051720210326
41597CB00032B/5551